Advances in Shannon's Sampling Theory

T0173987

Ahmed I. Zayed, Ph.D.
Department of Mathematics
University of Central Florida
Orlando, Florida

CRC Press
Taylor & Francis Group
Boca Raton London New York

CRC Press is an imprint of the
Taylor & Francis Group, an **informa** business

CRC Press
Taylor & Francis Group
6000 Broken Sound Parkway NW, Suite 300
Boca Raton, FL 33487-2742

First issued in paperback 2020

ISBN 13: 978-0-367-57986-9 (pbk)
ISBN 13: 978-0-8493-4293-6 (hbk)

Visit the Taylor & Francis Web site at
http://www.taylorandfrancis.com

and the CRC Press Web site at
http://www.crcpress.com

Library of Congress Card Number 93-19506

Library of Congress Cataloging-in-Publication Data

Zayed, Ahmed I.
Advances in Shannon's sampling theory / author, Ahmed I. Zayed.
p. cm.
Includes bibliographical references and index.
ISBN 0-8493-4293-7
1. Signal processing—Statistical methods. 2. Sampling (Statistics) 3. Engineering—Statistical methods. I. Title.
TK5102.9.Z39 1993
003'.54—dc20
DNLM/DLC
for Library of Congress 93-19506
CIP

To my mother
and
the memory of my father

PREFACE

Overview

It has now been almost forty five years since the publication of C. E. Shannon's work, "A Mathematical Theory of Communication," which has revolutionized the fundamentals of communication engineering and Information Theory. One of the basic mathematical ideas used in Shannon's theory is the sampling theorem for band-limited signals, commonly known in the western world as Shannon's sampling theorem. The publication of Shannon's work in 1949 spawned a profusion of research articles dealing with both the theory and applications of his sampling theorem.

What started as a theorem for reconstructing band-limited signals from their sample values at uniformly distributed set of points has now evolved into a branch of mathematical analysis, known as *Sampling Theory.* Most of the research in sampling theory has centered around extending Shannon's fundamental result in different directions, such as extending it to include non-equally spaced sampling points and other discrete data taken from the signals, as well as, extending it to multidimensional and more general types of signals. Error analysis and finding new and fast algorithms for implementation have also been the focus of intensive research.

Sampling theory has received considerable attention in the engineering community. The fact that most research articles on this subject have appeared in engineering journals and that the rudiments of the subject are covered in almost any engineering textbook on signal analysis attests to that.

Nevertheless, throughout its relatively short history, sampling theory has lived in the shadow of other more prominent and traditional areas of Mathematical Analysis, such as *Approximation Theory*, the *Theory of Fourier Series and Integrals*, and the *Theory of Entire Functions*, invariably borrowing ideas and techniques from these areas. Maybe that is why, despite its widespread use in physics and engineering, sampling theory has attracted only moderate attention in the mathematical community.

Although a fair amount of research in sampling theory has been conducted by mathematicians and has appeared in the literature, little of it has appeared in book form. The fundamentals of the subject are usually covered in only one or two chapters in most engineering textbooks on signal analysis, leaving out numerous results that are important to mathematicians. There is a very small number of books in the market that are dedicated almost entirely to sampling theory, the most recent of which are two books by R. J. Marks II, "Introduction to Shannon Sampling and Interpolation Theory"

(Springer-Verlag, 1991) and "Advanced Topics on Shannon Sampling and Interpolation Theory" (Springer-Verlag, 1992). The first is a graduate textbook for electrical engineering and the second is a collection of review and research articles by experts in the field.

Review articles collecting and tying together the most important findings in sampling theory have appeared from time to time, as is the case in most rapidly developing subjects. The first review article, of which we are aware, is A. Jerri's " The Shannon sampling theory-its various extensions and applications: a tutorial review," (*Proc. IEEE*, 65, 1977). Then came "A survey of the Whittaker-Shannon sampling theorem and some of its extensions" (*J. Math. Res. Exposition* 3, 1983) by P. Butzer," Five short stories about the cardinal series," (*Bull. Amer. Math. Soc.*, 12, 1985) by J. R. Higgins, and finally "The sampling theorem and linear prediction in signal analysis" (*Jber. d. Dt. Math. Verein.*, 90, 1988) by P. L. Butzer et al. The last article is basically a survey article of some of the main results obtained by Professor Butzer, his students, and colleagues at the Lehrstuhl A für Mathematik in Aachen, Germany.

In the last few years, with the surge of new techniques in mathematical analysis like the atomic decompositions of functions, wavelets, theory of frames and multiresolution analysis of Hilbert spaces, new ways to view and prove results in sampling theory have begun to emerge. Even more, sampling theory is now starting to take a new role and expand over a new horizon; it is no longer a co-dependent of other traditional branches of mathematics, but an equal partner. Newly discovered results in sampling theory have led to new discoveries in other areas of mathematics, such as *Special Functions*.

Because of all these recent developments, we felt that there was a need for a book to gather some of these scattered results on the subject and present them, in a relatively efficient and succinct way, as a unified theory.

Most of the results that will be presented in this monograph have never before been published in the form of a book.

Objectives

The purpose of this book is three-fold:

i) To give the reader an overview of the subject, its history and development.

ii) To provide a moderately concise survey of the present state of knowledge of the subject and to shed some light on new connections between sampling theory and other branches of mathematical analysis, such as the theory of boundary-value problems, the theory of frames and wavelets and multiresolution analysis of Hilbert spaces.

iii) To show how advances in sampling theory can lead to new discoveries in other branches of mathematical analysis, in particular, in the field of *Special Functions.*

These objectives will be achieved according to the following plan:

i) In Chapter 1, we shall give an overview of the Shannon sampling theorem and its history. For this book to be accessible to both engineers and mathematicians, we shall also introduce some of the engineering terminology that will be used throughout the book and their mathematical counterparts.

ii) In Chapter 2, we shall introduce the Shannon sampling theorem, which from that point on will be called (for reasons explained in Chapter 1), the Whittaker-Shannon-Kotel'nikov (WSK) sampling theorem. We shall then discuss the notion of band-limitedness and various extensions of the WSK theorem pertaining to different generalizations of this notion.

iii) Chapter 3, which is one of the longest chapters in the book, provides an overview of numerous generalizations and extensions of the WSK sampling theorem as well as a summary of some of the main results on error analysis.

Almost all the results in chapters 2 and 3 are well known. Therefore, in order to keep the monograph to a manageable size, we shall omit most of the proofs, but supply the reader with references instead.

iv) Chapter 4 is devoted entirely to the discussion of sampling theorems associated with Sturm-Liouville boundary-value problems following the general scheme of the Weiss-Kramer sampling theorem, which is a generalization of the WSK sampling theorem. This will be done for both regular and singular boundary-value problems.

v) Some of these results are first extended in Chapter 5 to general self-adjoint boundary-value problems associated with nth order differential operators and mixed boundary conditions, and then in Chapter 6 to non-self-adjoint boundary-value problems.

vi) The thrust of our work is demonstrated in Chapter 7, where some of the techniques developed in the previous chapters are applied to derive new summation formulae for infinite series. Although some of these formulae have just been recently discovered by others, we shall show that sampling theory can significantly simplify their proofs to just a few lines. The remaining formulae have been solely discovered and proved by using sampling theorems; no other simple proofs are known at the present time.

vii) Chapters 8 and 9 deal with multidimensional sampling theorems. In Chapter 8, some of the results of Chapter 4 are extended to higher dimensions to yield Kramer-type sampling theorems for multidimensional signals. The Feichtinger-Gröchenig sampling theory, which represents one of the most recent and promising approaches to irregular sampling, especially in multidimensions, is presented in Chapter 9.

viii) Finally, in Chapter 10, we shall present some very recent views and approaches to sampling theory, such as sampling theorems in the scope of the theory of frames, wavelets, and multiresolution analysis of Hilbert spaces.

One final word: this monograph is not meant to compete or overlap with other publications on the subject, but to supplement them. We strive to make it a welcome addition to the library of any individual interested in the field of sampling theory, whether a mathematician, scientist or an engineer.

Audience

This book is essentially a research monograph and contains original works; however, I have tried to write it for as wide an audience as I could. The first three chapters should be readily accessible to advanced students in their final year of undergraduate mathematics and to graduate students in applied fields with interest in sampling theory, in particular, to graduate students in electrical engineering with interest in signal analysis. Chapters 2 and 3 can serve as a handbook of sampling theorems.

Chapters 4 through 8, which comprise some of the author's own research, require some knowledge of special functions and the theory of boundary-value problems in one and several variables. Chapters 9 and 10 are, generally speaking, self-contained, yet may require some basic knowledge of functional analysis, Fourier series and integrals.

Chapter 10 can be used as a basis for teaching a crash course on wavelets and sampling theorems. It is an outgrowth of my lecture notes that I used in a one-semester graduate course on wavelets and sampling at the University of Central Florida.

Organization and Notation

Within each chapter, theorems are numbered consecutively by two digits, the first of which indicates the chapter in which the theorem is stated, for example, Theorem 3.7 means Theorem 7 in Chapter 3. Since corollaries are invariably corollaries of theorems, they will be numbered accordingly; for instance, Corollary 3.7.2 means Corollary 2 of Theorem 3.7. Lemmas and definitions, on the other hand, will be numbered by sections; for example, Lemma 3.2.5 means Lemma 5 of Section 2 in Chapter 3.

References are gathered at the end of each chapter. Sometimes, they may not refer to the original source, but to a more recent one where more references can be found. I have tried to document every item, yet since the list grows so rapidly, despite all efforts made to achieve completeness, it is very likely that I have overlooked some important contributions. I would like to take this opportunity to offer my sincere apologies to their authors.

Acknowledgments

It is my great pleasure to thank Professor Paul L. Butzer for his continuous support, especially in my early years of working on sampling theorems. I also would like to thank Professor Rowland Higgins for writing "Five short stories on the cardinal series." It was this article that spurred my interest in sampling theory.

It is with great pleasure that I acknowledge my gratitude to my friend and Ph.D. thesis advisor, Professor Gilbert G. Walter, for his constructive and sometimes penetrating criticism of the original version of the manuscript. Professor Abdul J. Jerri read the manuscript and made valuable comments and suggestions. I wish to express my sincere appreciation to him.

Many thanks are due to the staff of CRC Press, particularly, to Wayne Yuhasz, the Executive Editor of the Physical Science division, for their support and cooperation. Special thanks go to Dr. Daniel Zwillinger, an Advisory Editor to CRC Press, for recruiting me to write the book for this publishing company. I would like to thank Ms. June Wingler from the Division of Sponsored Research at the University of Central Florida for doing a superb job typing the manuscript.

Finally, this work would have been impossible to accomplish without the pleasant and stimulating atmosphere at the Mathematics Department, University of Central Florida, Orlando.

THE AUTHOR

Ahmed I. Zayed, Ph.D., is Professor of Mathematics at the University of Central Florida, Orlando. He held a similar position at the California Polytechnic State University (Cal Poly State University) in San Luis Obispo, California, until 1990 when he accepted his current position at the University of Central Florida.

Professor Zayed graduated in 1970 from Cairo University, Egypt, with a B.S. degree in pure and applied mathematics and obtained his Ph.D. degree in 1979 from the University of Wisconsin-Milwaukee.

He is a member of the American Mathematical Society, the Mathematical Association of America and the Society for Industrial and Applied Mathematics (SIAM). He is now serving as a member of the National Committee on the Teaching of Undergraduate Mathematics.

Among other awards, Professor Zayed has received a Meritorious Performance and Professional Promise Award and an Exceptional Merit Service Award.

Professor Zayed has presented over 20 invited lectures at international meetings, over 20 lectures at national and international meetings, and approximately 30 colloquium lectures at universities and research institutes in the United States, Canada, Japan, Europe, and the Middle East.

Professor Zayed has published over 50 research papers in different areas of Mathematics, including the theory of generalized functions and hyperfunctions, analytic function theory, special functions and orthogonal polynomials, differential equations and boundary-value problems, integral and function transforms, sampling theory, and wavelets. His current research interests include function transform techniques, sampling theory, and wavelets.

TABLE OF CONTENTS

1

INTRODUCTION AND A HISTORICAL OVERVIEW

1.1 A Historical Overview

One of the most important mathematical techniques used in communication engineering and information theory is *Sampling Theory*. Although most of its earliest applications were in these two fields, nowadays the applications of sampling theory permeate many branches of physics and engineering, such as signal analysis, image processing, radar, sonar, acoustics, optics, holography, meteorology, oceanography, crystallography, physical chemistry, medical imaging and many more. An extensive list of such applications containing over 800 references can be found in [12]; see also [16] for recent references.

Generally speaking, sampling theory can be used in any discipline where functions need to be reconstructed from sampled data, usually from the values of the functions or their derivatives at certain points. This naturally makes sampling theory an offspring of interpolation and approximation theory; yet, a closer look at its development reveals more intimate ties with the theory of entire functions. Admittedly, not many people may agree on what falls under the scope of sampling theory; however, it may seem fair to say that sampling theory is that branch of mathematics that deals with the reconstruction of members of a certain class of entire functions, commonly known as band-limited functions, from the values of either the functions themselves or some transformations thereof (such as their derivatives or their Hilbert transforms, etc.) at a discrete set of points.

Throughout its relatively short history, sampling theory has lived in the shadow of other more prominent and traditional areas of mathematical analysis, such as Approximation Theory, the Theory of Entire Functions, and the Theory of Fourier Series and Integrals, invariably borrowing ideas and techniques from these areas. Maybe that is why, despite its widespread use in physics and engineering, sampling theory has attracted only moderate attention in the mathematical community.

Unlike many other branches of applied mathematics, many of the fundamental mathematical results in sampling theory have been discovered as much by engineers as by mathematicians. In fact, most of the research articles have appeared in engineering literature.

Sampling theory as we know it now is about forty-five years old, but its mathematical roots can be traced back to the work of some great mathematicians, such as Poisson [19], Borel [1-3], Hadamard [10], La Vallée Poussin [25], and E. T. Whittaker [27]. It was even believed that the fundamental result in sampling theory was dated back to the work of Cauchy [5]; however, in his delightful article "Five short stories about the cardinal series," J. Higgins [11] contended that such a connection with the work of Cauchy could not be firmly substantiated.

The fundamental result in sampling theory, which we shall call *the sampling theorem* for the time being, states that if a signal $f(t)$ contains no frequencies higher than $W/2$ cycles per second, it is completely determined by giving its ordinates at a sequence of points spaced $1/W$ seconds apart, say $t_n = n/W, n = 0, \pm 1, \pm 2, \ldots$, and can be reconstructed from these ordinates, via the formula

$$f(t) = \sum_{n=-\infty}^{\infty} f\left(\frac{n}{W}\right) \frac{\sin \pi(Wt-n)}{\pi(Wt-n)}. \tag{1.1.1}$$

An important underlying engineering principle here is that all the information contained in such a signal is, in fact, contained in the sample values that are taken at equidistantly spaced instants. Moreover, the knowledge of the frequency bound determines the minimum rate at which the signal needs to be sampled in order to reconstruct it completely. This minimum rate is known as the Nyquist rate in reference to H. Nyquist who was the first to point out to its importance in connection with telegraphy [17].

There is almost a universal agreement that the sampling theorem was first discovered by a mathematician, E. T. Whittaker, in 1915 [27] in his study of the *Cardinal Functions (Series)* and was, further, developed by his second son J. M. Whittaker [28-30]. However, W. Ferrar claimed [9] (see also Higgins [11]) that another mathematician, F. J. W. Whipple, had actually discovered it five years before E. T. Whittaker, but did not publish his findings. If this is indeed the case, then this had set a precedent for another situation in sampling theory wherein a result was first discovered, announced but not published by one person, P. Weiss [26], and then it was rediscovered, modified and published by another person, H. Kramer, who justly or unjustly received all the credit [14].

In a recent article [4], P. Butzer and R. Stens drew the attention to a rather obscure paper [18] that was published in 1920 by the Japanese mathematician K. Ogura, in which he stated the sampling theorem in a form similar to its familiar form and sketched a simple and rigorous proof for it using the calculus of residues. This makes him, according to Butzer and Stens, the first mathematician ever to have done so. They have also stated that Ogura, however, had *erroneously* (sic) attributed his result to E. T. Whittaker. Their statement may appear a little bit perplexing at first since Ogura's paper was published five years after Whittaker's. To clarify this point and let the reader decide for himself/herself on who was the first to obtain the sampling theorem, we shall say here a few words about both Whittaker's and Ogura's results, then elaborate a little bit further on that in Chapter 2.

E. T. Whittaker began his paper by considering an arbitrary, but given, entire function $f(t)$ and then constructed the right hand side of (1.1.1), which he called the cardinal function, as an entire function that agrees with $f(t)$ at the sample points $\{t_n = n/W\}_{n=-\infty}^{\infty}$. Actually, he considered a more general type of sample points, namely, $\{\tau_n = a + nW\}_{n=-\infty}^{\infty}$, but it can be easily shown, by using a simple change of variables, that the two cardinal functions are equivalent. He, then, showed that the cardinal function is an entire function with no rapid oscillations, i.e., band-limited in our modern terminology. However, nowhere in his paper did E. T. Whittaker mention that the cardinal function in (1.1.1) was equal to $f(t)$. On the contrary, he said that there are many entire functions, which he called cotabular functions, that agree with f at the points $t_n = n/W$, $n = 0, \pm1, \pm2, \ldots$, but he singled out the cardinal function as the "*simplest*" cotabular function.

Ogura, on the other hand, started out by defining the cardinal function as an entire function $f(z)$ of exponential type that grows no faster than $\exp(\pi r \mid \sin\theta \mid)$ as z approaches infinity, where $z = re^{i\theta}$. He, then, said that one of the most important properties of the cardinal functions is that they can be constructed analytically from their values at the integers via (1.1.1) (with $W = 1$) as done in E. T. Whittaker's paper [27]! We shall return and discuss this controversy in more detail in Chapter 2, Section 1.

At any rate, the Whittakers' study was purely mathematical and clearly neither one of them had any specific applications in mind. They, of course, did not use the engineering words *signal* or *cycle* or *bounded frequencies*, but used their mathematical counterparts that we shall explain shortly. E. T. Whittaker's result was later rediscovered and introduced in information theory and communication engineering by C. E. Shannon in 1940, though it did not actually appear in the engineering literature until after World War II in 1949 [21]. In his two famous papers [21 and 22], "A mathematical theory of communication" and "Communication in the presence of noise," which won him several awards for his contribution in communication theory, Shannon acknowledged the work of E. T. Whittaker.

Shannon has also mentioned that other sets of data, such as samples of the signal, its derivatives and samples taken at non-uniformly distributed instants, can be used to reconstruct a signal with bounded frequencies.

In the late fifties, it became known in the western world that Shannon's result had been discovered independently in 1933 by a Russian engineer, V. Kotel'nikov [13], who applied it in communication engineering earlier than Shannon, and it was known by his name in the Russian and eastern European literature. Later, it was also reported that a Japanese engineer, I. Someya [24], had independently discovered this result about the same time Shannon did.

Although this fundamental result is commonly known in the western engineering circles as Shannon's sampling theorem, hereafter we shall call it the Whittaker-Shannon-Kotel'nikov sampling theorem or the WSK sampling theorem for short.

This brief history of the WSK sampling theorem is reminiscent of the history of the Radon transform, which enables one to reconstruct a suitable function of two variables from the knowledge of its integrals over all possible straight lines. The Radon transform was first discovered by a mathematician, J. Radon in 1917 [20], again with no specific applications in mind. His work passed unnoticed by mathematicians as well as physicists and engineers. Then it was later rediscovered over and over by many people in various fields of science and engineering. These rediscoveries ended in the seventies when a Russian, I. Shtein [23], and an American, A. Cormack [7], pointed out the original work of Radon. The 1979 Nobel Prize in physiology and medicine was shared by G. Hounsfield, a British engineer, and A. Cormack, a naturalized American physicist, for their work on Computerized Aided Tomography (CAT) in which they employed Radon's ideas. In their Nobel Prize addresses [6], they acknowledged the work of other pioneers in the field including Radon's.

Interestingly enough, the Radon transform and the WSK sampling theorem are related not only from a historical point of view, but from a mathematical one as well [8].

1.2 Introduction and Terminology

In interdisciplinary subjects, such as sampling theory, it is not uncommon for people in different disciplines to use different terminology causing occasional ambiguity and misunderstanding. To alleviate the potential for any such occurrence, we will start by mathematically defining the engineering terms that will be used throughout this monograph.

The most important term in communication engineering and signal analysis is the term *signal*. Herein, the word signal will always mean a *continuous (analog) signal* which is to be distinguished from a *discrete (digital) signal.*

Mathematically, a signal is nothing but a function $f(t)$, whether real or complex-valued. Throughout this monograph we will use the terms *signal* and *function* interchangeably. Therefore, phrases like bounded, or analytic or periodic signals should be self-explanatory. A multi-dimensional signal is a function of several variables. Physically, a signal $f(t)$ may represent the voltage difference at time t between two points in an electrical circuit, e.g., the voltage difference between the terminals of a microphone, or it may represent sound pressure or magnetic field strength. A digital signal, on the other hand, is just a sequence of numbers $\{f(n)\}_{n=-\infty}^{\infty}$. To see why signals are important, we need to know what a communication system is.

A communication system consists mainly of a transmitter, communication channel and a receiver. Its purpose is to send a message from the transmitter to the receiver. The message may consist of written or spoken words, or of pictures, or of sounds, etc. The transmitter changes this message into a signal that is actually sent over the communication channel to the receiver. The communication channel, which is merely the medium used to transmit the signal from the transmitter to the receiver, may be a wire as in telephone lines, or the atmosphere as in radio and television. The receiver is in a way an inverse transmitter, changing the transmitted signal back into a message. For example, in telephones the transmitter changes the sound pressure of the voice into a varying electrical current forming a signal that is sent through the telephone lines to the receiver which reverses the process.

In the process of being transmitted, the signal may acquire certain changes that may alter the content of the sender's original message by the time it arrives at the receiver. These unwanted changes may be distortions of sound, or static as in telephones and radios, or distortions in shape or shading of picture as in television. These changes fall into two main categories: distortions and noise. Distortion is a fixed operation applied to the signal and can, in principle, be corrected by applying the inverse operation, while noises involve statistical and unpredictable perturbations that cannot always be corrected.

A distortionless transmission of a signal through a circuit or a system means that the exact shape of the input signal is reproduced at the output regardless of whether or not the exact amplitude of the signal is preserved or the signal is delayed in time within reasonable limits. If the input and output signals are denoted by $f(t)$ and $g(t)$ respectively, then for distortionless transmission

$$g(t) = Lf(t) = Af(t - t_0),$$

where L is a linear, time-invariant operator.

By taking the Fourier transform of both sides of this equation, we obtain

$$G(\omega) = H(\omega)F(\omega),$$

where F, G are the Fourier transforms of f, g respectively and $H(\omega) = A\, e^{it_0\omega}$. $H(\omega)$ is called the system transfer function, or the system function for short, and its inverse Fourier transform $h(t)$ is called the impulse response of the system.

The word "filter" in electric engineering means a circuit or a system that has some frequency selective mechanism. Filters are usually characterized as low pass, high pass, band pass or band stop. The system transfer functions of the ideal types of these filters are:

Low pass filter $$H(\omega) = \begin{cases} A\, e^{it_0\omega}, & \text{for } |\omega| \le \omega_1 \\ 0, & \text{otherwise} \end{cases},$$

where ω_1 is the cut-off frequency;

High pass filter $$H(\omega) = \begin{cases} A\, e^{it_0\omega}, & \text{for } |\omega| \ge \omega_2 \\ 0, & \text{otherwise} \end{cases},$$

Band pass filter $$H(\omega) = \begin{cases} A\, e^{it_0\omega}, & \text{for } \omega_2 \le |\omega| \le \omega_1 \\ 0, & \text{otherwise} \end{cases},$$

Band stop filter $$H(\omega) = \begin{cases} 0, & \text{for } \omega_2 \le |\omega| \le \omega_1 \\ A\, e^{it_0\omega}, & \text{otherwise} \end{cases}.$$

By processing a signal $f(t)$ we mean operating on it in some fashion either to change its shape, configuration and properties or to extract some useful information. And in most cases this operation is required to be reversible. Mathematically, this amounts to applying some transformation (not necessarily linear) to the function $f(t)$ and requiring that the transformation be invertible. Sometimes for practical and economical reasons, only certain data extracted from the signal are transmitted and then used at the receiver to reconstruct the signal. This is where sampling theory comes into play.

By sampling a signal $f(t)$ at a discrete set of points, we are actually applying a linear transformation to it to convert it into a digital signal $\{f(t_n)\}_{n=-\infty}^{\infty}$. The WSK sampling theorem can now be viewed as a way to convert this digital signal back into the original signal, provided that the latter has bounded frequency content.

The energy of a signal f is defined by

$$E = \int_{-\infty}^{\infty} |f(t)|^2\, dt\,.$$

It is an important physical consideration to assume that signals have finite energies. That is

$$E = \int_{-\infty}^{\infty} |f(t)|^2 \, dt < \infty .$$

Therefore, the space of all finite-energy signals is the same as the familiar space $L^2(\mathfrak{R})$. For two finite energy signals f and g, the cross-correlation function $R_{f,g}(t)$ is defined as

$$R_{f,g}(t) = \frac{1}{\sqrt{2\pi}} \int_{-\infty}^{\infty} \bar{f}(\tau) g(t+\tau) \, d\tau ,$$

and the autocorrelation function of f is defined as

$$R_f(t) = R_{f,f}(t) = \frac{1}{\sqrt{2\pi}} \int_{-\infty}^{\infty} \bar{f}(\tau) f(t+\tau) \, d\tau ,$$

where $\bar{f}$ denotes the complex conjugate of f. The ambiguity function of f is defined as

$$A(t,\omega) = \frac{1}{E} \int_{-\infty}^{\infty} f\left(\tau + \frac{t}{2}\right) \bar{f}\left(\tau - \frac{t}{2}\right) e^{i\omega\tau} d\tau ,$$

in which the normalization constant has been chosen so that $A(0,0) = 1$.

The average power (energy) of a signal f over the interval (a,b) is defined as

$$\frac{1}{(b-a)} \int_a^b |f(t)|^2 \, dt ,$$

and its average power $\tilde{E}$ (over the whole real line $\mathfrak{R}$) is defined as

$$\tilde{E} = \lim_{T \to \infty} \frac{1}{2T} \int_{-T}^{T} |f(t)|^2 \, dt .$$

A signal is said to be of finite power if $\tilde{E}$ is finite. The Fourier transform

$$F(\omega) = \hat{f}(\omega) = \frac{1}{\sqrt{2\pi}} \int_{-\infty}^{\infty} f(t) e^{it\omega} dt ,$$

of a signal f is known as the *amplitude spectrum* of the signal and is of a special importance in engineering analysis. It represents in some sense the frequency content of the signal. Knowing the amplitude spectrum of the signal allows engineers to regard the signal as a sum of sinusoids of different frequencies via its inverse Fourier transform

$$f(t) = \frac{1}{\sqrt{2\pi}} \int_{-\infty}^{\infty} F(\omega) e^{-it\omega} d\omega .$$

In virtue of Parseval's relation, the energy of a signal can be given by

$$E = \int_{-\infty}^{\infty} |f(t)|^2 dt = \int_{-\infty}^{\infty} |F(\omega)|^2 d\omega .$$

It is easy to verify that

$$\hat{R}_{f,g}(\omega) = \bar{\hat{f}}(\omega)\hat{g}(\omega), \quad \text{and} \quad \hat{R}_f(\omega) = |\hat{f}(\omega)|^2 ;$$

hence

$$\int_{-\infty}^{\infty} \hat{R}_f(\omega) d\omega = E .$$

Moreover,

$$f\left(\tau + \frac{t}{2}\right) \bar{f}\left(\tau - \frac{t}{2}\right) = \frac{E}{2\pi} \int_{-\infty}^{\infty} A(t,\omega) e^{-i\tau\omega} d\omega ,$$

which, in view of Parseval's relation, leads to

$$\frac{1}{2\pi} \int_{-\infty}^{\infty} \int_{-\infty}^{\infty} |A(t,\omega)|^2 dt\, d\omega = 1 .$$

We may regard $f(t)$ and $F(\omega)$ as two representations of the same signal, one in the time domain and the other in the frequency domain. It is sometimes useful to view a signal in both the time and frequency domains simultaneously, but this requires more sophisticated mathematical techniques than the standard Fourier transform analysis. Some of these techniques, including the windowed Fourier transform, the Zak transform, the Gabor frames and wavelets, will be discussed in Chapter 10.

A finite-energy signal is said to be band-limited if its amplitude spectrum (its Fourier transform) vanishes outside an interval of the form $(-W, W)$, and the smallest such W is called the bandwidth of the signal. Clearly, such a signal has a finite-limit Fourier transform and can be written in the form

$$f(t) = \frac{1}{\sqrt{2\pi}} \int_{-W}^{W} F(\omega) e^{-it\omega} d\omega ,$$

with $F \in L^2(-W, W)$.

Band-limited signals are abundant in the physical world, for example a human voice usually has no frequencies higher than 8000 hertz (cycles per second), while conventional orchestral music has no frequencies higher than 20,000 hertz.

Analogously, a signal is said to be time-limited if $f(t)$ itself vanishes outside some interval of the form $(-T, T)$. It is a well-known fact that if a signal is band-limited, then its time representation $f(t)$ will trail indefinitely, even though it may become very small. It is, therefore, impossible to have a signal that is both band-limited and time-limited simultaneously, unless it is identically zero.

Nevertheless, it is possible to have a signal that is *almost* both band-limited and time-limited. In fact, most communication signals are of this type. To be more precise, if we denote the approximate time duration of the signal by T and its approximate bandwidth by W, then such signals are characterized by $2TW \geq 1$. This is known as the uncertainty principle in sampling theory, and on which we shall elaborate a little further at the end of Chapter 2. Shannon has indicated that any such signal can be determined everywhere to a high degree of accuracy by its values at $2TW$ sampling points spaced at time locations $1/2W$ apart.

It is also well known that band-limited signals are entire functions, and since most of our investigation will be dealing with such signals, we shall recall some basic definitions and facts pertaining to entire functions.

An entire function $f(z)$ is a function of complex variable analytic in the entire complex plane and consequently can be represented by a power series

$$f(z) = \sum_{n=0}^{\infty} a_n z^n ,$$

that converges everywhere, where $a_n = f^{(n)}(0)/n!$.

Let

$$M_f(r) = \max_{|z|=r} |f(z)| .$$

It follows from the maximum modulus principle that $M_f(r)$ is an increasing function of r.

An entire function $f(z)$ is said to be of finite order if there exists a positive constant k such that the inequality

$$M_f(r) < e^{r^k}$$

is valid for all sufficiently large values of r.

The greatest lower bound of such numbers k is called the order of the entire function. If ρ is the order of the entire function $f(z)$, then it can be shown [15] that

$$\rho = \limsup_{r \to \infty} \frac{\ln \ln M_f(r)}{\ln r}.$$

By the type σ of an entire function $f(z)$ of order ρ we mean the greatest lower bound of positive numbers C such that

$$M_f(r) < \exp(C r^\rho),$$

for sufficiently large r.

Similarly, it can be shown that

$$\sigma = \limsup_{r \to \infty} \frac{\ln M_f(r)}{r^\rho}.$$

An entire function of order one is called an entire function of exponential type. An example of an entire function of order m and type σ is the function $f(z) = \exp(\sigma z^m)$. The function, $f(z) = \sin \pi z$, is an entire function of exponential type and has type $\sigma = \pi$, while $f(z) = \cos\sqrt{z}$ has order $\rho = 1/2$ and type $\sigma = 1$.

The following important theorem will be used in subsequent chapters:

THEOREM 1.1 (Hadamard [15, p. 24])

Any entire function $f(z)$ of finite order ρ can be represented in the form

$$f(z) = z^m e^{P(z)} \prod_{n=1}^{\infty} G\left(\frac{z}{z_n}; p\right),$$

where z_n are the non zero roots of $f(z)$, $p (p \le \rho)$ is the smallest integer for which the series

$$\sum_{n=1}^{\infty} \frac{1}{|z_n|^{p+1}}$$

converges, $P(z)$ is a polynomial whose degree does not exceed ρ, m is the multiplicity of the zero at the origin and

$$G(u;p) = (1-u)\exp\left(u + \frac{u^2}{2} + \ldots + \frac{u^p}{p}\right), \quad G(u;0) = (1-u).$$

This theorem is known as Hadamard's factorization theorem for entire functions. The following representation for $\sin \pi z$ will also be needed

$$\sin \pi z = \pi z \prod_{n=1}^{\infty}\left(1 - \frac{z^2}{n^2}\right).$$

Most of the theoretical research in sampling theory has centered around generalizing and extending the WSK sampling theorem in different directions, such as extending it to include non-uniformly distributed sampling points, sampling of signals given by integral transforms other than the Fourier one, sampling of multidimensional signals, etc. Even the notion of band-limited signals has been the subject of intensive research that resulted in new and more general definitions of band-limited signals. The connection between the WSK sampling theorem and other branches of mathematical analysis has also been the focus of some research articles. On the practical side, the research has centered around finding new applications of the WSK sampling theorem and developing fast algorithms to reconstruct the signals.

In the next two chapters we shall review some of these generalizations and extensions of the WSK sampling theorem with two objectives in mind: first, to give the reader who is unfamiliar with the subject, a panoramic view of the field before getting involved in technical details in the following chapters; second, to set the stage for introducing some new results and show where they fit in the general framework of sampling theory.

The notations we shall adopt are mostly standard. The set of integers will be denoted by Z, the real numbers by $\Re$ and the complex numbers by **C**. The Fourier transform of $f(t)$ will be denoted by either $\hat{f}(\omega)$ or $F(\omega)$, and the convolution h of two functions f and g will be defined by

$$h(t) = (f * g)(t) = \frac{1}{\sqrt{2\pi}} \int_{-\infty}^{\infty} f(x) g(t-x)\, dx,$$

so that

$$\hat{h}(\omega) = (f * g)^{\wedge}(\omega) = \hat{f}(\omega)\hat{g}(\omega).$$

$\chi_A(t)$ will stand for the characteristic function of the set A and $C^k(I)$ for the set of all functions having k continuous derivatives on some given open interval I. If $I = \Re$, we write $C^k(\Re)$ or C^k for short.

References

1. E. Borel, Mémoire sur les séries divergentes, *Ann. École Norm. Sup.*, (3) 16 (1899), 9-131.

2. ______, Sur la recherche des singularités d'une fonction définie par un développement de Taylor, *C. R. Acad. Sci. Paris*, 127 (1898), 1001-1003.

3. ______, Sur l'interpolation, *C. R. Acad. Sci. Paris*, 124 (1897), 673-676.

4. P. L. Butzer and R. L. Stens, Sampling theory for not necessarily band-limited functions: A historical overview, *SIAM Review*, 34, 1 (1992), 40-53.

5. A. L. Cauchy, Mémoire sur diverses formules d'analyse, *C. R. Acad. Sci. Paris*, 12 (1841), 283-298.

6. A. M. Cormack, Nobel Prize Address, Dec. 8, 1979, Early two-dimensional reconstruction and recent topics stemming from it, *Med. Phys.*, 7 (1980), 227-282. Also, in *J. Computed Assisted Tomog.*, 4 (1980), 658-664.

7. ______, Reconstruction of densities from their projections with applications in radiological physics, *Phys. Med. Biol.*, 18 (1973), 195-207.

8. S. R. Deans, *The Radon Transform and Some of Its Applications*, John Wiley & Sons, New York (1983).

9. W. L. Ferrar, On the cardinal function of interpolation theory, *Proc. Roy. Soc. Edinburgh*, 46 (1926), 323-333.

10. J. Hadamard, La série de Taylor et son prolongement analytique, *Scientia*, 12 (1901), 1-100.

11. J. Higgins, Five short stories about the cardinal series, *Bull. Amer. Math. Soc.*, 12 (1985), 45-89.

12. A. J. Jerri, Part II: The sampling expansion—A detailed bibliography. Monograph, Clarkson University (1986).

13. V. Kotel'nikov, On the carrying capacity of the "ether" and wire in telecommunications, material for the first All-Union Conference on Questions of Communications, *Izd. Red. Upr. Svyazi RKKA*, Moscow, Russian (1933).

14. H. Kramer, A generalized sampling theorem, *J. Math. Phys.*, 38 (1959), 68-72.

15. B. Levin, *Distribution of Zeros of Entire Functions*, Translations of Mathematical Monographs Ser., Vol. 5, Amer. Math. Soc., Providence, Rhode Island (1964).

16. R. J. Marks II, Ed., *Advanced Topics on Shannon Sampling and Interpolation Theory*, Springer-Verlag, New York (1992).

17. H. Nyquist, Certain topics in telegraph transmission theory, *AIEE Trans.*, 47 (1928), 617-644.

18. K. Ogura, On a certain transcendental integral function in the theory of interpolation, *Tôhoku Math. J.*, 17 (1920), 64-72.

19. S. D. Poisson, Mémoire sur la maniere d'exprimer les fonctions, par des séries de quantités périodiques, et sur l'usage de cette transformation dans la résolution de différens problemes, *J. École Roy. Polytechnique*, 11 (1820), 417-489.

20. J. Radon, Über die Bestimmung von Funktionen durch ihre Integralwerte längs gewissen Mannigfaltigkeiten, Berichte Sächsische Akademie der Wissenschaften. *Leipzig, Math.-Phys. KL.*, 69 (1917), 262-267.

21. C. E. Shannon, Communication in the presence of noise, *Proc. IRE*, 137 (1949), 10-21.

22. ______, A mathematical theory of communication, *Bell System Tech. J.*, 27 (1948), 379-423.

23. I. N. Shtein, On the applications of the Radon transform in holographic interferometry, Radiotekh, *Elektron*, 17 (1972), 2436-2437.

24. I. Someya, *Waveform Transmission*, Shukyo, Tokyo (1949).

25. Ch. J. de la Vallée Poussin, Sur la convergence des formules d'interpolation entre ordonnées equidistantes, *Acad. Roy. Belg. Bull.*, C1. Sci. 1 (1908), 319-410.

26. P. Weiss, Sampling theorems associated with Sturm-Liouville systems, *Bull. Amer. Math. Soc.*, 63 (1957), 242.

27. E. T. Whittaker, On the functions which are represented by the expansion of the interpolation theory, *Proc. Roy. Soc. Edinburgh*, sec. A, 35 (1915), 181-194.

28. J. M. Whittaker, *Interpolatory Function Theory*, Cambridge University Press, Cambridge, England (1935).

29. ______, On the Fourier theory of the cardinal function, *Proc. Edinburgh Math. Soc.*, 1 (1929)b, 169-176.

30. ______, On the cardinal function of interpolation theory, *Proc. Edinburgh Math. Soc.*, 1 (1929)a, 41-46.

2

SHANNON SAMPLING THEOREM AND BAND-LIMITED SIGNALS

2.0 Introduction

The main goal of this chapter is to introduce Shannon's Sampling Theorem and some of its extensions pertaining to various classes of band-limited signals.

There was a time when almost every scholar working in the field of Sampling Theory agreed on what the term "band-limited signal" meant. The standard definition of band-limited signals at that time was the one introduced in Chapter 1.

Nowadays there does not seem to be a universal agreement, especially among mathematicians, on what a band-limited signal really is. This is not surprising since in the course of its evolution during the last four decades sampling theory has been enriched by several new mathematical discoveries, among them is the discovery of new classes of entire functions that resemble in their intrinsic properties the conventional class of band-limited signals; hence, they ought to be called band-limited signals as well.

In Section 1, we shall introduce the classical Shannon's sampling theorem as it applies to the conventional class of band-limited signals. In Section 2, we shall show how this conventional class of band-limited signals can be enlarged to contain more general types of signals, then we shall discuss the applicability of Shannon's theorem to these new types.

All the main results in this chapter are known; therefore, we shall omit most of the proofs, except for some short, but important ones.

2.1 Shannon Sampling Theorem and the Cardinal Series

2.1.A Shannon Sampling Theorem

The fundamental result in Sampling Theory is the celebrated Whittaker-Shannon-Kotel'nikov (WSK) sampling theorem, which in Shannon's original wording reads as follows [22]:

"If a function of time is limited to the band from 0 to W cycles per second, it is completely determined by giving its ordinates at a series of discrete points spaced $1/2W$ seconds apart in the manner indicated by the following result: If $f(t)$ has no frequencies over W cycles per second, then

$$f(t) = \sum_{n=-\infty}^{\infty} f\left(\frac{n}{2W}\right)\frac{\sin\pi(2Wt-n)}{\pi(2Wt-n)}\text{."}$$

Although this result is known in communication and electrical engineering literature as Shannon's sampling theorem, we shall call it from now on, for reasons explained in Chapter 1, the WSK sampling theorem. Shannon's result can be rephrased as follows.

THEOREM 2.1 (Whittaker-Shannon-Kotel'nikov)

If $f(t)$ is a signal (function) band-limited to $[-\sigma, \sigma]$, i.e.,

$$f(t) = \frac{1}{\sqrt{2\pi}}\int_{-\sigma}^{\sigma} F(\omega)e^{it\omega}d\omega, \tag{2.1.1}$$

for some $F \in L^2(-\sigma, \sigma)$, then it can be reconstructed from its sampled values at the points $t_k = k\pi/\sigma$, $k = 0, \pm 1, \pm 2, \ldots$, via the formula

$$f(t) = \sum_{k=-\infty}^{\infty} f(t_k)\frac{\sin\sigma(t-t_k)}{\sigma(t-t_k)}, t \in \Re \tag{2.1.2}$$

with the series being absolutely and uniformly convergent on compact sets.

We shall call the points $\{t_k\}_{k=-\infty}^{\infty}$ the sampling points and the functions

$$S_k(t) := \frac{\sin\sigma(t-t_k)}{\sigma(t-t_k)}, \tag{2.1.3}$$

the sampling functions. The sampling frequency σ/π in (2.1.2) is known as the Nyquist rate, which is the minimum rate at which the signal needs to be sampled in order to reconstruct it completely.

There are several proofs of this theorem (see [4 and 10] for some of them); the one we are about to present may not be the most elegant, but probably the most straightforward.

Proof. The idea is to extend $F(\omega)$ and $e^{it\omega}$ periodically to the entire real line $\Re$ as functions with period 2σ, then expand them in Fourier series. To simplify the notation, we shall not make a distinction between F and its periodic extension. We then have

$$F(\omega) = \sum_{n=-\infty}^{\infty} \hat{F}(n) e^{-in\omega\pi/\sigma},$$

where

$$\hat{F}(n) = \frac{1}{2\sigma} \int_{-\sigma}^{\sigma} F(\omega) e^{in\omega\pi/\sigma} d\omega,$$

and

$$e^{it\omega} = \sum_{n=-\infty}^{\infty} \frac{\sin\sigma(t-t_n)}{\sigma(t-t_n)} e^{in\omega\pi/\sigma}, \quad |\omega|\sigma.$$

By applying Parseval's equality to the right-hand side of (2.1.1), we obtain

$$f(t) = \frac{2\sigma}{\sqrt{2\pi}} \sum_{n=-\infty}^{\infty} \hat{F}(n) \frac{\sin\sigma(t-t_n)}{\sigma(t-t_n)}, \tag{2.1.4}$$

which, in view of the fact that

$$f(t_n) = \frac{2\sigma}{\sqrt{2\pi}} \hat{F}(n),$$

implies (2.1.2).

The convergence in (2.1.4) is understood to be in the sense of $L^2(\Re)$; however, the series in (2.1.2) is readily seen to converge absolutely and uniformly on any compact subset of $\Re$ when we apply the Cauchy-Schwarz inequality to (2.1.4). Note that because F belongs to $L^2(-\sigma, \sigma)$, $\Sigma_{n=-\infty}^{\infty} |F(n)|^2 < \infty$. ∎

The uniform convergence of the series (2.1.2) on compact subsets of the complex t-plane can be established just as easily.

If f is band-limited to an interval that is not symmetric about the origin, say to an interval of the form $[\omega_0 - \sigma, \omega_0 + \sigma]$, then one can easily show that the sampling series (2.1.2) will take the form

$$f(t) = \sum_{k=-\infty}^{\infty} f(t_k) \frac{\sin\sigma(t-t_k)}{\sigma(t-t_k)} \exp(i\omega_0(t-t_k)). \tag{2.1.5}$$

2.1.B The Cardinal Series and Whittaker's Cardinal Function

The sampling series on the right hand side of (2.1.2), which can be written in the form

$$\sin \sigma t \sum_{k=-\infty}^{\infty} f(t_k) \frac{(-1)^k}{(\sigma t - k\pi)},$$

is a special case of a more general type of series known as *cardinal series*, which are series of the form

$$\sin \sigma t \sum_{k=-\infty}^{\infty} C_k \frac{(-1)^k}{(\sigma t - k\pi)}.$$

Cardinal series were first studied in 1915 by E. T. Whittaker in [24], where he began with the assumption that $f(z)$ was a given entire function with values at the equidistant points $z_n = a + nW$, $n = 0, \pm 1, \pm 2, \ldots$, denoted by $f_n = f(a + nW)$. He then noted that there were many entire functions, which he called cotabular functions, coinciding with f at the points $\{z_n\}_{n=-\infty}^{\infty}$. For example, if g is an entire function satisfying $g(z_n) = 0$ for all n, then $h(z) = f(z) + g(z)$ is a function cotabular with f.

Now let us denote $(f_1 - f_0)$ by $\Delta f_{1/2}$, $(f_0 - f_{-1})$ by $\Delta f_{-1/2}$, $(\Delta f_{1/2} - \Delta f_{-1/2})$ by $\Delta^2 f_0$, etc. Whittaker raised the question of whether it was possible to construct analytically an entire function $C(z)$ that is both cotabular with f and representable by the Gregory-Newton (also known as the Gauss) series

$$f_0 + z\Delta f_{1/2} + \frac{z(z-1)}{2!}\Delta^2 f_0 + \frac{(z+1)z(z-1)}{3!}\Delta^3 f_{1/2} + \ldots, \tag{2.1.6}$$

which was known at that time that when it converged, it was *supposed* to converge to f. He then answered his question in the affirmative by constructing $C(z)$ as

$$C(z) = \sum_{n=-\infty}^{\infty} f(a + nW) \frac{\sin(\pi(z - a - nW)/W)}{(\pi(z - a - nW)/W)}. \tag{2.1.7}$$

Moreover, it was shown that $C(z)$, which Whittaker called the cardinal function of the cotabular set of f, is an entire function with no periodic constituents of period less than $2W$ (band-limited in our modern terminology). In the proof it was shown that if the cardinal series converges, then so does the series (2.1.6) and the two series converge to the same function. However, nowhere in this paper did Whittaker mention explicitly that $C(z)$ was equal to $f(z)$ or state any conditions under which they would be equal. Whittaker also said that among all the functions cotabular with f, $C(z)$ is the "*simplest.*"

Actually, E. T. Whittaker did not call the series (2.1.7) the cardinal series, but it was his son J. M. Whittaker who gave it this name [25-27]. It should be pointed out that E. T. Whittaker's paper [24] was not very accurate by today's standards as we shall see shortly. For a history of the cardinal series and some of their properties, we refer the reader to Higgins' article "*Five Short Stories About the Cardinal Series*" [9], in which he tried to unravel some mysteries surrounding the origin of the cardinal series.

Another mystery that Higgins would have probably liked to include in his article just came to light a few months ago in an article by P. Butzer and R. Stens [2], in which they gave credit to the Japanese mathematician K. Ogura for being the first mathematician ever to have stated and sketched a rigorous proof for the sampling theorem in the form we know it now. As promised in Chapter 1, we shall now shed some light on the connection between the work of E. T. Whittaker and K. Ogura.

In Whittaker's paper [24], two examples were given, in the first f assumed the values:

$$f(0)=0\,,\quad f(n)=(-1)^n/n\,,\quad f(-n)=(-1)^{n+1}/n\,,\quad n=1,2,3,\ldots\,,$$

and in the second f assumed the values:

$$f(a)=0\,,\quad f(a+w)=1\,,\quad f(a+2w)=1\,,\quad f(a+3w)=0\,,$$

$$f(a+4w)=-1\,,\quad f(a-w)=-1\,,\quad f(a-2w)=-1\,,\quad f(a-3w)=0\,,\quad f(a-4w)=1\,,$$

etc. The cardinal functions associated with these examples have been found to be

$$\frac{\cos\pi x}{x}-\frac{\sin\pi x}{\pi x^2}\quad\text{and}\quad\frac{2}{\sqrt{3}}\sin\left(\frac{\pi(x-a)}{3W}\right),$$

respectively. Whittaker noted that for his results to hold, the values $f(a+nw)$ of the entire function f need not go to zero as $|n|\to\infty$, as can be seen from his second example (cf. [24, p. 188]).

Capitalizing on this statement, Ogura started his article [17] by pointing to inaccuracies in Whittaker's paper, which he highlighted by showing that when f assumes the values:

$$f(0)=0\,,\quad f(n)=(-1)^{n+1}\,,\quad f(-n)=-f(n)\,,\quad n=1,2,3,\ldots\,,$$

the cardinal function cotabular with f is not entire since the defining cardinal series diverges at $z=1/2$. In fact, the cardinal series in this case can be *formally* given by

$$C(z)=\frac{\sin\pi z}{\pi}\sum_{n=1}^{\infty}\frac{2n}{n^2-z^2}\,.$$

To rectify this, Ogura, instead, defined the cardinal function $f(z)$ as an entire function of exponential type that grows no faster than $\exp(\pi r \mid \sin\theta \mid)$ as z approaches infinity, where $z = re^{i\theta}$. He, then, pointed out that one of the most important properties of the cardinal functions is that they can be constructed analytically from their values at the integers via (2.1.7) (with $W = 1$, $a = 0$), as was done in Whittaker's paper, i.e.,

$$f(z) = \sum_{n=-\infty}^{\infty} f(n) \frac{\sin \pi(z-n)}{\pi(z-n)}.$$

He also made the following observations: first, the series (2.1.6) may converge to an entire function, but this function may not in general be cardinal as in the case for $f(z) = e^z$, where

$$e^z = 1 + (e-1)\frac{z}{1!} + e^{-1}(e-1)^2 \frac{z(z-1)}{2!} + e^{-1}(e-1)^3 \frac{(z+1)z(z-1)}{3!} + \dots;$$

second, the function $f(z) = e^{az+b}$ is cardinal when and only when $a = i\beta$, β being real and $-\pi < \beta < \pi$.

The chief shortcoming in Whittaker's paper is the lack of a restriction on the growth of the entire function f, a fact of which Ogura, apparently, became aware. However, the two examples of cardinal functions given in Whittaker's paper, indeed, had the right sort of growth conditions and can be classified in our modern terminology as band-limited. These two functions belong, in the notation of Section 2.2, to the spaces $\mathbf{B}_\pi^2$ and $\mathbf{B}_{\pi/(3W)}^\infty$ of band-limited functions; see Theorems 2.2 and 2.3 below.

The work of E. T. Whittaker was later refined by his son J. M. Whittaker [25-28] who linked the cardinal functions to the finite-limit Fourier integral, hence bringing his result even closer to the present form of the sampling theorem. His two main results in this direction are (Theorems 16 and 17 in [25]) as follows.

His first result is that given any function $f(t)$ of the form

$$f(t) = \int_0^1 \cos(\pi t x)\, d\,\Phi(x) + \sin(\pi t x)\, d\,\Psi(x), \tag{2.1.8}$$

with Φ and Ψ being continuous functions on $[0, 1]$, then the series (2.1.7), with $W = 1$, $a = 0$, is $(C, 1)$ summable to f, that is

$$f(t) = \lim_{n\to\infty} \sum_{k=-n}^{n} \left(1 - \frac{|k|}{n+1}\right) f(k) \frac{\sin \pi(t-k)}{\pi(t-k)},$$

and if

$$\sum_{k=1}^{\infty} \frac{1}{k}(|a_k| + |a_{-k}|) < \infty ,$$

then the cardinal series

$$\sum_{k=-\infty}^{\infty} a_k \frac{\sin \pi (t-k)}{\pi (t-k)}$$

will converge absolutely to a function of the form (2.1.8). This result is weaker than the sampling theorem since every function band-limited to $[-\pi, \pi]$ can be written in the form (2.1.8) with

$$\Phi(x) = \sqrt{\frac{\pi}{2}} \int_{\pi}^{x} [f(\pi u) + f(-\pi u)]\, du ,$$

$$\Psi(x) = i\sqrt{\frac{\pi}{2}} \int_{-\pi}^{x} [f(\pi u) - f(-\pi u)]\, du ,$$

but the convergence of the cardinal series to $f(t)$ is replaced by $(C, 1)$ summability.

The second result of J. M. Whittaker is:

If

$$\sum_{k=2}^{\infty} (|a_k| + |a_{-k}|) \frac{\ln k}{k} < \infty ,$$

then the cardinal series defined by

$$C(t) = \sum_{k=-\infty}^{\infty} a_k \frac{\sin \pi (t-k)}{\pi (t-k)}$$

is absolutely convergent and for any real λ

$$C(t) = \sum_{k=-\infty}^{\infty} C(k+\lambda) \frac{\sin \pi (t-\lambda-k)}{\pi (t-\lambda-k)} ,$$

with the last series being absolutely convergent. This result was obtained earlier by Ferrar [7] under the stronger condition that

$$\sum_{n=-\infty}^{\infty} |a_n|^p < \infty , \quad \text{for some} \quad p > 1 .$$

One of the nicest properties of cardinal series is that if a cardinal series converges for a single noninteger value of t, it converges uniformly on any compact subset of the complex t-plane to an entire function of t.

2.1.C Oversampling

If a signal is band-limited to $[-\sigma,\sigma]$, it can also be considered as band-limited to $[-\delta,\delta]$ with $\delta>\sigma$. Thus, from the WSK sampling theorem, we have

$$f(t)=\sum_{k=-\infty}^{\infty} f(\eta_k)\frac{\sin\delta(t-\eta_k)}{\delta(t-\eta_k)}, \tag{2.1.9}$$

where $\eta_k = k\pi/\delta$.

But since

$$\hat{f}(\omega)=\hat{f}(\omega)\chi_{(-\sigma,\sigma)}(\omega),$$

we obtain by using (2.1.9) and the inverse Fourier transform

$$\begin{aligned} f(t) &= \left(f * \hat{\chi}_{(-\sigma,\sigma)}\right)(t)=\sqrt{2/\pi}\sum_{k=-\infty}^{\infty} f(\eta_k)(h_k * g)(t) \\ &= r\sum_{k=-\infty}^{\infty} f(\eta_k)\frac{\sin(\sigma t-rk\pi)}{(\sigma t-rk\pi)}, \end{aligned} \tag{2.1.10}$$

where

$$h_k(t)=\frac{\sin\delta(t-\eta_k)}{\delta(t-\eta_k)}, \quad g(t)=\frac{\sin\sigma t}{t}, \quad r=\frac{\sigma}{\delta}, \quad 0<r\le 1.$$

r is called the sampling rate parameter. When $r=1$, (2.1.10) reduces to (2.1.2). The sampling rate δ/π is larger than the Nyquist rate σ/π. This sampling procedure is known in engineering literature as oversampling. Oversampling is used frequently in practice since it can reduce interpolation noise levels due to noisy data [15, p. 112].

When a signal is oversampled, the sampled values become dependent. For example, when $t=\eta_m$ in (2.1.10) we obtain

$$f(\eta_m)=r\sum_{k=-\infty}^{\infty} f(\eta_k)\frac{\sin r\pi(m-k)}{r\pi(m-k)}=r\left\{f(\eta_m)+\sum_{\substack{k=-\infty \\ k\ne m}}^{\infty} f(\eta_k)\frac{\sin r\pi(m-k)}{r\pi(m-k)}\right\},$$

which implies

$$f(\eta_m) = \frac{r}{1-r} \sum_{\substack{k=-\infty \\ k \neq m}}^{\infty} f(\eta_k) \frac{\sin r\pi(m-k)}{r\pi(m-k)} .$$

The ramification of this last formula is that losing a single sampled value, when we oversample, will not affect the signal reconstruction since this missing value can be determined from the remaining ones. In fact, it can be shown that this phenomenon will hold even if we lose an arbitrarily large but finite number of sampled values [15]. And this is what makes oversampling a useful tool in applications.

2.2 More Band-Limited Functions

In this section we introduce more general classes of band-limited signals and discuss the applicability of the WSK sampling theorem to them.

For $\sigma \geq 0$ and $1 \leq p \leq \infty$, let $\mathbf{B}_\sigma^p$ denote the class of all entire functions f (on **C**) of exponential type at most σ belonging to $L^p(\mathfrak{R})$ when restricted to $\mathfrak{R}$; that is $f \in \mathbf{B}_\sigma^p$ if and only if

$$|f(z)| \leq \sup_{x \in \mathfrak{R}} |f(x)| \exp(\sigma |y|), \quad z \in \mathbf{C}$$

with

$$\int_{-\infty}^{\infty} |f(x)|^p dx < \infty \quad \text{if} \quad 1 \leq p < \infty, \quad \text{and} \quad \text{ess.}\sup_{x \in \mathfrak{R}} |f(x)| < \infty \quad \text{if} \quad p = \infty,$$

where $z = x + iy$.

By using the standard L^p-norm

$$\|f\|_p = \left(\int_{-\infty}^{\infty} |f(x)|^p dx \right)^{1/p} \quad \text{if} \quad 1 \leq p < \infty,$$

and

$$\|f\|_\infty = \text{ess.}\sup_{x \in \mathfrak{R}} |f(x)|, \quad \text{if} \quad p = \infty,$$

it can be shown that for $f \in \mathbf{B}_\sigma^p$, we have ([1, p. 102], [28, p. 99])

$$|f(x+iy)| \leq C \|f\|_p e^{\sigma |y|},$$

where C is a constant that depends only on p and σ.

The class $\mathbf{B}_\sigma^p(\sigma > 0)$ is nontrivial; however, the class $\mathbf{B}_0^p$ is trivial (it only contains the zero function), except for $p = \infty$.

For any $\alpha > 0$ and $f \in \mathbf{B}_\sigma^p$, we have [16, p. 123]

$$\|f\|_p \leq \sup_{t \in \Re} \left\{ \frac{\alpha}{\sqrt{2\pi}} \sum_{k=-\infty}^{\infty} | f(t - \alpha k) |^p \right\}^{1/p} \leq (1 + \alpha\sigma) \|f\|_p .$$

Unlike the classes $L^p(\Re)$, the classes $\mathbf{B}_\sigma^p$ behave very nicely under inclusion. In fact, the following inclusions hold

$$\mathbf{B}_\sigma^1 \subset \mathbf{B}_\sigma^p \subset \mathbf{B}_\sigma^q \subset \mathbf{B}_\sigma^\infty \quad (1 \leq p \leq q \leq \infty) . \tag{2.2.1}$$

This is in contrast with the classes $L^p(\Re)$, where $L^p(\Re)$ neither contains nor is contained in $L^q(\Re)$ for $1 \leq p,\ q \leq \infty,\ p \neq q$.

Another nice property of the class $\mathbf{B}_\sigma^p$ $(p \geq 1)$ is that it is closed under differentiation; indeed, if $f \in \mathbf{B}_\sigma^p$, then $f^{(n)} \in \mathbf{B}_\sigma^p$ with

$$\|f^{(n)}\|_p \leq \sigma^n \|f\|_p , \quad n = 0, 1, 2, \ldots . \text{ (cf. [16], p. 116)}$$

This is a striking difference between the class $\mathbf{B}_\sigma^p$ and the class $L^p(I)$ which is known to be not closed under differentiation, where I is any open interval.

If $\mathbf{B}$ is a class of entire functions such that for any $f \in \mathbf{B}$, f can be reconstructed from its sampled values at the points $\{t_k\}_{k=-\infty}^{\infty}$ via (2.1.2), we shall say that the WSK theorem is applicable to the class $\mathbf{B}$.

Having introduced the classes $\mathbf{B}_\sigma^p$ and some of their properties, we now turn our attention back to the class of band-limited functions to see how it can be enlarged.

Recall that our definition of band-limited signals stipulates that a band-limited signal f must have finite energy; hence, its Fourier transform F is in $L^2(-\sigma, \sigma)$, for some $\sigma > 0$.

According to the well-known Paley-Wiener theorem [18] (cf. [13])

$$f \in \mathbf{B}_\sigma^2 (\sigma > 0) \text{ if and only if } f(t) = \frac{1}{\sqrt{2\pi}} \int_{-\sigma}^{\sigma} F(\omega) e^{it\omega} d\omega ,$$

for some $F \in L^2(-\sigma, \sigma)$.

Therefore, it follows immediately that the class $\mathbf{B}_\sigma^2$, which is called the Paley-Wiener class of entire functions, is exactly the same as the class of all functions band-limited to $[-\sigma, \sigma]$.

It also follows from the WSK sampling theorem that if $f \in \mathbf{B}_\sigma^2$, then it can be reconstructed from its sampled values at the points $\{t_k\}_{k=-\infty}^{\infty}$ via (2.1.2). That is, the WSK sampling theorem is applicable to the class $\mathbf{B}_\sigma^2$, and hence in view of inclusions (2.2.1), it is also applicable to the class $\mathbf{B}_\sigma^p$, $1 \le p \le 2$.

The fact that the WSK sampling theorem is applicable to the class $\mathbf{B}_\sigma^p (1 \le p \le 2)$ can be also established directly as in the proof of Theorem 2.1 by using Hölder's inequality and the following modification of the Paley-Wiener theorem:

> A function $f \in L^p(\mathfrak{R})$, $1 \le p \le 2$, has an extension to the whole plane as an element of $\mathbf{B}_\sigma^p$ if and only if its Fourier transform F has support in $[-\sigma, \sigma]$, and if $f \in \mathbf{B}_\sigma^p$ with $1 < p \le 2$, then it has the representation (2.1.1) with $F \in L^q(\mathfrak{R})$, where $(1/p) + (1/q) = 1$. Moreover, $f \in B_\sigma^1$ if and only if it has the representation (2.1.1) with $F(\sigma) = F(-\sigma) = 0$ and the function obtained by extending F to be zero outside $(-\sigma, \sigma)$ has an absolutely convergent Fourier series on the interval $(-\sigma - \delta, \sigma + \delta)$, for some $\delta > 0$ [1, p. 106-107].

Our objective now is to enlarge the class of band-limited signals $\mathbf{B}_\sigma^2$ to include more general types of signals. There are a number of different ways to do so; however, we should first decide on which properties we would like to preserve for this new class of band-limited signals.

The most natural ones we require are:

A) That the Fourier transform F of a band-limited signal f have compact support; after all the term "band-limited" means that the frequency content of the signal f is bounded,

B) That the WSK sampling theorem be applicable to this new class of band-limited signals, since in practice we would like to be able to recover the signal from its sampled values.

Unfortunately, it turns out that the only class of functions that has these two properties is $\mathbf{B}_\sigma^2$. Nevertheless, we shall show that if condition (A) is relaxed a little, or in better words if it is interpreted appropriately, then we shall be able to preserve (B) as well. This will be indeed the case for the class $\mathbf{B}_\sigma^p$, $2 < p < \infty$.

Unlike $\mathbf{B}_\sigma^p$, $1 \le p \le 2$, members of the class $\mathbf{B}_\sigma^p$, $2 < p \le \infty$, do not necessarily have Fourier transforms in the classical sense. However, if we allow the Fourier transform to be taken in the generalized function (distributional) sense [8, 30], then members of the class $\mathbf{B}_\sigma^p$, $p > 2$, will have Fourier transforms with compact supports. For example, the function $f(t) = \cos \pi t$, which belongs to $\mathbf{B}_{\pi}^{\infty}$, has no Fourier transform in the classical sense, yet has one in the generalized function sense, namely,

$$F(\omega) = \sqrt{\pi/2}\ \left(\delta(\omega - \pi) + \delta(\omega + \pi)\right),$$

where $\delta(\omega)$ is the Dirac-delta function.

By a generalized function f we mean an element of the dual space A^* of some testing-function space A in the sense of Zemanian [30]. If A is the space of all infinitely differentiable functions with compact support, traditionally denoted by $\mathcal{D}$, then its dual $\mathcal{D}^*$ is usually called the space of Schwartz distributions.

The Fourier transform F of any f in B_{σ}^{∞} is a Schwartz distribution with support in $[-\sigma, \sigma]$. This is an immediate consequence of the following simplified version of a theorem of Schwartz, which is sometimes called the Paley-Wiener-Schwartz theorem [8, Vol. II, p. 162]:

> If an entire function $f(t)$ of first order of growth (exponential type) and of type $\leq \sigma$ does not grow more rapidly than $|x|^q$ as $|x| \to \infty$ for some q, then its Fourier transform is a Schwartz distribution with support in $[-\sigma, \sigma]$.

Another interpretation of the compactness of the support of the Fourier transform F of $f \in \mathbf{B}_{\sigma}^{\infty}$ has been given by Logan [14]. First, recall that [11, p. 125] for an appropriate function f,

$$F(\omega) = \lim_{T \to \infty} \frac{1}{\sqrt{2\pi}} \int_{-T}^{T} \left(1 - \frac{|t|}{T}\right) f(t) e^{it\omega} dt\ .$$

Logan has shown that if $f \in \mathbf{B}_{\sigma}^{\infty}$, then

$$\lim_{T \to \infty} \frac{1}{\sqrt{2\pi}} \int_{-T}^{T} \left(1 - \frac{|t|}{T}\right) f(t) e^{it\omega} dt = 0 \quad \text{for} \quad |\omega| > \sigma\ .$$

With these new interpretations of property (A), we can now prove property (B) for a larger class of functions than $\mathbf{B}_{\sigma}^{2}$, namely $\mathbf{B}_{\sigma}^{p}\ (2 < p < \infty)$.

THEOREM 2.2

Let $f \in \mathbf{B}_{\sigma}^{p}$, $1 \leq p < \infty$, $\sigma > 0$. Then f can be reconstructed from its sampled values at the points $\{t_k\}_{k=-\infty}^{\infty}$ via the formula

$$f(t) = \sum_{k=-\infty}^{\infty} f(t_k) \frac{\sin \sigma(t - t_k)}{\sigma(t - t_k)}, \quad t \in \Re$$

where the series converges absolutely and uniformly on compact sets.

The proof we provided for the WSK sampling theorem cannot be easily extended to Theorem 2.2 since functions in B_σ^p $(p > 2)$ cannot, in general, be written as Fourier integrals of functions with support in $[-\sigma, \sigma]$ as in (2.1.1); a new approach is needed. The idea of the following proof goes back to Butzer, Splettstösser and Stens [3].

Proof. Let

$$g(t) = \operatorname{sinc}(t) = \begin{cases} \dfrac{\sin \pi t}{\pi t} & \text{for} \quad t \neq 0 \\ 1 & \text{for} \quad t = 0 \end{cases} .$$

Since $g(k) = \begin{cases} 0 & \text{if} \quad k = \pm 1, \pm 2, \ldots \\ 1 & \text{if} \quad k = 0 \end{cases}$, it follows that

$$f(t) = \sum_{k=-\infty}^{\infty} f(t - t_k) g(k),$$

and hence the proof will be completed if we can show that

$$\sum_{k=-\infty}^{\infty} f(t - t_k) g(k) = \sum_{k=-\infty}^{\infty} f(t_k) g\left(\frac{\sigma}{\pi}(t - t_k)\right). \tag{2.2.2}$$

First, let us recall the following form of the Poisson summation formula:

$$\frac{1}{\sqrt{2\pi}} \sum_{k=-\infty}^{\infty} h\left(t + \frac{k\pi}{\sigma}\right) = \frac{1}{\sqrt{2\pi}} \sum_{k=-\infty}^{\infty} h(t + t_k) = \frac{\sigma}{\pi} \sum_{k=-\infty}^{\infty} \hat{h}(2k\sigma) e^{2ik\sigma t}, \tag{2.2.3}$$

provided that the series converge. A sufficient condition for that to hold is that $h \in L^1(\mathfrak{R})$ and $\hat{h}(\omega) = O((1 + |\omega|)^{-\alpha})$ as $|\omega| \to \infty$ for some $\alpha > 1$. Set

$$h_x(t) = f(t) g\left(\frac{\sigma}{\pi}(x - t)\right), \quad x \in \mathfrak{R}.$$

Since $f \in B_\sigma^p$ for $1 \le p < \infty$ and $g \in B_\pi^q$ for $1 < q \le \infty$, where $(1/p) + (1/q) = 1$, it follows that $h_x \in B_{2\sigma}^1$. By applying (2.2.3) with $t = 0$ to h_x, we get

$$\frac{1}{\sqrt{2\pi}} \sum_{k=-\infty}^{\infty} h_x(t_k) = \frac{1}{\sqrt{2\pi}} \sum_{k=-\infty}^{\infty} f(t_k) g\left(\frac{\sigma}{\pi}(x - t_k)\right) = \frac{\sigma}{\pi} \hat{h}_x(0), \tag{2.2.4}$$

since $\hat{h}_x(\omega) = 0$ for $|\omega| \geq 2\sigma$. But

$$\hat{h}_x(\omega) = \frac{1}{\sqrt{2\pi}} \int_{-\infty}^{\infty} f(t) g\left(\frac{\sigma}{\pi}(x-t)\right) e^{i\omega t} dt ,$$

hence

$$\hat{h}_x(0) = \frac{1}{\sqrt{2\pi}} \int_{-\infty}^{\infty} f(t) g\left(\frac{\sigma}{\pi}(x-t)\right) dt = \frac{1}{\sqrt{2\pi}} \int_{-\infty}^{\infty} f(x-t) g\left(\frac{\sigma}{\pi} t\right) dt . \tag{2.2.5}$$

By repeating the same argument for the function

$$\tilde{h}_x(t) = f(x-t) g\left(\frac{\sigma}{\pi} t\right),$$

we obtain

$$\frac{1}{\sqrt{2\pi}} \sum_{k=-\infty}^{\infty} \tilde{h}_x(t_k) = \frac{1}{\sqrt{2\pi}} \sum_{k=-\infty}^{\infty} f(x-t_k) g(k) = \frac{\sigma}{\pi} \hat{\tilde{h}}_x(0) , \tag{2.2.6}$$

where

$$\hat{\tilde{h}}_x(0) = \frac{1}{\sqrt{2\pi}} \int_{-\infty}^{\infty} f(x-t) g\left(\frac{\sigma}{\pi} t\right) dt = \hat{h}_x(0) . \tag{2.2.7}$$

The last equality follows from (2.2.5). By combining (2.2.4) through (2.2.7) we obtain (2.2.2). ▮

Having shown that properties (A) and (B) hold for the class $\mathbf{B}_\sigma^p$, $1 \leq p < \infty$, $\sigma > 0$, we may now define $\mathbf{B}_\sigma^p$ as a class of band-limited functions. The extension of Theorem 2.2 to the case $p = \infty$ is impossible in view of the counter-example provided by the function $f(t) = \sin \pi t$, which is in $\mathbf{B}_\pi^\infty$, but whose values at the points $\{t_k\}_{k=-\infty}^{\infty}$ are all zeros. However, if we restrict ourselves to the class $\mathbf{B}_{\sigma'}^\infty$, with $0 < \sigma' < \sigma$, then the extension is possible [19].

We may then define the class of band-limited functions with bandwidth at most σ as

$$\left(\bigcup_{1 \leq p < \infty} \mathbf{B}_\sigma^p \right) \cup \left(\bigcup_{0 < \sigma' < \sigma} \mathbf{B}_{\sigma'}^\infty \right) .$$

To obtain a WSK-type sampling theorem for the class $\mathbf{B}_\sigma^\infty$, a price has to be paid in terms of the sampling rate; a sampling rate σ'/π faster than the Nyquist rate σ/π $(0 < \sigma < \sigma')$ is needed.

THEOREM 2.3 [19, 20]

Let $f \in \mathbf{B}_\sigma^\infty$, $\sigma > 0$, then

$$f(t) = \sum_{n=-\infty}^{\infty} f(\eta_n) \frac{\sin\sigma'(t-\eta_n)}{\sigma'(t-\eta_n)},$$

where $\eta_n = n\pi/\sigma'$ and $0 < \sigma < \sigma'$. The series converges uniformly on any compact subset of the complex t-plane.

One may be tempted to define a band-limited function as merely an entire function whose Fourier transform has compact support in some general sense, e.g., in the generalized function sense. The pitfall of this is that property (B), i.e., the applicability of the WSK sampling theorem, may no longer hold as can be seen from the aforementioned example $f(t) = \sin\pi t$ or $f(t) = -ite^{i\pi t}$ whose Fourier transform is $F(\omega) = \sqrt{2\pi}\,\delta'(\omega - \pi)$. In the former case the sampling series (2.1.2) is identically zero and in the latter it is divergent since $f(t_k) = O(k)$ as $|k| \to \infty$.

More generally, the converse of the Paley-Wiener-Schwartz theorem cited above asserts that [8, Vol. II, p. 131] if F is a generalized function with compact support, then its inverse Fourier transform $f(z)$ is an entire function of exponential type with $f(x) = O(x^p)$ as $|x| \to \infty$ for some p. Thus, the series (2.1.2) is, in general, divergent whenever $p \geq 1$.

Attempts have been made to extend the WSK sampling theorem to such functions by either using summability methods, or by taking the convergence in (2.1.2) to be in the generalized function sense, or by introducing a convergence factor in the sampling series. One of the earliest results in this direction is due to Campbell [6], who showed that if $F(w)$ is a Schwartz distribution with support in the open interval $\{w: |w| < (1-q)\sigma$, for some q with $0 < q < 1\}$ and $f(t)$ is its inverse Fourier transform, then

$$f(t) = \sum_{k=-\infty}^{\infty} f(t_k) \frac{\sin\sigma(t-t_k)}{\sigma(t-t_k)} S(q\sigma[t-t_k]),$$

where

$$S(y) = \frac{\left\{\int_{-1}^{1} \exp[(1-x^2)^{-1} - ixy]\,dx\right\}}{\left\{\int_{-1}^{1} \exp[(1-x^2)^{-1}]\,dx\right\}},$$

$t_k = (k\pi)/\sigma$, and the series converges pointwise.

The presence of the S factor in the above series makes it lack the familiar form suggested by the WSK sampling theorem. This is an undesirable feature shared in general by sampling series expansions of entire functions that are unbounded on the real axis.

Surprisingly, it turns out that in some cases we can get rid of the summability and the convergence factors; hence, restoring the familiar form of the WSK sampling series, if we merely sample the signal at a rate faster than the Nyquist rate as indicated by Theorem 2.3. In other words, if we relax condition (B) by, for example, not insisting on sampling at the Nyquist rate, we can also enlarge the class of σ-band-limited functions. This usually requires a new definition of the bandwidth.

A larger class H of band-limited functions was introduced by M. Zakai in [29]. We shall say that f is a band-limited function in the sense of Zakai with bandwidth σ if it is an entire function of exponential type satisfying

$$|f(z)| \leq B e^{A|z|} \quad \text{and} \quad \int_{-\infty}^{\infty} \frac{|f(x)|^2}{1+x^2}\, dx < \infty ,$$

for some positive numbers A and B.

The bandwidth σ of f is defined as the smallest W so that the Fourier transform of $(f(z)-f(0))/z$ vanishes almost everywhere outside $(-W, W)$. The bandwidth σ is, in general, less than or equal to A.

Clearly, if $f \in \mathbf{B}_\sigma^\infty$, then it is band-limited in the sense of Zakai, i.e., $\mathbf{B}_\sigma^\infty \subset H$. More generally, if f is band-limited in the sense of Zakai, then

$$g(z) = \frac{f(z)-f(0)}{z}$$

is band-limited in the conventional sense, i.e., $g \in \mathbf{B}_\sigma^2$ for some $\sigma > 0$.

Zakai's definition of band-limited functions encompasses functions like

$$Si(t) = \int_0^t \frac{\sin x}{x} dx ,$$

which can be written as a Fourier transform of a function F with compact support, namely,

$$Si(t) = \frac{1}{2i} \int_{-1}^{1} \frac{1}{\omega} e^{i\omega t} d\omega .$$

The function, $F(\omega) = (1/\omega)\chi_{(-1,1)}$, however, is not in $L^p(-1,1)$ for any $p \geq 1$. Zakai has obtained the following sampling theorem for the class H using sampling rate σ'/π larger than the Nyquist rate σ/π.

THEOREM 2.4 (Zakai [25])

If f is a band-limited function in the sense of Zakai with bandwidth σ, then

$$f(t) = \sum_{n=-\infty}^{\infty} f(\eta_n) \frac{\sin \sigma'(t-\eta_n)}{\sigma'(t-\eta_n)},$$

where $\eta_n = (n\pi)/\sigma'$ and $0 < \sigma < \sigma'$. The series converges uniformly on any compact subset of the complex t-plane.

We may occasionally call this theorem the Whittaker-Shannon-Kotel'nikov-Zakai sampling theorem. A more refined class of band-limited functions was also first introduced by Zakai in [29], but was further developed by S. Cambanis and E. Masry [5]. Initially, Zakai defined the class $H(\sigma,\delta)$ of band-limited functions as the class of all functions $f(t)$ satisfying

$$\int_{-\infty}^{\infty} \frac{|f(t)|^2}{1+t^2} dt < \infty \quad \text{and} \quad f(x) = \int_{-\infty}^{\infty} f(t) g(x-t)\, dt ,$$

for almost all x, where

$$g(t) = g(t;\sigma,\delta) = \sqrt{\frac{2}{\pi}} \frac{2}{\delta t^2} \sin\left[\left(\sigma + \frac{\delta}{2}\right)t\right] \sin\left(\frac{\delta t}{2}\right),$$

which is the inverse Fourier transform of

$$G(\omega) = G(\omega;\sigma,\delta) = \begin{cases} 1 & \text{for } |\omega| \leq \sigma, \\ 1 - \dfrac{|\omega| - \sigma}{\delta} & \text{for } \sigma < |\omega| \leq \sigma + \delta, \\ 0 & \text{for } \sigma + \delta < |\omega| . \end{cases}$$

Cambanis and Masry showed that the class $H(\sigma,\delta)$ is, in fact, independent of δ and hence they denoted it by $H(\sigma)$. Following their notation, we shall call functions in the class $H(\sigma)$ σ-band-limited or band-limited to σ in the sense of Zakai.

Let us denote the bandwidth of $f \in H(\sigma)$ by $\sigma_0(f)$, then it is reasonable to define the bandwidth σ_0 of the class $H(\sigma)$ as

$$\sigma_0 = \sup_{f \in H(\sigma)} \sigma_0(f).$$

The following facts were shown in [5]:

i) $f \in H(\sigma)$ if and only if $f(t) = f(0) + tg(t)$, where g is band-limited in the conventional sense, i.e., $g \in B_\sigma^2$,

ii) $\sigma_0 = \sigma$,

iii) for all $f \in H(\sigma)$ and $0 < \tau < \pi/\sigma$, we have the sampling series expansion

$$f(t) = \sum_{n=-\infty}^{\infty} f(n\tau) \frac{\sin[(\pi/\tau)(t - n\tau)]}{[(\pi/\tau)(t - n\tau)]},$$

or

$$f(t) = \sum_{n=-\infty}^{\infty} f(\eta_n) \frac{\sin \sigma'(t - \eta_n)}{\sigma'(t - \eta_n)},$$

where $\eta_n = (n\pi)/\sigma'$, and $0 < \sigma < \sigma'$. The convergence is uniform on any compact subset of the t-plane.

The class $H(\sigma)$ has also been generalized by A. J. Lee [12] as follows: Let H_σ^k be the class of all entire functions of exponential type satisfying

$$\int_{-\infty}^{\infty} \frac{|f(x)|^2}{(1+x^2)^k} dx < \infty,$$

and $|f(z)| \le C(1 + |z|)^k \exp(\sigma |\operatorname{Im} z|)$ for some constant C depending on f and any complex number z. Lee showed, among other things, that if

$$\int_{-\infty}^{\infty} \frac{|f(x)|^2}{(1+x^2)^k} dx < \infty,$$

then $f \in H_\sigma^k$ is equivalent to either of the following conditions:

i) $f(t) = \sum_{j=0}^{k-1} [f^{(j)}(0)/j!] t^j + (t^k/k!) g(t)$, where $g \in B_\sigma^2$, or

ii) the Fourier transform of f has support in $[-\sigma, \sigma]$.

The class H_σ^0 is the same as the class B_σ^2 of conventional band-limited functions and the class H_σ^1 is the same as Zakai's class of σ band-limited functions $H(\sigma)$.

The class $H_\sigma^\infty = \bigcup_{k=0}^{\infty} H_\sigma^k$ consists of all functions f that are temperate distributions having Fourier transforms with support in $[-\sigma, \sigma]$. Moreover, $f \in H_\sigma^\infty$ is such that

$$\int_{-\infty}^{\infty} \frac{|f(x)|^2}{(1+x^2)^k}\, dx < \infty$$

if and only if the order of its distributional Fourier transform is less than or equal to k.

The following sampling theorem for the class H_σ^k has been obtained by Lee [12], (cf. [20]):

THEOREM 2.5 (Lee [12])

Let $f \in H_\sigma^k$. Then if

$$0 < \tau < \frac{\pi}{\sigma} \quad \text{and} \quad 0 < \beta < \frac{\pi}{k}\left(\frac{1}{\tau} - \frac{\sigma}{\pi}\right),$$

we have

$$f(t) = \sum_{n=-\infty}^{\infty} f(n\tau) \frac{\sin[(\pi/\tau)(t-n\tau)]\sin^k[\beta(t-n\tau)]}{[(\pi/\tau)(t-n\tau)][\beta(t-n\tau)]^k},$$

and the convergence is uniform on any compact subset of the t-plane.

In closing, we should point out that there is an inherited paradox in the definition of band-limited signals that stems from property (A). On the one hand, by definition, a band-limited signal is an entire function whose Fourier transform has compact support, whether in the classical or in the generalized functions sense, but on the other hand, as an entire function, the signal itself cannot have compact support, unless it is identically zero. That is, it cannot be both band-limited and time-limited, unless it is identically zero. This is in contradiction with the physical world where signals are known to be both band-limited and time-limited.

A possible resolution of this apparent paradox was provided by Slepian [23] in 1976 in his philosophical discussion of the role of mathematical models in exact sciences. One way to express this impossibility of confining the signal and its amplitude spectrum simultaneously to a bounded region is through an analogue of the Heisenberg uncertainty principle of quantum mechanics [19, p. 273]. If we use, as a measure of the duration of $f(t)$ and $F(\omega)$, the numbers T and W defined by

$$T^2 = \frac{1}{E} \int_{-\infty}^{\infty} t^2 \, |f(t)|^2 \, dt \, , \quad W^2 = \frac{1}{E} \int_{-\infty}^{\infty} \omega^2 \, |F(\omega)|^2 \, d\omega \, ,$$

where

$$E = \int_{-\infty}^{\infty} |f(t)|^2 \, dt = \int_{-\infty}^{\infty} |F(\omega)|^2 \, d\omega \, ,$$

and assume that $tf^2(t) \to 0$ as $|t| \to \infty$, then

$$1/2 \le TW \, . \tag{2.2.8}$$

To prove (2.2.8), we use integration by parts to deduce that

$$\int_{-\infty}^{\infty} tf(t) \frac{df}{dt} dt = 1/2 \int_{-\infty}^{\infty} t df^2 = \frac{1}{2} \left\{ tf^2(t) \, |_{-\infty}^{\infty} - \int_{-\infty}^{\infty} f^2(t) \, dt \right\} ,$$

which, in view of the assumption on f, yields

$$\int_{-\infty}^{\infty} tf(t) \frac{df}{dt} dt = -\frac{1}{2} E \, . \tag{2.2.9}$$

By applying the Cauchy-Schwarz inequality, we obtain

$$\left| \int_{-\infty}^{\infty} tf(t) \frac{df}{dt} dt \right|^2 \le \left(\int_{-\infty}^{\infty} t^2 \, |f(t)|^2 \, dt \right) \left(\int_{-\infty}^{\infty} \left| \frac{df}{dt} \right|^2 dt \right) \le E^2 T^2 W^2 \, ,$$

which, when combined with (2.2.9), yields (2.2.8). Here we have used the fact that the Fourier transform of df/dt is $i \omega F(\omega)$.

References

1. R. P. Boas, *Entire Functions*, Academic Press, New York (1954).

2. P. L. Butzer and R. L. Stens, Sampling theory for not necessarily band-limited functions: A historical review, *SIAM Review*, 34, 1 (1992), 40-53.

3. P. L. Butzer, W. Splettstösser and R. L. Stens, The sampling theorem and linear prediction in signal analysis, Jahresbar. *Deutsch. Math. Verein.*, 90 (1988), 1-70.

4. P. L. Butzer, A survey of the Whittaker-Shannon sampling theorem and some of its extensions, *J. Math. Res. Exposition*, 3 (1983), 185-212.

5. S. Cambanis and E. Masry, Zakai's class of band-limited functions and processes: its characterization and properties, *SIAM J. Appl. Math.*, 30 (1976), 10-21.

6. L. Campbell, Sampling theorem for the Fourier transform of a distribution with bounded support, *SIAM J. Appl. Math.*, 16 (1968), 626-636.

7. W. L. Ferrar, On the consistency of cardinal function interpolation, *Proc. Roy. Soc.*, Edinburgh, 47 (1927), 230-242.

8. I. M. Gelfand and G. E. Shilov, *Generalized Functions*, Vol. I, II, Academic Press, New York (1968).

9. J. Higgins, Five short stories about the cardinal series, *Bull. Amer. Math. Soc.* 12 (1985), 45-89.

10. A. J. Jerri, The Shannon sampling theorem—its various extensions and applications: A tutorial review, *Proc. IEEE* (11) 65 (1977), 1565-1596.

11. Y. Katznelson, *An Introduction to Harmonic Analysis*, Dover Publ., New York (1976).

12. A. J. Lee, Characterization of band-limited functions and processes, *Inform. Control*, 31 (1976), 258-271.

13. N. Levinson, *Gap and Density Theorem*, Amer. Math. Soc. Colloquium Publ. Ser., Vol. 26, Amer. Math. Soc., Providence, Rhode Island (1940).

14. B. F. Logan, Jr., Information in the zero crossings of the bandpass signals, *Bell Systems Tech. J.*, 56 (1977), 487-510.

15. R. J. Marks II, *Introduction to Shannon Sampling and Interpolation Theory*, Springer-Verlag, New York (1991).

16. S. M. Nikol'skii, *Approximation of Functions of Several Variables and Imbedding Theorems*, Springer-Verlag, New York (1975).

17. K. Ogura, On a certain transcendental integral function in the theory of interpolation, *Tôhoku Math. J.*, 17 (1920), 64-72.

18. R. Paley and N. Wiener, *Fourier Transforms in the Complex Domain*, Amer. Math. Soc. Colloquium Publ. Ser., Vol. 19, Amer. Math. Soc., Providence, Rhode Island (1934).

19. A. Papoulis, *Signal Analysis*, McGraw-Hill, New York (1977).

20. K. Seip, An irregular sampling theorem for functions band limited in a generalized sense, *SIAM J. Appl. Math.*, 47, 5 (1987), 1112-1116.

21. C. E. Shannon, Communication in the presence of noise, *Proc. IRE*, 137 (1949), 10-21.

22. ______, A mathematical theory of communication, *Bell System Tech. J.*, 27 (1948), 379-423.

23. D. Slepian, On bandwidth, *Proc. IEEE*, 64 (1976), 292-300.

24. E. T. Whittaker, On the functions which are represented by the expansion of the interpolation theory, *Proc. Roy. Soc. Edinburgh*, Sec. A, 35 (1915), 181-194.

25. J. M. Whittaker, *Interpolatory Function Theory*, Cambridge University Press, Cambridge, England (1935).

26. ______, On the Fourier theory of the cardinal function, *Proc. Edinburgh Math. Soc.*, 1 (1929b), 169-176.

27. ______, On the cardinal function of interpolation theory, *Proc. Edinburgh Math. Soc.*, 1 (1929a), 41-46.

28. R. M. Young, *An Introduction to Nonharmonic Analysis*, Academic Press, New York, 1980.

29. M. Zakai, Band-limited functions and the sampling theorem, *Inform. Control*, 8 (1965), 143-158.

30. A. H. Zemanian, *Generalized Integral Transforms*, Dover Publ., New York (1987).

3

GENERALIZATIONS OF SHANNON SAMPLING THEOREM

3.0 Introduction

Although the Shannon (Whittaker-Shannon-Kotel'nikov [WSK]) sampling theorem has been employed successfully in many applications in various disciplines, there have been occasions where it ceased to be such a useful tool to reconstruct certain types of signals. More general sampling theorems were needed for different reconstruction problems. A great number of research papers have been published in the last four decades dealing with various aspects of the WSK sampling theorem, in particular, its extensions, generalizations and applications.

The primary goal of this chapter is to give an overview of some of the main generalizations and extensions of the WSK sampling theorem and its relationships with some classical, as well as, new results in mathematical analysis.

The secondary goal is to pave the way for introducing, in the subsequent chapters, some recent results on sampling theorems. To appreciate some of these results, one needs to know where they fit in the general scheme of things in Sampling Theory.

To achieve these goals in a perspicuous and concise way, we shall confine ourselves to certain aspects of the WSK sampling theorem and its generalizations with more emphasis on non-uniform sampling, sampling of integral transforms other than the Fourier one, and sampling of multidimensional signals.

In so doing, we are not attempting to duplicate or compete with other excellent and comprehensive review articles by P. Butzer [14], P. Butzer et al. [16], R. Higgins [34] and A. Jerri [42] but trying to provide a new perspective.

Most of the results presented in this chapter are well known and several of them can be found in at least one of the aforementioned references. Therefore, most of the proofs will be omitted, but references will be provided; only short proofs of some important theorems will be given. However, some new yet relevant results will also be briefly discussed, in particular in Section 3.2, but their proofs will be postponed until later chapters.

3.1 Non-Uniform Sampling

The sampling points in the WSK sampling theorem are uniformly spaced; the sampling procedure associated with it is known as uniform sampling. Non-uniform sampling occurs as frequently in practice and sometimes is known to give better results. In fact, uniform sampling with either some of the sampling points missing or with time jitters can be regarded as non-uniform sampling. In some situations it is more natural to sample a signal non-uniformly. For example, when a signal is varying rapidly it is more appropriate to sample it at a higher rate than when it is varying slowly.

Reconstruction of signals from non-uniformly distributed sampling points has been used to a great extent in many fields, such as Radio Astronomy [33], Computed Tomography (CT), Magnetic Resonance Imaging (MRI) [29, 71, 90], optical and electronic imaging systems [8, 9, 28, 29, 73, 99-101] and many others [2, 42, 77, 86].

There are two main techniques used to reconstruct signals from non-uniformly and uniformly collected data; one is based on using the values of the signal, and perhaps some of its derivatives, at predetermined instants, such as in the WSK sampling theorem. The other is based on using instants, where the signal assumes predetermined values such as zero.

We will digress momentarily to briefly discuss the second technique before addressing the problem of non-uniform sampling.

Let us recall our original definition of a band-limited signal. A signal (function) $f(t)$ is band-limited with bandwidth σ if it is the Fourier transform of a function $F \in L^2(-\sigma,\sigma)$ with support in $[-\sigma,\sigma]$, i.e.,

$$f(t) = \frac{1}{\sqrt{2\pi}} \int_{-\sigma}^{\sigma} F(w) e^{itw} dw , \quad F \in L^2(-\sigma,\sigma) .$$

We have seen from the Paley-Wiener theorem that $f(t)$ is an entire function of exponential type and its type is at most equal to σ.

Although entire functions are not completely determined by the location of their zeros as can be seen from the Hadamard factorization theorem for entire functions (Theorem 1.1), surprisingly band-limited functions are— up to a constant. This is an easy consequence of a less known result of

Titchmarsh [105], which states that if

$$f(z)=\frac{1}{\sqrt{2\pi}}\int_a^b F(w)e^{zw}dw\,,\quad F\in L^1(a,b)\,,$$

then

$$f(z)=f(0)\exp\left\{\frac{1}{2}(a+b)z\right\}\prod_{n=1}^{\infty}\left(1-\frac{z}{z_n}\right),$$

where z is a complex number, $\{z_n\}$ are the zeros of f (which, of course, are not necessarily uniformly distributed) and the product is conditionally convergent.

Therefore, a signal that is band-limited to the interval $[a,b]$ is uniquely determined by its zeros up to an exponential factor that depends on the spectral end points. In particular, if a signal is band-limited to an interval symmetric about the origin, say to $[-\sigma,\sigma]$, then it is completely determined by its zeros—up to a constant—and can be reconstructed from these zeros via the formula

$$f(z)=f(0)\prod_{n=1}^{\infty}\left(1-\frac{z}{z_n}\right). \tag{3.1.1}$$

Here we are, of course, assuming that $f(0)\neq 0$; otherwise we will have

$$f(z)=Az^m\prod_{n=1}^{\infty}\left(1-\frac{z}{z_n}\right),$$

where m denotes the order of the zero at $z=0$.

It is easy to show that a signal having either a finite number or uncountably many zeros cannot be band-limited, but it is more difficult to show that the complex zeros of a band-limited function f tend to cluster near the real axis and that the density of the zeros is proportional to its bandwidth according to the formula

$$N(r)\sim\frac{b-a}{\pi}r\,,$$

where $N(r)$ is the number of zeros of f that lie inside the disc $|z|\leq r$. That is the zeros occur, in the limit, at the Nyquist rate [105].

Reconstructing signals from their zeros was first introduced in engineering by Bond and Cahn [10], and further developed by Voelcker [110] and Sekey [92].

Clearly, to reconstruct a band-limited signal from its zeros, one needs to know all of them, whether real or complex. The real zeros are known in engineering terminology as the zero-crossings of the signal; they can be detected and used to reconstruct the signal. If all the zeros of a band-limited signal are real, then it can be reconstructed (up to a constant) from its zero-crossings. Such a signal is called a real-zero signal, commonly known in engineering literature as an RZ signal, and the process of reconstructing it from its zero crossings is known as real-zero interpolation (RZI). Typical examples of such a signal in modern communication systems are 2-level facsimile and clipped speech [10 and 52].

An important class of signals, from both mathematical and engineering points of view, is the class of band-pass signals, which is the class of all bounded signals whose Fourier transforms are confined to $[a, b]$ and $[-b, -a]$. Logan [55] has proved that a band-pass signal of less than one octave in bandwidth and which has no common zeros with its Hilbert transform, except real zeros of order one, is uniquely determined (up to a constant) and can be reconstructed from its zero crossings.

Reconstructing signals from their zero crossings has important practical applications, especially in the field of Image Processing [108]. For example, in situations where an image is blurred or distorted, but in such a way that the zero crossings are preserved, it is possible to recover the image from its distorted version. More generally, if a band-limited signal f undergoes a nonlinear transformation T, i.e., $g(t) = T\{f(t)\}$, it may be possible to recover f from the knowledge of the output signal g if T is a zero-crossing preserving transformation, e.g., $T(z) = z^a$ for some non-negative rational number a. For more details on this, the reader may consult [59] where the case $a = 1/n$ has also been treated; see also [58].

Since nonlinear transformations, in general, increase the bandwidth of the input signal and hence, increase the sampling rate, it may be, at least theoretically, more advantageous to recover the signal from its zero crossings using RZI techniques than from its sampled values using sampling theorems.

Admittedly, RZI techniques rely on the fact that all the zeros of the signal are real, which would limit their applications. Nevertheless, there are several techniques available to convert general band-limited signals to RZ signals, that is converting complex zeros to real zeros.

Extensive work has been done on identifying and developing conditions under which this conversion is realizable [1 and 110]. Some methods suggested by Voelcker are modulation and differentiation. The latter, which is assumed to take place at the transmitter, requires integration at the receiver; hence, the transmitted signal is determined—up to an additive constant. Most band-limited signals encountered in practice do not, however, yield easily to these conversion techniques.

Fortuitously, there are other methods available to reconstruct band-limited signals from non-uniformly given data; the most widely known one, which is a generalization of the WSK sampling theorem, is due to Paley and Wiener [69]. In this generalization of the WSK sampling theorem the equidistantly spaced sampling points $\{t_k\}_{k\in\mathbf{Z}}$ in (2.1.2) are replaced by certain nonequidistantly spaced ones with more general sampling functions $S_k^{**}(t)$ than $S_k(t)$ of (2.1.3), where **Z** is the set of integers. This important result can be stated as follows:

THEOREM 3.1 (Paley-Wiener [69])

Let $\{t_k\}_{k\in\mathbf{Z}}$ be a sequence of reals such that

$$D := \sup_{k\in\mathbf{Z}} \left| t_k - \frac{k\pi}{\sigma} \right| < \frac{\pi}{4\sigma}, \tag{3.1.2}$$

and let $G(t)$ be the entire function defined by

$$G(t) = (t - t_0)\prod_{k=1}^{\infty}\left(1 - \frac{t}{t_k}\right)\left(1 - \frac{t}{t_{-k}}\right). \tag{3.1.3}$$

Then, for any $f \in \mathbf{B}_\sigma^2$,

$$f(t) = \sum_{k=-\infty}^{\infty} f(t_k) S_k^{**}(t) \quad (t \in \mathfrak{R}), \tag{3.1.4}$$

where

$$S_k^{**}(t) := \frac{G(t)}{G'(t_k)(t - t_k)}, \tag{3.1.5}$$

and the series on the right hand side of (3.1.4) converges uniformly on compact subset of $\mathfrak{R}$.

Clearly, when $t_k = (k\pi)/\sigma = -t_{-k}$, $G(t)$ becomes $\sin(\sigma t)/\sigma$ and (3.1.4), (3.1.5) reduce to (2.1.2), (2.1.3). The sampling series (3.1.4) can be regarded as an extension of the classical Lagrange interpolation formula to $\mathfrak{R}$ for functions of exponential type. Unlike the classical Lagrange interpolation formula, formula (3.1.4) contains infinitely many terms. Therefore, we will call (3.1.4) a Lagrange-type interpolation series and S_k^{**} Lagrange-type sampling (interpolating) functions.

We will refer to this theorem as the Paley-Wiener interpolation theorem for band-limited functions; however, in some literature it is referred to as the Paley-Wiener-Levinson theorem. The constant $\pi/(4\sigma)$ in (3.1.2) is best possible, in the sense that if $D = \pi/(4\sigma)$ the result does not hold in general [69]. The original constant obtained by Paley and Wiener in (3.1.2) was

$1/(\sigma\pi)$ and it was later sharpened to its present value by N. Levinson and M. Kadec.

Levinson [51, p. 48] showed that if $\{t_k\}_{k\in Z}$ is a sequence of real numbers satisfying (3.1.2), then

i) the set $\{e^{it_k x}\}_{k\in Z}$ is complete in $L^2(-\sigma,\sigma)$, i.e., if

$$\int_{-\sigma}^{\sigma} f(x)e^{it_k x}dx = 0 \quad \text{for all } k ,$$

then f is identically zero; hence $\{e^{it_k x}\}_{k\in Z}$ possesses a unique biorthonormal set $\{h_k(x)\}_{k\in Z}$, i.e.,

$$\int_{-\sigma}^{\sigma} h_k(x)e^{it_m x}dx = \begin{cases} 0 & \text{if} \quad k \neq m \\ 1 & \text{if} \quad k = m \end{cases} \quad ; \tag{3.1.6}$$

cf. [119], p. 29.

ii)
$$\frac{G(t)}{(t-t_k)G'(t_k)} = \int_{-\sigma}^{\sigma} h_k(x)e^{itx}dx , \tag{3.1.7}$$

where G is defined by (3.1.3). Moreover, $G(t)/(t-t_k) \in L^2(\Re)$ for any k.

iii) For any $F \in L^2(-\sigma,\sigma)$,

$$F(x) = \sum_{k=-\infty}^{\infty} F_n h_n(x) , \tag{3.1.8}$$

in the mean, as well as, uniformly on any closed subinterval of $(-\sigma,\sigma)$,

where
$$F_n = \int_{-\sigma}^{\sigma} F(x)e^{it_n x}dx . \tag{3.1.9}$$

iv) These results are not, in general, true if $D \geq \pi/(4\sigma)$.

It is easy to see that equation (3.1.9) is a consequence of (3.1.8) and (3.1.6). Theorem 3.1 follows from Levinson's results by taking the Fourier transform of (3.1.8) and using (3.1.7), (3.1.9) to obtain

$$f(t) = \sum_{k=-\infty}^{\infty} f(t_k)S_k^{**}(t) ,$$

in which

$$f(t) = \int_{-\sigma}^{\sigma} F(x) e^{itx} dx ,$$

and the series being convergent in the mean. To show the uniform convergence, we note that for any $f \in B_\sigma^2$,

$$f(t) = \lim_{A \to \infty} \int_{-A}^{A} f(x) \frac{\sin \sigma(t-x)}{\pi(t-x)} dx .$$

Therefore, if we set

$$f_N(t) = \sum_{k=-N}^{N} f(t_k) S_k^{**}(t),$$

and use the Cauchy-Schwarz inequality, we obtain

$$|f(t) - f_N(t)| \le \|f - f_N\|_2 \left\| \frac{\sin \sigma(t-x)}{\pi(t-x)} \right\|_2 = \|f - f_N\|_2 \sqrt{\frac{\sigma}{\pi}} \to 0$$

as $N \to \infty$, uniformly in t.

Kadec [119, p. 42] sharpened Levinson theorem by showing that if the sequence $\{t_k\}_{k \in Z}$ satisfies (3.1.2), then the set $\{e^{it_k x}\}_{k \in Z}$ is a Riesz basis for $L^2(-\sigma, \sigma)$, and that this result is not true if $D \ge \pi/(4\sigma)$. The constant $d = \pi/(4\sigma)$ is known as Kadec's constant.

We refer the reader to Chapter 10 for the definition of Riesz basis and other related concepts, but for the time being we only remark that a Riesz basis for $L^2(-\sigma, \sigma)$ possesses the above-mentioned properties (i) and (iii). However, the converse is not true, in the sense that any set in $L^2(-\sigma, \sigma)$ having properties (i) and (iii) is not necessarily a Riesz basis.

The sampling functions $S_k^{**}(t)$ in (3.1.5) are obviously band-limited themselves, i.e., $S_k^{**}(t) \in B_\sigma^2$ for all k. In fact, Levinson showed that if

$$D < \frac{(p-1)\pi}{2p\sigma}, \quad 1 < p \le 2,$$

then (i) through (iv) are valid in $L^p(-\sigma, \sigma)$, in particular, $S_k^{**}(t) \in B_\sigma^p$.

It also follows from Kadec's theorem that the expansion (3.1.4) is unique and if any term from the sequence $\{t_k\}_{k \in Z}$ is deleted, (3.1.4) is not in general valid for any $f \in B_\sigma^2$. The restriction that D is strictly smaller than $\pi/4\sigma$ can be relaxed considerably if we confine our attention to functions in $B_{\sigma-\delta}^2$, for some δ with $0 < \delta < \sigma$; see [116, Theorem 2].

Definition 3.1.1. The class $\mathbf{B}_\sigma^p$ is said to possess a stable sampling expansion with respect to a class of sampling sequences $\left\{T : T = \{t_n\}_{n=-\infty}^{\infty}\right\}$ if there exists a positive constant C, independent of f and T, such that the relation

$$\int_{-\infty}^{\infty} |f(t)|^p \, dt \le C \sum_{n=-\infty}^{\infty} |f(t_n)|^p$$

is valid for each sampling sequence T and each $f \in \mathbf{B}_\sigma^p$.

This condition guarantees that small errors in the samples will cause correspondingly small errors in the reconstructed signal.

The class $\mathbf{B}_\sigma^2$ possesses a stable sampling expansion with respect to the class of sampling sequences defined by (3.1.2) [116]. If $D = 0$ in (3.1.2), Theorem 3.1 reduces to the WSK sampling theorem, the constant C in Definition 3.1 becomes 1 and the inequality becomes an equality in virtue of Parseval's relation

$$\int_{-\infty}^{\infty} |f(t)|^2 \, dt = \sum_{n=-\infty}^{\infty} |f(t_n)|^2 .$$

Similar results can be obtained for the class $\mathbf{B}_\sigma^p$ with $1 < p \le 2$.

There are also analogs of Theorems 2.3 and 2.5 for non-uniform sampling.

THEOREM 3.2 (Seip [91])

Let $\{t_n\}_{n=-\infty}^{\infty}$ and $G(t)$ be as in Theorem 3.1, and let δ be any positive real number such that $0 < \delta < \sigma$. Then, for any $f \in \mathbf{B}_{\sigma-\delta}^{\infty}$, we have

$$f(t) = \sum_{n=-\infty}^{\infty} f(t_n) \frac{G(t)}{G'(t_n)(t - t_n)},$$

uniformly on any compact subset of the complex t-plane.

THEOREM 3.3 (Seip [91])

Let $\{t_n\}_{n=-\infty}^{\infty}$, $G(t)$ and δ be as in Theorem 3.2, and β be a real number such that $0 < \beta < \delta/k$. Then, for any $f \in H_{\sigma-\delta}^{k}$ we have

$$f(t) = \sum_{n=-\infty}^{\infty} f(t_n) \frac{G(t)}{G'(t_n)(t - t_n)} \frac{\sin^k \beta(t - t_n)}{\beta^k (t - t_n)^k},$$

uniformly on any compact subset of the complex t-plane.

Recently, M. Rawn [85] derived an analogous result to the Paley-Wiener interpolation theorem for band-limited functions, but for functions that are Hankel transforms of order $2k$ of functions with compact support.

THEOREM 3.4 (Rawn [85])

Let $\{t_n\}$ be a sequence of real numbers satisfying $|t_n - (n - 1/4)| \le d' < 1/4$, $n = k+1, \ldots;$

$$G(t) = \prod_{n=k+1}^{\infty}\left(1 - \left(\frac{t}{t_n}\right)^2\right),$$

and let f be any function J_{2k}-Bessel band-limited to $(0, \pi)$, i.e.,

$$f(t) = \int_0^{\pi} wH(w) J_{2k}(wt)\, dw, \quad k = 0, 1, 2, \ldots \tag{3.1.10}$$

with $\sqrt{w}\, H(w) \in L^2(0, \pi)$. Then

$$f(t) = \sum_{n=k+1}^{\infty} f(t_n)\left(\frac{t}{t_n}\right)^{2k} \frac{(2t_n)G(t)}{(t^2 - t_n^2)G'(t_n)}, \quad n = k+1, k+2, \ldots .$$

Moreover, the class of functions

$$B_{\pi,2k} = \{f : R \to R \text{ such that there exists } H(w)$$
$$\text{with } \sqrt{w}\, H(w) \in L^2(0, \pi) \text{ satisfying (3.1.10)}\}$$

has a stable expansion with respect to the class of sampling points $\{t_n\}$ previously defined.

For more on non-uniform sampling, see [6, 31, 60, 61, 118].

3.2 Sampling Theorems for Other Integral Transforms and Representations of Band-Limited Signals—Kramer's Sampling Theorem

Since band-limited signals are represented by Fourier integrals of functions with compact support, we can say that the WSK sampling theorem provides a sampling series expansion for the reconstruction of certain Fourier transforms taken over finite intervals. If a signal is represented by another type of integral transform, the WSK sampling theorem is no longer applicable; a more general sampling theorem for the reconstruction of integral transforms of types other than the Fourier one is needed.

Reconstructing integral transforms other than the Fourier one from some of their sampled values occurs frequently in some physical applications. One such integral transform is the Hankel transform, which has been proved to be a useful tool in the analysis of several optical problems such as the Fresnel and Fraunhofer diffraction [73, 85]. In such problems, digital reconstruction would lend itself directly to the sampling series expansion of the Hankel transform. Another application of the Hankel transform can be seen in view of the fact that, when circular symmetry is assumed, an m-dimensional Fourier transform can be reduced to a one-dimensional $J_{(m/2)-1}$-Hankel transform, where $J_p(z)$ is the Bessel function of the first kind and order p [73, 95 and 40, p. 842]. Therefore, in solving various m-dimensional optical and signal analysis problems, where circular symmetry can be assumed, one may use the Hankel transform instead of m-dimensional Fourier transform. For image and signal reconstruction in these problems, the sampling series expansion of the Hankel transform will be needed.

The idea of reconstructing integral transforms from a discrete set of data stems from the work of J. M. Whittaker himself [114, p. 71], who had suggested a sampling series expansion for an integral transform taken over a finite interval with the Bessel function, instead of the exponential function, as its kernel. Nevertheless, the primary result in this direction is attributed to H. Kramer [50] and is known as Kramer's sampling theorem although, again, its main idea actually goes back to P. Weiss [113], who had first announced his results in the Bulletin of the American Mathematical Society in 1957, but never published the proofs. Weiss announced that he had obtained sampling series expansions for finite limit integral transforms whose kernels are associated with second-order regular Sturm-Liouville boundary-value problems. Two years later, Kramer published his work (including the proofs) in which he extended Weiss' results to integral transforms whose kernels are generated from boundary-value problems associated with nth order differential operators and illustrated his method by deriving a sampling series expansion for an integral transform with the Bessel function as its kernel. This integral transform is indeed a Hankel transform taken over a finite interval.

Since Kramer's sampling theorem will be the focus of forthcoming discussions we shall state it with a proof. When Kramer's theorem is associated with Sturm-Liouville boundary-value problems, we shall call it the Weiss-Kramer sampling theorem.[1]

THEOREM 3.5 (Kramer [50])

Let there exist a function $K(x,t)$ continuous in t such that $K(x,t) \in L^2(I)$ for every real number t. Assume that there exists a sequence of real numbers

[1] Professor A. Jerri suggested calling Theorem 3.5 the Whittaker-Weiss-Kramer (WWK) sampling theorem.

$\{t_k\}_{k=-\infty}^{\infty}$ such that $\{K(x,t_k)\}_{k=-\infty}^{\infty}$ is a complete orthogonal family of functions in $L^2(I)$. Then for any $f(t)$ of the form

$$f(t) = \int_a^b F(x)K(x,t)\,dx\,, \tag{3.2.1}$$

where $F(x) \in L^2(I)$ and I is a finite closed interval $[a,b]$, we have the sampling series representation

$$f(t) = \sum_{k=-\infty}^{\infty} f(t_k)S_k^*(t)\,, \tag{3.2.2}$$

with

$$S_k^*(t) = \frac{\int_a^b K(x,t)\overline{K(x,t_k)}\,dx}{\int_a^b |K(x,t_k)|^2\,dx}. \tag{3.2.3}$$

Proof. Because $F \in L^2(I)$, we have

$$F(x) = \sum_{k=-\infty}^{\infty} \hat{F}(k)\overline{K(x,t_k)}\,,$$

where

$$\hat{F}(k) = \frac{\int_a^b F(x)K(x,t_k)\,dx}{\int_a^b |K(x,t_k)|^2 dx}.$$

Similarly,

$$K(x,t) = \sum_{k=-\infty}^{\infty} S_k^*(t)K(x,t_k)\,.$$

An application of Parseval's equality to (3.2.1), together with the fact that

$$f(t_k) = \int_a^b F(x)K(x,t_k)\,dx\,,$$

completes the proof. ▌

To see how this theorem generalizes the WSK sampling theorem, just take

$$I = [-\pi, \pi], \quad K(x,t) = e^{ixt} \quad \text{and} \quad t_k = k,$$

then it easily follows that $S_k^*(t) = (\sin \pi(t-k))/\pi(t-k)$. Hence, if

$$f(t) = \int_{-\pi}^{\pi} F(x) e^{ixt} dx, \quad \text{with} \quad F(x) \in L^2(-\pi, \pi),$$

then

$$f(t) = \sum_{k=-\infty}^{\infty} f(k) \frac{\sin \pi(t-k)}{\pi(t-k)},$$

which is the WSK sampling theorem.

Yet, it is not really clear how the function $K(x,t)$ and the sequence $E = \{t_k\}_{-\infty}^{\infty}$ in Theorem 3.5 can be found. Kramer has indicated that if the regular self-adjoint boundary-value problem

$$Ly = ty, \quad x \in I \tag{3.2.4}$$

$$U_j(y) = 0; \quad j = 1, 2, \ldots, n, \tag{3.2.5}$$

possesses a function $\phi(x,t)$ that generates the eigenfunctions of the problem $\{\phi_k(x)\}$ when the parameter t is replaced by the eigenvalues $\{t_k\}$, i.e., $\phi_k(x) = \phi(x, t_k)$, where $I = [a,b]$ is some finite closed interval,

$$L = p_0(x)\frac{d^n}{dx^n} + p_1(x)\frac{d^{n-1}}{dx^{n-1}} + \ldots + p_n(x), \tag{3.2.6}$$

$p_k(x)$ being complex-valued function with $n-k$ continuous derivatives, $k = 0, 1, \ldots, n$, $p_0(x) \neq 0$ for any $x \in I$ and $U_j(y) = 0$, $j = 1, \ldots, n$, are linearly independent homogeneous self-adjoint boundary conditions independent of t, then we can choose $\phi(x,t)$ as $K(x,t)$ and the eigenvalues $\{t_k\}$ as the sampling points E. For more details on the boundary-value problem (3.2.4) and (3.2.5), see Chapter 5.

To illustrate the idea, let us consider the following regular Sturm-Liouville boundary-value problem:

$$-y'' = ty, \quad x \in [0, \pi] \tag{3.2.7}$$

$$y'(0) = 0 = y'(\pi). \tag{3.2.8}$$

The function $\phi(x,t)$ that generates the eigenfunctions $\{\cos kx\}$ is $\cos\sqrt{t}\,x$, where $t_k = k^2$ are the eigenvalues.

It is easily seen that

$$S_k^*(t) = \frac{2\sqrt{t}\sin\pi(\sqrt{t}-k)}{\pi(t-k^2)} \quad \text{if} \quad k = 1,2,\ldots, \quad \text{and} \quad S_0^*(t) = \frac{\sin\pi\sqrt{t}}{\pi\sqrt{t}}.$$

Therefore, the Weiss-Kramer theorem now takes the form:

If

$$f(t) = \int_0^\pi F(x)\cos\sqrt{t}\,x\,dx\,, \tag{3.2.9}$$

for some $F \in L^2(0,\pi)$, then

$$f(t) = f(0)\frac{\sin\pi\sqrt{t}}{\pi\sqrt{t}} + \sum_{k=1}^{\infty} f(k^2)\frac{2\sqrt{t}\sin\pi(\sqrt{t}-k)}{\pi(t-k^2)}. \tag{3.2.10}$$

If we replace t by t^2 and set $\tilde{f}(t) = f(t^2)$, we obtain

$$\tilde{f}(t) = \tilde{f}(0)\frac{\sin\pi t}{\pi t} + 2\sum_{k=1}^{\infty} \tilde{f}(k)\frac{t\sin\pi(t-k)}{\pi(t^2-k^2)},$$

which, in turn, can be put in a more familiar form

$$\tilde{f}(0) = \tilde{f}(0)\frac{\sin\pi t}{\pi t} + \sum_{k=-\infty}^{\infty}{}' \tilde{f}(k)\frac{\sin\pi(t-k)}{\pi(t-k)}.$$

The prime on the summation sign indicates that the term corresponding to $k = 0$ is omitted. We shall show in Chapter 4 that these last three series converge uniformly on compact subsets of $\Re$.

We should emphasize that the kernel $K(x,t)$ does not always arise from boundary-value problems; one such example was given by Kak [47] in which he derived the Walsh sampling theorem as a special case of Kramer's theorem. The Walsh sampling theorem was derived originally by F. Pichler [83] using a different approach (cf. [111]).

Kramer did not spell out the conditions on the differential operator (3.2.6) or the boundary condition (3.2.5) under which the regular boundary-value problem (3.2.4) and (3.2.5) would have one function $\phi(x,t)$ generating all the eigenfunctions, when the eigenvalue parameter t is replaced by the eigenvalues t_n, $n = 0,1,2,\ldots$

As it turns out, Kramer's assumption that the eigenfunctions are all generated by one single function $\phi(x,t)$ is very restrictive, especially when a general boundary-value problem like (3.2.4) and (3.2.5) is considered. In fact, this assumption does not hold if the order n of the differential operator (3.2.6) is greater than 2, except in some very special cases.

Worse yet, even if $n = 2$ this assumption is not always valid under general boundary-conditions, as seen from the following simple example:

$$-y'' = ty\,, \quad x \in [0,\pi] \tag{3.2.11}$$

$$y(0) = y(\pi) \quad \text{and} \quad y'(0) = y'(\pi)\,. \tag{3.2.12}$$

The eigenfunctions of this problem are $\sin 2nx$ and $\cos 2nx$; $n = 0, 1, 2, \ldots$, which are not generated by one single real-valued function.

Nevertheless, one may be tempted to derive a sampling series expansion associated with this problem. Since such sampling expansions do not seem to exist in the literature we shall try to derive one not only to fill a gap in the literature, but also to show some of the complications and subtleties associated with this type of boundary-value problems. This will also show how difficult it is, if not even impossible, to apply Kramer's argument to a general boundary-value problem such as (3.2.4) and (3.2.5).

The obvious choice for the function $\phi(x,t)$ that generates the eigenfunctions of problem (3.2.11) and (3.2.12), when the parameter t is replaced by the eigenvalues $t_n = 4n^2$, $n = 0, 1, 2, \ldots$, is

$$\phi(x,t) = A\cos\sqrt{t}\,x + B\sin\sqrt{t}\,x\,.$$

Therefore, if

$$f(t) = \int_0^\pi F(x)(A\cos\sqrt{t}\,x + B\sin\sqrt{t}\,x)dx\,, \tag{3.2.13}$$

for some $F \in L^2(0,\pi)$, we obtain after some calculations,

$$f(t) = \frac{f(0)}{\pi}\left\{\frac{\sin(\pi\sqrt{t})}{\sqrt{t}} + \frac{B}{A}\,\frac{2\sin^2\left(\frac{\pi}{2}\sqrt{t}\right)}{\sqrt{t}}\right\} + \sum_{k=1}^{\infty} a_{2k}\left\{A\,\frac{\sqrt{t}\sin(\pi\sqrt{t})}{(t-4k^2)} + B\,\frac{2\sqrt{t}\sin^2\left(\frac{\pi}{2}\sqrt{t}\right)}{(t-4k^2)}\right\}$$

$$+ \sum_{k=1}^{\infty} b_{2k}\left\{A\,\frac{(-4k)\sin^2\left(\frac{\pi}{2}\sqrt{t}\right)}{(t-k^2)} + B\,\frac{(2k)\sin(\pi\sqrt{t})}{(t-4k^2)}\right\}, \tag{3.2.14}$$

where a_k and b_k are the Fourier cosine and sine coefficients of $F(x)$ given by

$$a_k = \frac{2}{\pi}\int_0^{\pi} F(x)\cos kx\,dx\ ,$$

$$b_k = \frac{2}{\pi}\int_0^{\pi} F(x)\sin kx\,dx\ .$$

The last series is *not* a sampling series expansion of $f(t)$ since the coefficients a_k and b_k cannot be uniquely expressed in terms of the sampled values of f at the eigenvalues. In fact,

$$f(4k^2) = \frac{\pi}{2}(A\,a_{2k} + B\,b_{2k})\,. \tag{3.2.15}$$

Therefore, no Kramer-type sampling theorem exists for the boundary-value problem (3.2.11), (3.2.12). Nevertheless, if we denote the Hilbert transform (or the conjugate) of f by $\tilde{f}$, where

$$\tilde{f}(t) = \int_0^{\pi} F(x)(A\sin\sqrt{t}\,x - B\cos\sqrt{t}\,x)\,dx\ ,$$

we obtain

$$\tilde{f}(4k^2) = \frac{\pi}{2}(A\,b_{2k} - B\,a_{2k})\,,$$

which, when combined with (3.2.15), yields

$$a_{2k} = \frac{2\left(f(4k^2) - r\tilde{f}(4k^2)\right)}{\pi A(1+r^2)}\,, \quad b_{2k} = \frac{2\left(rf(4k^2) + \tilde{f}(4k^2)\right)}{\pi A(1+r^2)}\,,$$

where $r = B/A$.

Upon substituting these in (3.2.14) and carrying out some calculations, we obtain

$$f(t) = \frac{f(0)}{\pi\sqrt{t}}\left\{\sin(\pi\sqrt{t}) + 2r\sin^2\left(\frac{\pi}{2}\sqrt{t}\right)\right\}$$

$$+\frac{2}{\pi(1+r^2)}\sum_{k=1}^{\infty}\frac{1}{(t-4k^2)}\left\{f(4k^2)\left[(\sqrt{t}+2kr^2)\sin(\pi\sqrt{t}) + 2r(\sqrt{t}-2k)\sin^2\left(\frac{\pi}{2}\sqrt{t}\right)\right]\right.$$

$$\left.+\tilde{f}(4k^2)\left[r(2k-\sqrt{t})\sin(\pi\sqrt{t}) - 2(r^2\sqrt{t}+2k)\sin^2\left(\frac{\pi}{2}\sqrt{t}\right)\right]\right\}. \qquad (3.2.16)$$

One interesting observation here is this: although (3.2.13) reduces to (3.2.9) when $r = 0$, the sampling series (3.2.16) associated with (3.2.13) does not reduce to that associated with (3.2.9), namely (3.2.10). Indeed, when $r = 0$, (3.2.16) renders the following interesting sampling series

$$f(t) = f(0)\frac{\sin(\pi\sqrt{t})}{\pi\sqrt{t}} + \frac{2}{\pi}\sum_{k=1}^{\infty}\frac{1}{(t-4k^2)}\left\{f(4k^2)\sqrt{t}\sin(\pi\sqrt{t}) - 4k\tilde{f}(4k^2)\sin^2\left(\frac{\pi}{2}\sqrt{t}\right)\right\}. \qquad (3.2.17)$$

This indicates that the sampling series associated with the boundary-value problems (3.2.7), (3.2.8) and (3.2.11), (3.2.12) are essentially different.

Notice that the sampling rate in (3.2.17) is smaller than that in (3.2.10), which is advantageous from a practical point of view since fewer samples are needed to reconstruct f; however, the price we pay for this is that we need to sample $\tilde{f}$ as well. More on this phenomenon will be discussed in Section 3.6.

Reconstructing f from the samples of its Hilbert transform $\tilde{f}$ is not new (see for example sec. 3.2 in [16], [28, p. 76] and [102]); however, what is new here is its connection with the boundary-value problem (3.2.11), (3.2.12).

To sum it all up, Kramer's assumption does not appear to be the right sort of assumption for his sampling series expansion. We shall introduce more appropriate assumptions in Chapters 4 through 6.

In practice, however, almost all the known examples are produced by the regular Sturm-Liouville boundary-value problem:

$$-y'' + q(x)y = ty, \quad x \in I = [a,b] \qquad (3.2.18)$$

$$y(a)\cos\alpha + y'(a)\sin\alpha = 0, \qquad (3.2.19)$$

$$y(b)\cos\beta + y'(b)\sin\beta = 0, \qquad (3.2.20)$$

where q is a continuous function on I.

The existence of $\phi(x,t)$ in this case is always guaranteed; just take $\phi(x,t)$ to be a solution of (3.2.18) and (3.2.19) (or (3.2.18) and (3.2.20)). For, the eigenvalues $\{t_k\}$ are the zeros of the function $\phi(b,t)\cos\beta+\phi'(b,t)\sin\beta$ and the eigenfunctions may be taken as $\phi_k(x)=\phi(x,t_k)$.

This is in contrast with the boundary-value problem (3.2.11) and (3.2.12). In general, if either the boundary conditions are of mixed type, i.e., they involve the values of y and its derivatives at the endpoints $x=a$, $x=b$, or if the differential operator L is of order $n>2$, then the existence of $\phi(x,\lambda)$ is not always guaranteed.

The sampling series expansions associated with the regular Sturm-Liouville boundary-value problem (3.2.18) through (3.2.20) is essentially Weiss' result, except Weiss considered the differential equation

$$(p(x)y')'+(tr(x)-q(x))y=0, \quad a\le x\le b, \tag{3.2.21}$$

where $p(x)$, $r(x)>0$, $q(x)\ge 0$ and $p(x)$ has a continuous first derivative. But, (3.2.21) can be transformed back into (3.2.18) via the substitutions

$$z=\int\left[\frac{r(x)}{p(x)}\right]^{1/2}dx, \quad u=[r(x)p(x)]^{1/4}y$$

to yield

$$u''-Q(z)u=tu,$$

in which

$$Q(z)=\frac{\theta''(z)}{\theta(z)}+\frac{q(x)}{r(x)} \quad \text{and} \quad \theta(z)=[r(x)p(x)]^{1/4}; \quad \text{cf. [104], p. 22.}$$

To show that his sampling theorem implies the WSK sampling theorem, Kramer considered the 1st order boundary-value problem:

$$-iy'=ty, \quad -\sigma\le x\le\sigma$$

with

$$y(-\sigma)=y(\sigma).$$

The eigenvalues and eigenfunctions of this problem are $t_k = k\pi/\sigma$ and $\phi_k(x) = e^{it_k x}$, $k = 0, \pm 1, \pm 2, \ldots$. Hence, the Sampling Theorem 3.5 takes the form:

If

$$f(t) = \int_{-\sigma}^{\sigma} F(x) e^{itx} dx ,$$

for some $F \in L^2(-\sigma, \sigma)$, then $S_k^*(t) = \sin\sigma(t - t_k)/\sigma(t - t_k)$ and

$$f(t) = \sum_{k=-\infty}^{\infty} f(t_k) \frac{\sin\sigma(t - t_k)}{\sigma(t - t_k)} .$$

Finally, Kramer concluded his paper by considering the boundary-value problem:

$$y'' - \frac{(\nu^2 - 1/4)}{x^2} y = ty , \quad 0 \le x \le 1 , \tag{3.2.22}$$

with

$$y(0) = 0 \quad \text{and} \quad y(1) = 0 , \tag{3.2.23}$$

to deduce the sampling series expansion

$$f(t) = \sum_{k=1}^{\infty} f(t_k) \frac{2\sqrt{t_k} J_\nu(\sqrt{t})}{(t_k - t) J_\nu'(\sqrt{t_k})} \tag{3.2.24}$$

for

$$f(t) = \int_0^1 F(x) \sqrt{x} J_\nu(x\sqrt{t}) dx , \quad \text{with} \quad \sqrt{x} F(x) \in L^2[0, 1]$$

where $J_\nu(z)$ is the Bessel function of the 1st kind and order ν, and t_k is the kth non-null zero of $J_\nu(\sqrt{z})$.

A close inspection of this innocuous example reveals that the boundary-value problem (3.2.22) and (3.2.23) does not really satisfy Kramer's hypotheses because of the singularity at $x = 0$ in the differential equation. Nevertheless, Kramer was able to derive the sampling series expansion (3.2.24). Therefore, it is fair to say that Kramer knew—although he never mentioned it explicitly—that his theorem could be extended to some singular boundary-value problems. Few years later, Kramer-type sampling series associated with other singular Sturm-Liouville boundary-value problems, such as boundary-value problems involving the Legendre and Laguerre differential equations, were discovered; see [43, 44, 93, and 121].

This raises the question of whether the Weiss-Kramer theorem could be extended to singular boundary-value problems. Since in sampling series expansions associated with boundary-value problems the samples are taken at the eigenvalues of the problem, and since singular problems, generally speaking, have continuous spectra, there is no hope to have an affirmative answer.

Well, is it then safe to say that the Weiss-Kramer sampling theorem holds for singular boundary-value problems with discrete spectra? Alas, the answer is still negative. A counter example is provided by the singular Sturm-Liouville problem

$$y'' - x^2 y = -ty\,, \quad -\infty < x < \infty$$

$$\lim_{|x| \to \infty} |y(x)| < \infty\,.$$

The spectrum is discrete; in fact, the eigenvalues and eigenfunctions are $t_k = 2k+1$ and $\phi_k(x) = \exp(-x^2/2) H_k(x)$, $k = 0, 1, \ldots$ respectively, where $H_k(x)$ is the Hermite polynomial of degree k. Kramer's assumption that the eigenfunctions are all generated by one single function seems to hold in this case as well. In fact, the parabolic cylindrical function $\phi(x,t)$ is such a function. This appears to be in contradiction with Kramer's theorem; yet a careful analysis of the problem reveals that there is no contradiction because Kramer's proof does not necessarily hold for singular problems. The reason that Theorem 3.5 does not hold in this case is that the function $\phi(x,t)$ that generates the eigenfunctions is not in $L^2(-\infty,\infty)$. We shall discuss this example in more detail in Chapter 4.

The questions of whether the Weiss-Kramer sampling theorem gives essentially new results and whether it is an improvement over the WSK sampling theorem or not, were partially answered by Campbell [20], who showed that if the kernel $K(x,t)$ arises from either the regular Sturm-Liouville boundary-value problem (3.2.18) through (3.2.20) or from the first order boundary-value problem

$$-iy' + q(x)y = ty\,, \quad x \in [-\sigma, \sigma]$$

$$y(-\sigma) = y(\sigma) e^{i\phi}\,, \quad -\pi < \phi \le \pi\,,$$

then any function that can be represented by Kramer's sampling series (3.2.2), can also be represented by the WSK sampling series (2.1.2). In short, if Kramer's sampling theorem applies to $f(t)$, then so does the WSK theorem; see also [45]. Recently, this close connection between the WSK and Kramer sampling series has been studied more thoroughly in [120, 122], where it has been shown that both series are indeed special cases of Lagrange-type interpolation series.

Campbell also conjectured that his result might be true if the kernel $K(x,t)$ arises from self-adjoint boundary-value problems associated with nth order differential operators with $n > 2$. He then proceeded to show that his result was still valid even in two cases where the kernel arose from singular Sturm-Liouville problems, namely boundary-value problems associated with the Bessel and Legendre differential equations.

A. Jerri [41, 43 and 44] also derived Kramer-type sampling series associated with singular boundary-value problems involving the associated Legendre, Gegenbauer, Chebyshev, Laguerre and Hermite differential equations; see also [93 and 120].

Nevertheless, the study of sampling theorems associated with general singular Sturm-Liouville boundary-value problems was confined to specific examples until it was fully developed in 1991 [100]. This will be the main topic of Chapter 4.

3.3 Multidimensional Sampling

One of the most natural and practical extensions of the WSK sampling theorem is its extension to higher dimensions. Multidimensional signals are abundant in nature. A black and white picture can be thought of as a two-dimensional signal taking values between zero and one, where black areas for example are described by zero, white areas by one, and intermediate grey areas by numbers between these two limits. Similarly, a colored T.V. signal can be thought of as a multidimensional signal or equivalently as a function of several variables.

For reconstruction of multidimensional signals, multidimensional sampling theorems are needed. E. Parzen [78] was the first to generalize the WSK sampling theorem to band-limited signals in higher dimensions; however, he never attempted to publish his result as a paper because he felt that it was a straightforward generalization.

Definition 3.3.1. Let $G \subset \Re^N$ $(N \geq 1)$ be a bounded set symmetric with respect to the origin, then a function $f(t_1, \ldots, t_N)$ of N variables is said to be band-limited to G if there exists a function $F(\omega_1, \ldots, \omega_N)$ with support in G such that

$$f(\underline{t}) = \frac{1}{(\sqrt{2\pi})^N} \int_G F(\underline{\omega}) e^{i(\underline{t}\cdot\underline{\omega})} d\underline{\omega},$$

with

$$\int_G |F(\underline{\omega})|^2 d\underline{\omega} < \infty,$$

where $\underline{t} = (t_1, \ldots, t_N)$, $\underline{\omega} = (\omega_1, \ldots, \omega_N)$, $\underline{t} \cdot \underline{\omega} = \sum_{i=1}^{N} t_i \omega_i$, and $d\underline{\omega}$ means that the integration is with respect to the N-dimensional Lebesgue measure.

Functions that are band-limited to N-dimensional rectangles symmetric about the origin are uniquely determined by their sampled values at an N-dimensional sampling lattice; moreover, they can be reconstructed from the knowledge of these samples as can be seen from Parzen's theorem.

THEOREM 3.6 (Parzen [78])

Let $f(t_1, \ldots, t_N)$ be a function band-limited to the N-dimensional rectangle $B = \prod_{i=1}^{N} (-\sigma_i, \sigma_i)$, $\sigma_i > 0$, $i = 1, \ldots, N$, i.e., its Fourier transform $F(\omega_1, \ldots, \omega_N)$ is such that

$$\int_{-\sigma_1}^{\sigma_1} \cdots \int_{-\sigma_N}^{\sigma_N} |F(\omega_1, \ldots, \omega_N)|^2 \, d\omega_1 \ldots d\omega_N < \infty ,$$

and

$$F(\omega_1, \ldots, \omega_N) = 0 , \quad \text{for} \quad |\omega_k| > \sigma_k > 0 , \quad k = 1, 2, \ldots, N ,$$

then

$$f(t_1, \ldots, t_N) = \sum_{k_1=-\infty}^{\infty} \cdots \sum_{k_N=-\infty}^{\infty} f\left(\frac{\pi k_1}{\sigma_1}, \ldots, \frac{\pi k_N}{\sigma_N}\right) \frac{\sin(\sigma_1 t_1 - k_1 \pi)}{(\sigma_1 t_1 - k_1 \pi)} \cdots \frac{\sin(\sigma_N t_N - k_N \pi)}{(\sigma_N t_N - k_N \pi)} .$$

The proof mimics that of the WSK sampling theorem.

Parzen's result can, in turn, be generalized to an N-dimensional analogue of the Paley-Wiener interpolation theorem for band-limited signals as follows:

THEOREM 3.7 (Paley-Wiener-Parzen)

Let G_i be the entire function in t_i defined by

$$G_i(t_i) = (t_i - t_{i,0}) \prod_{k_i=1}^{\infty} \left(1 - \frac{t_i}{t_{i,k_i}}\right)\left(1 - \frac{t_i}{t_{i,-k_i}}\right), \quad i = 1, 2, \ldots, N ,$$

where t_{i,k_i} are real numbers satisfying the estimate

$$\sup_{k_i \in \mathbb{Z}} \left| t_{i,k_i} - \frac{k_i \pi}{\sigma_i} \right| < \frac{\pi}{4\sigma_i}, \quad i = 1, 2, \ldots, N .$$

Then, for any signal in the separable form $f(t_1,\ldots,t_N) = f_1(t_1)\ldots f_N(t_N)$ that is band-limited to the N-dimensional rectangle $B = \prod_{i=1}^{N}[-\sigma_i,\sigma_i]$, we have

$$f(t_1,\ldots,t_N) = \sum_{k_1=-\infty}^{\infty}\cdots\sum_{k_N=-\infty}^{\infty} f\left(t_{1,k_1},\ldots,t_{N,k_N}\right)\frac{G_1(t_1)}{G_1'(t_{1,k_1})(t_1-t_{1,k_1})}\cdots\frac{G_N(t_N)}{G_N'(t_{N,k_N})(t_N-t_{N,k_N})}.$$

If the signal is not in the above form, then the theorem is no longer true, and no simple explicit sampling series expansion is known in the general case. However, for $N = 2$, we have

THEOREM 3.8 (Butzer and Hinsen [17])

Let $\{x_n\}_{n=-\infty}^{\infty}$, $\{y_{n,m}\}_{m=-\infty}^{\infty}$ be two sequences of real numbers satisfying

$$\sup_{n\in Z}\left|x_n - \frac{n\pi}{a}\right| < \frac{\pi}{4a} \quad \text{and} \quad \sup_{m\in Z}\left|y_{n,m} - \frac{m\pi}{b}\right| < \frac{\pi}{4b} \quad \text{for each } n.$$

Then for any function $f(x,y)$ band-limited (in the conventional sense) to the rectangle $B = I\times J$ where $I = [-a,a]$ and $J = [-b,b]$, we have

$$f(x,y) = \sum_{n=-\infty}^{\infty}\sum_{m=-\infty}^{\infty} f(x_n, y_{n,m})\Psi_{n,m}(x,y), \tag{3.3.1}$$

where

$$\Psi_{n,m}(x,y) = \frac{G(x)}{(x-x_n)G'(x_n)}\,\frac{G_n(y)}{(y-y_{n,m})G_n'(y_{n,m})} = \Psi_n(x)\Psi_{n,m}(y),$$

$$G(x) = (x-x_0)\Pi_{n=1}^{\infty}\left(1-\frac{x}{x_n}\right)\left(1-\frac{x}{x_{-n}}\right)$$

and

$$G_n(y) = (y-y_{n,0})\Pi_{m=1}^{\infty}\left(1-\frac{y}{y_{n,m}}\right)\left(1-\frac{y}{y_{n,-m}}\right).$$

The series (3.3.1) converges to $f(x,y)$ uniformly on each compact subset of the complex plane.

Proof. Since f is band-limited to B, there exists an entire function of two complex variables $F(z_1, z_2)$ [67, p. 132] satisfying

$$|F(z_1, z_2)| \leq C \exp(a|Im z_1| + b|Im z_2|)$$

such that $F(x, y) = f(x, y)$ for all $(x, y) \in \Re^2$. Now, it is easy to see that for each fixed y, $f(x, y)$ is band-limited to I as a function of x. Therefore, by Theorem 3.1

$$f(x, y) = \sum_{n=-\infty}^{\infty} f(x_n, y) \Psi_n(x). \tag{3.3.2}$$

But again, each of the terms $f(x_n, y)$ is band-limited to J as a function of y; hence another application of Theorem 3.1 yields

$$f(x_n, y) = \sum_{m=-\infty}^{\infty} f(x_n, y_{n,m}) \Psi_{n,m}(y). \tag{3.3.3}$$

By combining (3.3.2) and (3.3.3), we obtain (3.3.1). The uniform convergence follows from iterative applications of Theorem 3.1. ▮

One of the shortcomings of this approach is that all the sampling points must lie on straight lines parallel to the y-axis. However, these lines are not necessarily equally spaced and the points on each line need not be uniformly distributed; see also [65].

Except for some notational difficulties, this theorem can be extended to higher dimensions and to functions that are not necessarily band-limited in the conventional sense.

For more detailed treatments of sampling theorems in N dimensions ($N \geq 2$), we refer the reader to the work of Petersen [79]; Petersen and Middleton [80, 81] and Prosser [84].

Sampling series expansions for radially symmetric functions that are band-limited to the unit sphere in $\Re^N$ have also been obtained. Expectedly, because of the radial symmetry, the sampling series expansions for such functions involve the Bessel function of order $(N/2) - 1$ [40, 73 and 95].

Sampling series expansions for functions band-limited to other regions that are symmetric about the origin in $\Re^2$, such as parallelograms and discs, have been derived using different techniques. For example, for functions band-limited to a disc centered at the origin, Blazek [8, 9], Stark [100], and Stark and Sarna [101] used sampling in polar coordinates; see also [56].

For functions band-limited to a general region in $\Re^N$, no general sampling theorem was available until recently. This was a gap in the theory of sampling of band-limited signals because of a theorem of Plancherel and Polya. But before we state this theorem, let us introduce the space $\mathbf{B}^p_\Omega(\mathfrak{R}^N)$:

$$\mathbf{B}^p_\Omega(\mathfrak{R}^N) = \{f : f \text{ is a complex-valued tempered distribution, } \operatorname{supp} \hat{f} \subset \Omega, \|f\|_p < \infty\},$$

$$\|f\|_p = \left(\int_{\mathbf{R}^N} |f(\underline{x})|^p \, d\underline{x} \right)^{1/p}, \quad 1 \le p < \infty.$$

In virtue of an N-dimensional version of the Paley-Wiener-Schwartz theorem ([27, Vol. II, p. 131], see also Section 2.2), members of the space $\mathbf{B}^p_\Omega(\mathfrak{R}^N)$ are entire functions of exponential type. Therefore, it appears not only natural but also consistent with the definition of band-limited signals introduced in Section 2.2 to define the space $\mathbf{B}^p_\Omega(\mathfrak{R}^N)$ $(1 \le p < \infty)$ as the space of signals band-limited to Ω. As a special case, we obtain for $N = 1$, $\Omega = (-\sigma, \sigma)$, $1 \le p \le 2$, that $\mathbf{B}^p_\Omega(\mathfrak{R}^1) = \mathbf{B}^p_\sigma$. Now we can state the Plancherel-Polya theorem.

THEOREM 3.9 (Plancherel-Polya [106])

If $f \in \mathbf{B}^p_\Omega(\mathfrak{R}^N)$ and $\{t_k\}_{k=1}^\infty$ is an appropriate set of points in $\mathfrak{R}^N$, e.g., lattice points where the length of the mesh is sufficiently small, then there exist two positive constants C_1 and C_2 such that

$$C_1 \|f\|_p \le \left(\sum_{k=1}^\infty |f(t_k)|^p \right)^{1/p} \le C_2 \|f\|_p .$$

The theorem essentially says that under suitable conditions on the sampling points $\{t_k\}$, the discrete l^p-norm of the sampled values $\{f(t_k)\}_{k=1}^\infty$ defines an equivalent norm on the band-limited $L^p(\mathfrak{R}^N)$ functions. Thus, a band-limited signal f is uniquely determined by its sampled values $\{f(t_k)\}_{k=1}^\infty$ taken at an appropriate set of points. Unfortunately, Theorem 3.9 does not offer any clues to how f can be reconstructed from its sampled values. Thus, despite its important theoretical ramifications, this theorem has no practical implementations.

The most commonly used technique to tackle this problem is to enclose the spectral support of the band-limited signal $f(\underline{t})$, the region Ω in $\mathfrak{R}^N$ over which the Fourier transform $F(\underline{\omega})$ of f is nonvanishing, in a suitable set Γ called a period cell. The latter is any region that, when translated by the column vectors of a periodicity matrix (for the definition of a periodicity matrix, see [57, p. 178], will fill the entire space without gaps. By covering the entire space with copies of Γ, we obtain a periodic extension of $F(\underline{\omega})$ to $\mathfrak{R}^N$, which can be expanded in a Fourier series. Lastly, by taking the inverse Fourier transform of this series, the sampling series expansion of $f(\underline{t})$ is obtained. This is in essence a generalization of the proof of Theorem 2.1.

The choice of Γ is arbitrary, but it must be in such a way that the replications of the spectral support Ω do not overlap and hence alias. The geometry of the sampling lattice depends on the choice of Γ and a sensible choice of the latter will be the one that minimizes the sampling rate. The lowest rate at which a multidimensional band-limited signal can be sampled without aliasing is the Nyquist rate. Unlike in one-dimensional sampling, in multidimensional sampling there may be more than one sampling geometry that can achieve the Nyquist rate.

Although rectangular sampling, which corresponds to evaluating the function at sampling points located at the corners of a hypercube, is the most obvious generalization of one-dimensional sampling, it is not always the most efficient for some applications. It has been shown in [62 and 63] that one type of non-rectangular sampling, namely hexagonal sampling, and its higher dimensional generalizations yield a lower sampling density and more efficient signal processing algorithms for some applications, such as phased array antennas [63].

New and promising results concerning the reconstruction of band-limited multidimensional signals have recently been reported by H. Feichtinger and K. Gröchenig [23-25] and K. Gröchenig [30]. We shall discuss some of these results in more details in Chapter 9, where we shall also derive, by using the Green's function in several variables, a Kramer-type sampling theorem for integral transforms of functions supported in a general simply connected bounded domain in $\Re^N$.

We conclude this section by noting that the reconstruction problem of multidimensional signals from their zero crossings is much more difficult than that for one dimensional signals since the zero crossings of a multidimensional signal are, in general, contours rather than isolated points. Although much less research has been devoted to this problem than in the one-dimensional case some interesting results have been obtained in [21, 68, 70 and 87], where in the latter Logan's one-dimensional result for bandpass signals (cf. Section 3.1) has been extended to 2 dimensions.

In closing, we should emphasize that although the topic of multidimensional sampling is very wide and full of deep and profound results, we have resisted the temptation to discuss and elaborate on its different aspects and instead concentrated more on those that are related to our subsequent presentation.

3.4 Sampling Theorems and Generalized Functions

Let us recall that if

$$f(t) = \frac{1}{\sqrt{2\pi}} \int_{-\sigma}^{\sigma} F(\omega) e^{it\omega} d\omega ,$$

for some integrable function F, then

$$f(t) = \sum_{k=-\infty}^{\infty} f(t_k) \frac{\sin \sigma (t - t_k)}{\sigma (t - t_k)} .$$

If the integrability condition is dropped or if the function F is replaced by a generalized function, then the sampling series may not converge.

For example, if $F(\omega) = \delta'(\omega + a)$, $-\sigma < a < \sigma$, then $f(t) = (it/\sqrt{2\pi}) e^{-iat}$. Therefore, by substituting this in the sampling series (2.1.2), we obtain a divergent series due to the fact that $f(k\pi/\sigma) = O(k)$ as $|k| \to \infty$. Whereas, if $F(\omega) = \delta(\omega + a)$; $-\sigma < a < \sigma$, then $f(t) = (1/\sqrt{2\pi}) e^{-iat}$, and another substitution into the sampling series (2.1.2) yields,

$$e^{-iat} = \sum_{k=-\infty}^{\infty} (e^{-ik\pi a/\sigma}) \frac{\sin \sigma (t - t_k)}{\sigma (t - t_k)} ,$$

which is correct since it is easily verified that the series on the right-hand side is just the Fourier series expansion of the function on the left-hand side when regarded as a function in a.

The question of whether the WSK sampling theorem is valid when $F(\omega)$ is a generalized function with support in $(-\sigma, \sigma)$ was first investigated by L. Campbell in [19]. Before stating Campbell's result, let us recall the following facts from the theory of Generalized Functions [27 and 123].

Consider the vector spaces

$$E = \{\phi : \phi \in C^{\infty}\} ,$$

$$\mathcal{D} = \{\phi : \phi \in C^{\infty}, \operatorname{supp} \phi \subset [-a, a] \text{ for some } 0 < a < \infty\} ,$$

$$K(\sigma) = \{\phi : \phi \in C^{\infty}, \operatorname{supp} \phi \subset [-\sigma, \sigma], \text{for some fixed } \sigma > 0\} ,$$

$$S = \left\{\phi : \phi \in C^{\infty}, \lim_{|x| \to \infty} (1 + |x|)^p |\phi^{(q)}(x)| = 0; p, q = 0, 1, 2, \ldots\right\} ,$$

and

$$Z(\sigma) = \{\hat{\phi} : \hat{\phi} \text{ is the Fourier transform of some } \phi \in K(\sigma)\} .$$

Provided with appropriate topologies, these vector spaces become testing-function spaces in the sense of Zemanian [123]. Let E^*, $\mathcal{D}^*$, $K^*(\sigma)$, S^*, $Z^*(\sigma)$ denote their dual spaces respectively. The space S is known as the Schwartz space of C^∞ rapidly decreasing functions and its dual S^* is the space of tempered distributions. $\mathcal{D}^*$ is the space of Schwartz distributions.

It is known that if F is a Schwartz distribution with compact support, then it is an element of E^*, and since $e^{itw} \in E$ for any fixed t, we can define the Fourier transform of F by

$$f(t) = \frac{1}{\sqrt{2\pi}} \langle F(w), e^{itw} \rangle ,$$

where $\langle F, \phi \rangle$ denotes the number that the functional F assigns to the element ϕ. It can be shown [27, p. 131] that if $\operatorname{supp} F \subset [-\sigma, \sigma]$, then f is an entire function of exponential type satisfying

$$|f(t)| \le C \exp((\sigma + \varepsilon) |t|), \quad t \in \mathbb{C}$$

and for each nonnegative integer m

$$|f(t)| \le C(1 + |t|)^m, \quad t \in \Re .$$

We can now state two of the main sampling theorems related to generalized functions.

THEOREM 3.10 (Campbell [19])

Let $F(w)$ be a distribution with support contained in the open interval $\{\omega : |\omega| < (1-q)\sigma, 0 < q < 1\}$. Let $f(t)$ be its inverse Fourier transform. Then

$$f(t) = \sum_{k=-\infty}^{\infty} f(t_k) \frac{\sin \sigma(t - t_k)}{\sigma(t - t_k)} S(q\sigma[t - t_k]),$$

where

$$S(y) = \frac{\left\{ \int_{-1}^{1} \exp[(1-x^2)^{-1} - ixy] dx \right\}}{\left\{ \int_{-1}^{1} \exp[(1-x^2)^{-1}] dx \right\}},$$

$t_k = (k\pi)/\sigma$, and the series converges pointwise.

Campbell's sampling series lacks the familiar appearance of the WSK sampling theorem because of the presence of the S factors that were introduced to speed up the convergence of the series. If the notion of convergence is relaxed from pointwise to convergence in the generalized function sense, then an analogue of the WSK sampling theorem can be obtained for a larger class of generalized functions.

THEOREM 3.11 (Pfaffelhuber [82])

Let $f \in S^*$ be such that its Fourier transform F has compact support in $(-\sigma, \sigma)$. Then

$$f(t) = \sum_{k=-\infty}^{\infty} f(t_k) \frac{\sin \sigma(t - t_k)}{\sigma(t - t_k)},$$

where the series converges in the sense of $Z^*(\sigma)$.

3.5 Sampling Theorems for Other Types of Signals

3.5.A Time-Limited Signals (Nonband-Limited Signals)

We have seen at the end of Section 2.2 that a signal cannot be both band-limited and time-limited simultaneously. It follows immediately that for a time-limited signal $f(t)$, $\sigma = \infty$; therefore, the sampling series (2.1.2) is not defined. Nevertheless, since the series converges for any $0 < \sigma < \infty$ it may be reasonable to anticipate that it will converge to $f(t)$ as $\sigma \to \infty$; that is

$$f(t) = \lim_{\sigma \to \infty} \sum_{k=-\infty}^{\infty} f\left(\frac{k\pi}{\sigma}\right) \frac{\sin(\sigma t - k\pi)}{(\sigma t - k\pi)}. \tag{3.5.1}$$

According to Butzer and co-workers [15 and 16], De la Vallée Poussin [109] was the first to deal with series of the form given in (3.5.1), and hence he was also the first to consider sampling expansions of nonband-limited functions. In fact, his work, which began in 1908, was concerned with a slightly different but related problem that can be formulated as follows: Given a bounded function f on some finite interval $[a, b]$, a set of points $\{t_k = k\pi/m,\ k = 0, \pm1, \pm2, \ldots$ and $m = n$ or $m = n + 1/2, n = 1, 2, 3, \ldots\}$ and

$$F_m(t) = \sum_{t_k \in [a,b)} f(t_k) \frac{\sin m(t - t_k)}{m(t - t_k)}, \tag{3.5.2}$$

where $f(t_k)$ is assumed to be zero if $t_k \notin [a, b)$, what is the behavior of $F_m(t)$ as $m \to \infty$? In particular, does $F_m(t)$ converge to f as $m \to \infty$?

Unlike the summation in (3.5.1), the one in (3.5.2) contains only a finite number of terms. However, if f is regarded as a time-limited function, i.e., $f(t) = 0$ whenever $t \notin [a,b)$, then (3.5.2) can be rewritten as

$$F_m(t) = \sum_{k=-\infty}^{\infty} f(t_k) \frac{\sin m(t - t_k)}{m(t - t_k)}.$$

One of the differences between this series and the one in (3.5.1) is that the parameter m is discrete whereas σ is assumed to be continuous. At any rate, De la Vallée Poussin proved that the behavior of F_m as $m \to \infty$ is similar to that of the partial sums of the Fourier series of f, namely, for a bounded and Riemann integrable function f on $[a,b]$, if f is continuous at $t_0 \in (a,b)$ and of bounded variation in $[t_0 - \varepsilon, t_0 + \varepsilon]$ for some $\varepsilon > 0$, then

$$\lim_{m \to \infty} F_m(t_0) = f(t_0);$$

moreover, if f is continuous and of bounded variation on $[a,b]$, then

$$f(t) = \lim_{m \to \infty} F_m(t) = \lim_{m \to \infty} \sum_{k=-\infty}^{\infty} f(t_k) \frac{\sin m(t - t_k)}{m(t - t_k)}$$

uniformly on any subinterval $[c,d] \subset (a,b)$.

This last formula can be thought of as a sampling series for a time-limited function or equivalently a nonband-limited function. A generalization of this result was formulated by M. Theis [103] in 1919 as follows: Let f be a continuous function on $[a,b]$ that vanishes outside that interval, then

$$f(t) = \lim_{m \to \infty} \sum_{k=-\infty}^{\infty} f(t_k) \left\{ \frac{\sin m(t - t_k)}{m(t - t_k)} \right\}^2,$$

uniformly on any subinterval $[c,d] \subset (a,b)$.

The condition that f is of bounded variation is not needed here as is the case in the theory of Fourier series.

Interest in sampling theorems for nonband-limited functions was revived in the early sixties with the publication of P. Weiss' paper [112], but this time the research has taken a different approach, mostly in the form of studying the aliasing error and using properties of the Fourier transform of f.

The aliasing error $(R_\sigma f)$ is defined by

$$(R_\sigma f)(t) = f(t) - \sum_{k=-\infty}^{\infty} f\left(\frac{k\pi}{\sigma}\right) \frac{\sin(\sigma t - k\pi)}{(\sigma t - k\pi)}.$$

For a nonband-limited signal, the following questions naturally arise: under what conditions does $\lim_{\sigma\to\infty}(R_\sigma f) = f$?; for a given f, how large is the aliasing error $(R_\sigma f)$? Answers to these questions and other related ones have been the focus of extensive research; see [16, 97 and 98] for more details. Since aliasing errors will be investigated further in Section 3.8, here we will only give one simple answer in the form of

THEOREM 3.12 (Butzer et al. [16])

If $f \in L^2(\mathfrak{R}) \cap C(\mathfrak{R})$ and $\hat{f} = F \in L^1(\mathfrak{R})$, then (3.5.1) holds uniformly in t and

$$\|(R_\sigma f)\|_\infty \le \sqrt{2/\pi} \int_{|\omega|>\sigma} |F(\omega)|\, d\omega . \tag{3.5.3}$$

Moreover, if $(i\omega)^n F(\omega) \in L^1(\mathfrak{R})$ for some nonnegative integer n, then

$$|f^{(n)}(t) - (R_\sigma f)^{(n)}(t)| \le \sqrt{2/\pi} \int_{|\omega|\ge\sigma} |\omega|^n |F(\omega)|\, d\omega .$$

When f is band-limited, the bound in (3.5.3) implies that the aliasing error is zero.

Splettstösser [96] has obtained some related results for multidimensional non-band-limited signals and multidimensional aliasing errors.

It is worth noting that sampling series expansions for some non-band-limited signals represented by integrals other than the Fourier one have also been obtained. For example, A. Papoulis [72] gave a sampling series expansion for a signal represented by the infinite limit Hilbert transform, and similarly did A. Jerri [41] for an infinite limit Laguerre and Hermite transforms. More recently, A. Zayed [120] has obtained a sampling series expansion, analogous to the WSK sampling series, for signals represented by integral transforms taken over infinite intervals with kernel functions generated from singular Sturm-Liouville boundary-value problems. This sampling series expansion has a very surprising feature: it does look almost exactly like the WSK sampling series, except that the summation is one sided, i.e., it extends only from zero to infinity, but unlike the WSK sampling series, it does converge to a non-band-limited signal; see Chapter 4 for more details.

3.5.B Band-Pass Signals

Recall that a band-pass signal is a signal that contains no high and no low frequencies; that is, mathematically, it is a bounded function f whose Fourier transform F is supported in $[-\omega_0-\sigma, -\omega_0+\sigma] \cup [\omega_0-\sigma, \omega_0+\sigma]$, $\omega_0 - \sigma > 0$.

Though not very efficient, reconstructing such a signal by using the WSK sampling theorem is possible by regarding it as a signal band-limited to $[-\omega_0-\sigma, \omega_0+\sigma]$. More efficient sampling series expansions have been obtained by using different methods [11, 72 and 49]. Goldman [28, p. 75] suggested the following approach. Set

$$g(t) = \sqrt{2/\pi} \int_{\omega_0-\sigma}^{\omega_0+\sigma} F(\omega) e^{it\omega} d\omega ,$$

where

$$f(t) = \frac{1}{\sqrt{2\pi}} \left\{ \int_{-\omega_0-\sigma}^{-\omega_0+\sigma} F(\omega) e^{it\omega} d\omega + \int_{\omega_0-\sigma}^{\omega_0+\sigma} F(\omega) e^{it\omega} d\omega \right\}.$$

Then, it is readily seen that $g(t) = f(t) - i\tilde{f}(t)$, where $\tilde{f}$ is the Hilbert transform of f, which can be given by

$$\tilde{f}(t) = \sqrt{2/\pi} \left\{ \int_{\omega_0-\sigma}^{\omega_0+\sigma} (b(\omega) \cos \omega t - a(\omega) \sin \omega t)\, d\omega \right\},$$

and $F(\omega) = a(\omega) - ib(\omega)$.

By noting that $f(t) = Re\, g(t)$ and that $g(t)$ has a shifted sampling series of the type given by (2.1.5), we can, with some straightforward calculations, derive the sampling series expansion

$$f(t) = \sum_{k=-\infty}^{\infty} \{ f(t_k) \cos \omega_0 (t - t_k) + \tilde{f}(t_k) \sin \omega_0 (t - t_k) \} \frac{\sin \sigma (t - t_k)}{\sigma (t - t_k)}, \quad (3.5.4)$$

which, as (3.2.17), shows that f can be reconstructed from the samples of its Hilbert transform.

Sharma and Mehta [94] derived a Kramer-type sampling theorem for band-pass signals. Their results can be stated as follows:

THEOREM 3.13 (Sharma and Mehta [94])

Let $g(\omega)$ be a complex-valued function such that $g \in L^1(\mathfrak{R})$, and $K(t, \omega)$ be a continuous complex function of time such that $|K(t,\omega)| = |K(t,-\omega)|$.

Let f be a real-valued signal that is band-limited to the band-pass region $B = [-\omega_0 - \sigma, -\omega_0 + \sigma] \cup [\omega_0 - \sigma, \omega_0 + \sigma]$. If $\{K(t_k, \omega)\}$ is an orthogonal family on $L^2(\omega_0 - \sigma, \omega_0 + \sigma)$ and

$$f(t) = \int_B g(\omega) K(t,\omega)\, d\omega ,$$

then

$$f(t) = \sum_{k=-\infty}^{\infty} f(t_k) L_k(t),$$

where

$$L_k(t) = \frac{2\int_B K(t,\omega)\overline{K(t_k,\omega)}\, d\omega}{\int_B |K(t_k,\omega)|^2\, d\omega},$$

and $t_k = (k\pi)/\sigma$. The proof of this theorem is straightforward and mimics those of Theorem 3.5 and (3.5.4).

3.5.C Finite Power Signals

A signal $f(t)$ is said to be of finite power if its average power (over the whole real line) $\tilde{E}$ is finite, i.e.,

$$\tilde{E} = \lim_{T\to\infty} \frac{1}{2T} \int_{-T}^{T} |f(t)|^2 dt < \infty . \tag{3.5.5}$$

Clearly, any finite-energy signal f ($f \in L^2(\Re)$) is of finite power. In fact the class of all finite power signals contains $L^p(\Re)$ for any $p \geq 2$ since, in view of Hölder's inequality, we have

$$\frac{1}{2T}\int_{-T}^{T} |f(t)|^2 dt \leq \frac{1}{(2T)}\left(\int_{-T}^{T} |f(t)|^{2p} dt\right)^{1/p} \left(\int_{-T}^{T} 1^q dt\right)^{1/q}$$

$$= \frac{1}{(2T)^{1-1/q}}\left(\int_{-T}^{T} |f(t)|^r dt\right)^{2/r},$$

where $1 \leq p,q < \infty$, $(1/p)+(1/q) = 1$, $r = 2p$.

The result now follows by taking the limit of both sides of the above inequality as $T \to \infty$. The case $p = \infty$ is trivial and is left to the reader. For any constant A, the average power $\tilde{E}$ can also be written in the form

$$\tilde{E} = \lim_{T\to\infty} \frac{1}{2T} \int_{-T+A}^{T+A} |f(t)|^2 dt = \lim_{T\to\infty} \frac{1}{2T} \int_{-T}^{T} |f(t+A)|^2 dt . \qquad (3.5.6)$$

For, if we write $\tilde{E} = \tilde{E}_1 + \tilde{E}_2$, where

$$\tilde{E}_1 = \lim_{T\to\infty} \frac{1}{2T} \int_{-T}^{0} |f(t)|^2 dt \quad \text{and} \quad \tilde{E}_2 = \lim_{T\to\infty} \frac{1}{2T} \int_{0}^{T} |f(t)|^2 dt ,$$

then

$$\lim_{T\to\infty} \frac{1}{2T} \int_{-T+A}^{T+A} |f(t)|^2 dt = \lim_{T\to\infty} \left(\frac{1}{2T} \int_{-T+A}^{0} |f(t)|^2 dt + \frac{1}{2T} \int_{0}^{T+A} |f(t)|^2 dt \right)$$

$$= \lim_{T\to\infty} \left(\frac{2(T-A)}{2T} \frac{1}{2(T-A)} \int_{-T+A}^{0} |f(t)|^2 dt + \frac{2(T+A)}{2T} \frac{1}{2(T+A)} \int_{0}^{T+A} |f(t)|^2 dt \right)$$

$$= \tilde{E}_1 + \tilde{E}_2 = \tilde{E} .$$

From this, we similarly obtain, by replacing T by $T+A$ and $2A$ by a, that

$$\tilde{E} = \lim_{T\to\infty} \frac{1}{2T} \int_{-T}^{T+a} |f(t)|^2 dt . \qquad (3.5.7)$$

Finite power signals, in general, have no Fourier transform or finite energy. They are, however, spectrally analyzed in terms of their power spectrum which is defined by

$$S(\omega) = \lim_{T\to\infty} \frac{1}{2T} \left| \frac{1}{\sqrt{2\pi}} \int_{-T}^{T} f(t) e^{it\omega} dt \right|^2 . \qquad (3.5.8)$$

We will show that

$$\int_{-\infty}^{\infty} S(\omega)\, d\omega = \frac{\tilde{E}}{\sqrt{2\pi}} . \qquad (3.5.9)$$

The integral in (3.5.9) will be taken, in general, in the sense of distributions. Let $f_T(t) = f(t)\chi_{[-T,T]}$ and $F_T(\omega)$ be its Fourier transform; hence

$$S(\omega) = \lim_{T\to\infty} S_T(\omega),$$

where

$$S_T(\omega) = \frac{1}{2T}|F_T(\omega)|^2.$$

It is easy to see that when f is real, the inverse Fourier transform of S_T is the autocorrelation function $R_T(t)$ of f_T, which is defined by

$$R_T(t) = \frac{1}{(2T)}\frac{1}{(2\pi)}\int_{-\infty}^{\infty} f_T(\tau)f_T(t+\tau)\,d\tau. \qquad (3.5.10)$$

For $t > 0$, $R_T(t)$ can be written in the form

$$R_T(t) = \frac{1}{(2T)}\frac{1}{(2\pi)}\int_{-T}^{T-t} f(\tau)f(t+\tau)\,d\tau. \qquad (3.5.11)$$

For a finite power signal, the autocorrelation function $R(t)$ is defined by

$$R(t) = \lim_{T\to\infty}\frac{1}{(2T)}\frac{1}{(2\pi)}\int_{-T}^{T} f(\tau)f(t+\tau)\,d\tau.$$

To show that this definition makes sense, we apply the Cauchy-Schwarz inequality to obtain

$$\left|\frac{1}{2T}\int_{-T}^{T} f(\tau)f(t+\tau)\,d\tau\right|^2 \le \left(\frac{1}{2T}\int_{-T}^{T}|f(\tau)|^2 d\tau\right)\left(\frac{1}{2T}\int_{-T}^{T}|f(t+\tau)|^2 d\tau\right),$$

which, when we take the limit as $T\to\infty$ and use (3.5.6), implies that

$$|R(t)| \le R(0) = \frac{1}{(2\pi)}\bar{E}. \qquad (3.5.12)$$

As in the proof of (3.5.6), we can also show that

$$R(t) = \lim_{T\to\infty}\frac{1}{(2T)}\frac{1}{(2\pi)}\int_{-T+A}^{T+A} f(\tau)f(t+\tau)\,d\tau;$$

hence, $R(t)$ is even. For, we have

$$R(-t) = \lim_{T\to\infty} \frac{1}{(2T)} \frac{1}{(2\pi)} \int_{-T}^{T} f(\tau) f(-t+\tau)\, d\tau$$

$$= \lim_{T\to\infty} \frac{1}{(2T)} \frac{1}{(2\pi)} \int_{-T-t}^{T-t} f(\tau+t) f(\tau)\, d\tau$$

$$= R(t).$$

Clearly, $R(t), R_T(t)$ are bounded; hence they are tempered distributions, and

$$\lim_{T\to\infty} R_T(t) = R(t). \tag{3.5.13}$$

If the convergence in (3.5.13) is in the sense of S^* (the space of tempered distributions), e.g., uniform convergence on compact subsets of $\Re$, then since the Fourier transform is a continuous transformation on S^*, it follows by taking the Fourier transform of both sides of (3.5.13) that

$$\lim_{T\to\infty} S_T(\omega) = S(\omega) = \hat{R}(\omega), \tag{3.5.14}$$

in the sense of tempered distributions. Since R is even, we may formally write

$$S(\omega) = \frac{1}{\sqrt{2\pi}} \int_{-\infty}^{\infty} R(t) \cos\omega t\, dt, \tag{3.5.15}$$

and

$$R(t) = \frac{1}{\sqrt{2\pi}} \int_{-\infty}^{\infty} S(\omega) \cos\omega t\, d\omega. \tag{3.5.16}$$

In particular, by (3.5.12),

$$\int_{-\infty}^{\infty} S(\omega)\, d\omega = \sqrt{2\pi}\, R(0) = \frac{\tilde{E}}{\sqrt{2\pi}}. \tag{3.5.17}$$

For example, if $f(t)$ is constant, say $f(t) = C$, then

$$R(t) = \lim_{T\to\infty} \frac{1}{(2T)} \frac{C^2}{(2\pi)} \int_{-T}^{T} d\tau = \frac{C^2}{2\pi}.$$

Therefore,

$$S(\omega) = \frac{1}{\sqrt{2\pi}} C^2 \delta(\omega),$$

and

$$\int_{-\infty}^{\infty} S(\omega)\, d\omega = \frac{C^2}{\sqrt{2\pi}} = \frac{\tilde{E}}{\sqrt{2\pi}}.$$

For finite power signals, the WSK sampling theorem does not, in general hold; however, as we shall see shortly it holds in the sense that

$$\lim_{T\to\infty} \frac{1}{2T} \int_{-T}^{T} \left| f(t+\tau) - \sum_{-\infty}^{\infty} f(t+nT) \frac{\sin \omega_0(\tau - nT)}{\omega_0(\tau - nT)} \right|^2 dt = 0,$$

where f is a signal whose power spectrum is band-limited in the sense

$$S_f(\omega) = 0 \quad \text{for} \quad |\omega| \geq \omega_0 > 0. \tag{3.5.18}$$

We shall prove this result because the original proof, which is due to Papoulis [75], was not rigorous.

First, let h be integrable and f be of finite power such that

$$g(t) = \int_{-\infty}^{\infty} f(t-\tau) h(\tau)\, d\tau$$

is defined. Let $R_{g,T}$ and $R_{f,T}$ be the autocorrelation functions of g_T and f_T respectively. Then,

$$R_{g,T}(t) = \frac{1}{2T} \int_{-T}^{T} g(\tau) g(t+\tau)\, d\tau$$

$$= \frac{1}{2T} \int_{-T}^{T} d\tau \int_{-\infty}^{\infty} f(\tau - x) h(x)\, dx \int_{-\infty}^{\infty} f(t+\tau-y) h(y)\, dy$$

$$= \int_{-\infty}^{\infty} h(x)\, dx \int_{-\infty}^{\infty} h(y)\, dy \left(\frac{1}{2T} \int_{-T}^{T} f(\tau - x) f(t+\tau-y)\, d\tau \right),$$

and hence

$$R_g(t) = \lim_{T\to\infty} R_{g,T}(t) = \int_{-\infty}^{\infty} h(x)dx \int_{-\infty}^{\infty} h(y)dy\, R_f(t+x-y)$$

$$= (R_f * h * \tilde{h})(t), \tag{3.5.19}$$

where $\tilde{h}(x) = h(-x)$. Since R_f is bounded and h is integrable, it follows by taking the Fourier transform of both sides of (3.5.19) and using (3.5.14) that

$$S_g(\omega) = S_f(\omega)|H(\omega)|^2,$$

in the sense of tempered distributions. In view of (3.5.18) and (3.5.17), we have

$$\frac{1}{\sqrt{2\pi}}\tilde{E}_g = \int_{-\omega_0}^{\omega_0} S_f(\omega)|H(\omega)|^2 d\omega,$$

in which the integral converges in the classical sense. Hence,

$$\tilde{E}_g \le \max_{|\omega|\le\omega_0} |H(\omega)|^2 \tilde{E}_f. \tag{3.5.20}$$

THEOREM 3.14 (Papoulis [75])

Let f be a finite power signal with power spectrum $S_f(\omega)$ band-limited to $[-\omega_0, \omega_0]$, i.e.,

$$S_f(\omega) = 0 \quad \text{for} \quad |\omega| \ge \omega_0 > 0.$$

Then, for any real τ

$$\lim_{N\to\infty}\lim_{T\to\infty} \frac{1}{2T}\int_{-T}^{T}\left| f(t+\tau) - \sum_{n=-N}^{N} f(t+nT)\frac{\sin\omega_0(\tau-nT)}{\omega_0(\tau-nT)}\right|^2 dt = 0. \tag{3.5.21}$$

Proof. Let

$$e_N(t+\tau) = f(t+\tau) - \sum_{n=-N}^{N} f(t+nT)\frac{\sin\omega_0(\tau-nT)}{\omega_0(\tau-nT)}.$$

It is easy to verify that for fixed τ, e_N is of finite power and

$$e_N(t+\tau) = \int_{-\infty}^{\infty} f(t+\tau-x)\, h_N(x)\, dx\,, \tag{3.5.22}$$

where

$$h_N(x) = \delta(x) - \sum_{n=-N}^{N} \delta(x-\tau+nT)\, \frac{\sin\omega_0(\tau-nT)}{\omega_0(\tau-nT)}\,,$$

and that the Fourier transform $H_N(\omega)$ of h_N is given by

$$H_N(\omega) = \frac{1}{\sqrt{2\pi}}\left(1 - \sum_{n=-N}^{N} e^{-i\omega(\tau-nT)}\, \frac{\sin\omega_0(\tau-nT)}{\omega_0(\tau-nT)}\right).$$

From (3.5.22) and (3.5.20), it follows that

$$\begin{aligned}
\tilde{E}_{e_N} &= \lim_{T\to\infty} \frac{1}{2T} \int_{-T}^{T} \left| f(t+\tau) - \sum_{n=-N}^{N} f(t+nT)\, \frac{\sin\omega_0(\tau-nT)}{\omega_0(\tau-nT)} \right|^2 dt \\
&\le \max_{|\omega|\le\omega_0} |H_N(\omega)|^2 \tilde{E}_f\,.
\end{aligned} \tag{3.5.23}$$

From the observation that

$$e^{i\omega\tau} = \sum_{-\infty}^{\infty} e^{inT\omega}\, \frac{\sin\omega_0(\tau-nT)}{\omega_0(\tau-nT)}\,, \qquad |\omega| \le \omega_0$$

uniformly on $[-\omega_0, \omega_0]$, it follows that

$$\lim_{N\to\infty}\ \max_{|\omega|\le\omega_0} |H_N(\omega)|^2 = 0\,, \tag{3.5.24}$$

and finally by combining (3.5.24) and (3.5.23) we obtain (3.5.21). ∎

3.6 Sampling by Using More General Types of Data and Sampling Functions

3.6.A Sampling by Using Other Types of Data

Up until now, we have been discussing different sampling series expansions for the reconstruction of signals from their sampled values at a discrete set of points. Reconstructing signals from other types of data is also conceivable. When Shannon introduced his sampling theorem, he also noted that a band-limited signal could be reconstructed from the values of the signal and its first derivative at every other sampling point. This means that one can reconstruct a band-limited signal by sampling it at half the Nyquist rate, provided that at each sampling point two sample values are taken, one from the signal and the other from the signal's derivative. He then extended his statement to higher derivatives by saying that a band-limited signal could also be reconstructed from the values of the signal, its first and second derivatives at every other third sampling point, and so on. Some years later, these statements were mathematically formulated, proved and even generalized by other people, including A. Papoulis [72], D. Linden [53], D. Linden and N. Abramson [54], L. Fogel [26], and D. Jagerman and L. Fogel [39]. The last two authors also gave a few applications where the sampled derivatives were needed. The most important of these applications is the one in the field of air traffic control, wherein the aircraft estimated velocity, as well as position, is used to determine a continuous course plot of the air-path with half the sampling rate. One of the earliest main results on sampling with the values of the function and its derivative is due to Jagerman and Fogel.

THEOREM 3.15 (Jagerman and Fogel [39])

Let $f \in B_\sigma^2$, then

$$f(t) = \sum_{k=-\infty}^{\infty} \left\{ f\left(\frac{2k\pi}{\sigma}\right) + \left(t - \frac{2k\pi}{\sigma}\right) f'\left(\frac{2k\pi}{\sigma}\right) \right\} \left\{ \frac{\sin\frac{\sigma}{2}\left(t - \frac{2k\pi}{\sigma}\right)}{\frac{\sigma}{2}\left(t - \frac{2k\pi}{\sigma}\right)} \right\}^2 ,$$

or

$$f(t) = \sum_{k=-\infty}^{\infty} \{f(kT) + (t - kT) f'(kT)\} \operatorname{sinc}^2\left(\frac{\sigma}{2\pi}(t - kT)\right),$$

where $T = 2\pi/\sigma$ and the series converges uniformly on any compact subset of $\mathfrak{R}$.

This simultaneous sampling of the signal and its first derivative reduces the sampling rate by a half, i.e., from σ/π to $\sigma/2\pi$. More generally, sampling by using the values of the signal and its first p derivatives reduces the sampling rate by $1/(p+1)$. This will be seen from a generalization of Theorem 3.15, where a band-limited-signal is reconstructed by using sample values of the first p derivatives of the signal. Instead of stating and proving this generalization of Theorem 3.15 now, we will derive it later as a special case of a more general theorem of Papoulis.

Papoulis first addressed the following question: Given a system (or a filter) with system function $H(\omega)$ and a σ-band-limited signal f, reconstruct the input signal f from the values $g(k\pi/\sigma)$ of the output signal

$$g(t) = \frac{1}{\sqrt{2\pi}} \int_{-\sigma}^{\sigma} F(\omega) H(\omega) e^{it\omega} d\omega . \qquad (3.6.1)$$

Surprisingly, the answer to this question is easy:

$$f(t) = \sum_{k=-\infty}^{\infty} g(kT) y(t - kT) \quad \text{where} \quad T = \frac{\pi}{\sigma},$$

and

$$y(t) = \frac{1}{2\sigma} \int_{-\sigma}^{\sigma} \frac{e^{it\omega}}{H(\omega)} d\omega .$$

The proof is straightforward. Let us write

$$f(t) = \frac{1}{\sqrt{2\pi}} \int_{-\sigma}^{\sigma} F(\omega) e^{it\omega} d\omega = \frac{1}{\sqrt{2\pi}} \int_{-\sigma}^{\sigma} F(\omega) H(\omega) \frac{e^{it\omega}}{H(\omega)} d\omega$$

$$= \frac{1}{\sqrt{2\pi}} \int_{-\sigma}^{\sigma} F(\omega) H(\omega) \sum_{k=-\infty}^{\infty} b_k(t) e^{ikT\omega} d\omega$$

$$= \sum_{k=-\infty}^{\infty} b_k(t) \frac{1}{\sqrt{2\pi}} \int_{-\sigma}^{\sigma} F(\omega) H(\omega) e^{ikT\omega} d\omega = \sum_{k=-\infty}^{\infty} g(kT) b_k(t),$$

where $b_k(t)$ are the Fourier coefficients of the function $e^{it\omega}/H(\omega)$ when expanded in a Fourier series in the interval $(-\sigma, \sigma)$, i.e.,

$$\frac{e^{it\omega}}{H(\omega)} = \sum_{k=-\infty}^{\infty} b_k(t) e^{ikT\omega}, \quad |\omega| < \sigma .$$

It is evident that

$$b_k(t) = \frac{1}{2\sigma} \int_{-\sigma}^{\sigma} \frac{e^{it\omega}}{H(\omega)} e^{-ikT\omega} d\omega = y(t - kT),$$

which completes the proof. Interchanging the summation and the integration signs is permissible if $H(\omega)$ is a nice function, e.g., is differentiable and different from zero in the interval $(-\sigma, \sigma)$.

An upper bound for the output signal g can be derived by applying the Cauchy-Schwarz inequality to (3.6.1) to obtain

$$|g(t)|^2 \le \frac{E}{2\pi} \int_{-\sigma}^{\sigma} |H(\omega)|^2 d\omega,$$

where $E = \int_{-\sigma}^{\sigma} |F(\omega)|^2 d\omega < \infty$ is the total energy of f. For $H(\omega) = (-i\omega)^n$, it is easy to see that $g(t) = f^{(n)}(t)$ and hence

$$|f^{(n)}(t)| \le \sigma^n \sqrt{\frac{E\sigma}{(2n+1)\pi}}. \qquad (3.6.2)$$

As a special case of (3.6.2), we have

$$|f(t)| \le \sqrt{\frac{E\sigma}{\pi}}, \qquad (3.6.3)$$

where the upper bound is attained for

$$f(t) = \sqrt{\frac{E\pi}{\sigma}} \frac{\sin\sigma(t-t_0)}{\pi(t-t_0)}$$

at $t = t_0$.

Papoulis then considered a more general problem: suppose that a band-limited signal f is fed simultaneously into p different systems (filters) with system functions $H_1(\omega), \ldots, H_p(\omega)$, reconstruct f from the samples $g_1(kT), \ldots, g_p(kT)$ of the output signals

$$g_k(t) = \frac{1}{\sqrt{2\pi}} \int_{-\sigma}^{\sigma} F(\omega) H_k(\omega) e^{it\omega} d\omega, \quad k = 1, \ldots, p \qquad (3.6.4)$$

where $T = p\pi/\sigma$.

The answer is provided in the following theorem whose proof is certainly not a trivial generalization of the case $p = 1$.

THEOREM 3.16 (Papoulis [72])

Let f, $H_k(\omega)$ and $g_k(t)$ be given as above. Then

$$f(t) = \sum_{i=1}^{p} \sum_{k=-\infty}^{\infty} g_i(kT) y_i(t-kT), \qquad (3.6.5)$$

where

$$y_k(t) = \frac{1}{c} \int_{-\sigma}^{-\sigma+c} Y_k(\omega,t) e^{it\omega} d\omega, \quad k = 1,\ldots,p, \tag{3.6.6}$$

$c = 2\sigma/p = 2\pi/T$ and $Y_1(\omega,t),\ldots,Y_p(\omega,t)$, if they exist, are solutions of the following system of equations

$$\sum_{k=1}^{p} H_k[\omega + (m-1)c] Y_k(\omega,t) = e^{i(m-1)ct}, \quad m = 1,\ldots,p, \tag{3.6.7}$$

in which t is arbitrary and ω is in the interval $(-\sigma, -\sigma + c)$.

Proof. First, let us observe that since $H_k[\omega + (m-1)c] (k = 1,\ldots,p)$ are independent of t and the right hand sides of (3.6.7) are periodic functions in t with period $T = 2\pi/c$, then $Y_k(\omega,t)$ are periodic functions in t with the same prescribed period. Therefore, from (3.6.6) and the periodicity of $Y_k(\omega,t)$, we obtain

$$y_n(t-kT) = \frac{1}{c} \int_{-\sigma}^{-\sigma+c} Y_n(\omega, t-kT) e^{i(t-kT)\omega} d\omega$$

$$= \frac{1}{c} \int_{-\sigma}^{-\sigma+c} Y_n(\omega,t) e^{it\omega} e^{-ikT\omega} d\omega, \quad n = 1,\ldots,p.$$

The last integral shows that $y_n(t-kT)$ is the kth Fourier coefficient of the function $Y_n(\omega,t)e^{it\omega}$ when expanded in a Fourier series in the interval $(-\sigma, -\sigma+c)$. Hence,

$$Y_n(\omega,t)e^{it\omega} = \sum_{k=-\infty}^{\infty} y_n(t-kT) e^{ikT\omega}. \tag{3.6.8}$$

Next, we show that for every ω in the interval $(-\sigma,\sigma)$ and every t we have

$$e^{it\omega} = \sum_{n=1}^{p} \sum_{k=-\infty}^{\infty} H_n(\omega) y_n(t-kT) e^{ikT\omega}. \tag{3.6.9}$$

Multiplying the first equation in (3.6.7) by $e^{it\omega}$ and using (3.6.8) yield that (3.6.9) is valid for every ω in the interval $(-\sigma, -\sigma+c)$. Multiplying the second equation in (3.6.7) by $e^{it\omega}$ and using (3.6.8) once more yield that

$$e^{it(\omega+c)} = \sum_{n=1}^{p} \sum_{k=-\infty}^{\infty} H_n(\omega+c) y_n(t-kT) e^{ikT\omega}$$

is valid for every ω in the interval $(-\sigma, -\sigma + c)$. But as ω varies in the interval $(-\sigma, -\sigma + c)$, $\omega + c$ varies in the interval $(-\sigma + c, -\sigma + 2c)$. Hence, (3.6.9) is valid in that interval, and by continuing in this fashion, we can easily show that it is also valid in the interval $(-\sigma + (p-1)c, -\sigma + pc) = (\sigma - c, \sigma)$. Hence, by putting all these steps together, we obtain that (3.6.9) is valid in the interval $(-\sigma, \sigma)$. Finally, by replacing the function $e^{it\omega}$ with the right hand side of (3.6.9) in

$$f(t) = \frac{1}{\sqrt{2\pi}} \int_{-\sigma}^{\sigma} F(\omega) e^{it\omega} d\omega ,$$

and using (3.6.4), we obtain (3.6.5). ▮

The sampling expansion (3.6.5) holds if the system of equations (3.6.7) has a unique solution, which is equivalent to saying that the determinant of the system is different from zero for every ω in the interval $(-\sigma, -\sigma + c)$.

As an example, let us take $p = 2$, $H_1(\omega) = 1$ and $H_2(\omega) = i\omega$; hence, $c = \sigma$, $T = 2\pi/\sigma$, $g_1(t) = f(t)$ and $g_2(t) = f'(t)$. The system of equations (3.6.7) now takes the simple form

$$Y_1(\omega, t) + i\omega Y_2(\omega, t) = 1$$

$$Y_1(\omega, t) + i(\omega + \sigma) Y_2(\omega, t) = e^{i\sigma t} ,$$

which upon solving gives

$$Y_1(\omega, t) = 1 - \frac{\omega}{\sigma}(e^{it\sigma} - 1) \quad \text{and} \quad Y_2(\omega, t) = \frac{1}{i\sigma}(e^{it\sigma} - 1) .$$

By substituting this into (3.6.6) and calculating the integrals, we obtain

$$y_1(t) = \left(\frac{\sin(\sigma t/2)}{(\sigma t/2)} \right)^2 = \operatorname{sinc}^2\!\left(\frac{\sigma t}{2\pi} \right) ,$$

and

$$y_2(t) = t \left(\frac{\sin(\sigma t/2)}{(\sigma t/2)} \right)^2 = t \operatorname{sinc}^2\!\left(\frac{\sigma t}{2\pi} \right) .$$

The sampling series (3.6.5) now reduces to the one given in Theorem 3.15.

Whereas if we take $p = 3$, $H_n(\omega) = (i\omega)^{n-1}$, $n = 1, 2, 3$; then, $T = 3\pi/\sigma$, $c = 2\sigma/3$, and $g_1(t) = f(t)$, $g_2(t) = f'(t)$, $g_3(t) = f''(t)$. The system of equations

(3.6.7) becomes

$$Y_1(\omega,t) + i\omega Y_2(\omega,t) - \omega^2 Y_3(\omega,t) = 1$$

$$Y_1(\omega,t) + i(\omega+c)Y_2(\omega,t) - (\omega+c)^2 Y_3(\omega,t) = e^{ict}$$

$$Y_1(\omega,t) + i(\omega+2c)Y_2(\omega,t) - (\omega+2c)^2 Y_3(\omega,t) = e^{2ict}\,.$$

Upon solving this system, substituting the solutions into (3.6.6) and calculating the integrals, we obtain

$$y_1(t) = \operatorname{sinc}^2\left(\frac{\sigma t}{3\pi}\right), \quad y_2(t) = t\operatorname{sinc}^3\left(\frac{\sigma t}{3\pi}\right), \quad y_3(t) = \frac{t^2}{2}\operatorname{sinc}^3\left(\frac{\sigma t}{3\pi}\right).$$

The sampling series (3.6.5) now gives

$$f(t) = \sum_{k=-\infty}^{\infty}\left(f\left(\frac{3k\pi}{\sigma}\right) + \left(t-\frac{3k\pi}{\sigma}\right)f'\left(\frac{3k\pi}{\sigma}\right) + \frac{1}{2}\left(t-\frac{3k\pi}{\sigma}\right)^2 f''\left(\frac{3k\pi}{\sigma}\right)\right)\operatorname{sinc}^3\left(\frac{\sigma t}{3\pi}-k\right).$$

Having seen the pattern, the reader should be able to derive the abovementioned generalization of Theorem 3.15 that gives a sampling series expansion for reconstructing a band-limited signal f from its sample values as well as the sample values of its first $(p-1)$ derivatives, namely

$$f(t) = \sum_{n=0}^{p-1}\sum_{k=-\infty}^{\infty}\left(\frac{1}{n!}\left(t-\frac{kp\pi}{\sigma}\right)^n f^{(n)}\left(\frac{kp\pi}{\sigma}\right)\right)\operatorname{sinc}^p\left(\frac{\sigma t}{p\pi}-k\right),$$

or

$$f(t) = \sum_{n=0}^{p-1}\sum_{k=-\infty}^{\infty}\left(\frac{1}{n!}(t-kT)^n f^{(n)}(kT)\right)\operatorname{sinc}^p\left(\frac{\sigma}{p\pi}(t-kT)\right). \qquad (3.6.10)$$

Note that for $H_n(\omega) = (i\omega)^{n-1}$, $n = 1,\ldots,p$, the determinant of the system (3.6.7) (the Vandermonde determinant) is different from zero for every ω in the interval $(-\sigma, -\sigma+c)$.

We should also note that signals represented by integrals other than the Fourier one, may also be reconstructed by using sampling series expansions involving the sample values of the signals and their derivatives; see [46].

In view of Theorem 3.16, (3.2.17) and (3.5.4), it is evident that the reconstruction of a band-limited signal using data other than the sample values of the signal and its derivatives is also possible.

We conclude our discussion by noting that some of the ideas mentioned in this section have been extended to multidimensional signals as well. For example, Petersen and Middleton [80] derived a sampling series expansion that involves sample values of the amplitude and the gradient of an

N-dimensional stochastic field. Their idea was extended further by Montgomery [64] who derived, for multidimensional signals, sampling series expansions that involve the values of the signals and their partial derivatives up to order $p \geq 1$.

3.6.B Sampling by Using Other Types of Sampling Functions

The sampling functions $S_k(t) = \sin\sigma(t - t_k)/\sigma(t - t_k)$, where $t_k = k\,\pi/\sigma$, $k = 0, \pm 1, \pm 2, \ldots$, in the WSK sampling theorem are translates of the function $S_0(t) = \sin\sigma t/\sigma t$. Hence, if we define the translation operator T by $T_y(f(t)) = f(t - y)$, then the WSK sampling series can be written in the form

$$f(t) = \sum_{k=-\infty}^{\infty} f(t_k) T_{t_k}(S_0(t)).$$

It is of special interest to know if there are more general sampling series expansions for band-limited functions, wherein the sampling functions are generated from one single function by translations, i.e., series expansions of the form

$$f(t) = \sum_{k=-\infty}^{\infty} f(\alpha_k) T_{\beta_k}(\phi(t)) = \sum_{k=-\infty}^{\infty} f(\alpha_k)\,\phi(t - \beta_k), \tag{3.6.11}$$

where $\{\alpha_k\}_{k \in Z}$ and $\{\beta_k\}_{k \in Z}$ are two given sequences of real numbers.

More precisely, let us consider the following problem: given two sequences of real numbers $\{\alpha_k\}_{k \in Z}$ and $\{\beta_k\}_{k \in Z}$, determine a function ϕ for which (3.6.11) holds for any function f band-limited to, say, $[-\sigma, \sigma]$.

Another related problem is this: if f is band limited to $[-\sigma, \sigma]$, ϕ is arbitrary but given and g is a function defined by

$$g(t) = \sum_{k=-\infty}^{\infty} f(\alpha_k)\,\phi(t - \beta_k), \tag{3.6.12}$$

under what conditions can we recover f from g?

Answers to these questions can be easily provided if we assume that $\alpha_k = \beta_k = \lambda k$ for some $0 < \lambda < 2\pi/\sigma$. Under this assumption, (3.6.11) becomes

$$f(t) = \sum_{k=-\infty}^{\infty} f(\lambda k)\,\phi(t - \lambda k), \tag{3.6.13}$$

where f is assumed to be band-limited to $[-\sigma, \sigma]$. By taking the Fourier transform of (3.6.13) and employing the Poisson summation formula (cf. (2.2.3)), we obtain

$$\hat{f}(\omega) = \left(\sum_{k=-\infty}^{\infty} f(\lambda k) e^{i\lambda k\omega}\right)\hat{\phi}(\omega) = \frac{\sqrt{2}\,\pi}{\lambda}\left(\sum_{k=-\infty}^{\infty} \hat{f}\left(\omega + \frac{2k\pi}{\lambda}\right)\right)\hat{\phi}(\omega). \tag{3.6.14}$$

Finding a ϕ to satisfy (3.6.13) is equivalent to finding a $\hat{\phi}$ to satisfy (3.6.14). The latter is easy to find, just take

$$\hat{\phi}(\omega) = \begin{cases} \dfrac{\lambda}{\sqrt{2\pi}} & \text{if } |\omega| < \sigma_1 \\ 0 & \text{if } |\omega| \geq \sigma_2 \end{cases},$$

where $0 < \sigma \leq \sigma_1 \leq \sigma_2$, provided that $0 < \lambda \leq 2\pi/(\sigma + \sigma_2)$. For, all the translates of $\hat{f}(\omega)$ will have supports disjoint from that of $\hat{\phi}$, except for $k = 0$, i.e., $\hat{f}(\omega + 2k\pi/\lambda)\hat{\phi}(\omega) = 0$, $k = \pm 1, \pm 2, \dots$. In the special case where $\sigma = \sigma_1 = \sigma_2$ and $\lambda = \pi/\sigma$,

$$\phi(t) = \frac{\sin \sigma t}{\sigma t}.$$

However, if we define

$$\hat{\phi}(\omega) = \begin{cases} \dfrac{\lambda}{\sqrt{2\pi}} & \text{if } |\omega| \leq \sigma_1 \\ \dfrac{\lambda(\omega - \sigma_2)}{\sqrt{2\pi}(\sigma_1 - \sigma_2)} & \text{if } \sigma_1 \leq |\omega| \leq \sigma_2 \\ 0 & \text{if } |\omega| \geq \sigma_2, \end{cases}$$

then

$$f(t) = \sum_{k=-\infty}^{\infty} f(\lambda k)\phi(t - \lambda k),$$

where

$$\phi(t) = \frac{2\lambda}{\pi} \frac{\sin\left(\frac{\sigma_2+\sigma_1}{2}\right)t \, \sin\left(\frac{\sigma_2-\sigma_1}{2}\right)t}{(\sigma_2 - \sigma_1)t^2}, \quad 0 < \lambda \leq \frac{2\pi}{\sigma + \sigma_2}.$$

Again, if $\sigma_1 = \sigma_2 = \sigma$, and $\lambda = \pi/\sigma$, then $\phi(t) = \sin \sigma t/\sigma t$ and (3.6.13) reduces to the WSK sampling series (2.1.2).

As for recovering f from g in (3.6.12), we repeat the same argument to obtain

$$\hat{f}(\omega) = \frac{\lambda}{\sqrt{2\pi}} \frac{\hat{g}(\omega)}{\hat{\phi}(\omega)} \chi_{(-\sigma,\sigma)}(\omega)$$

for $0 < \lambda \leq \pi/\sigma$, provided that $\hat{\phi} \neq 0$ in $(-\sigma, \sigma)$.

For general α_k and β_k the problems are more difficult and need a different approach since Poisson's summation formula is no longer valid. One such approach will be given in Chapter 9 when we discuss the Feichtinger-Gröchenig sampling theory. The question of whether there are counterparts of the Poisson summation formula for integral transforms other than the Fourier one was raised by P. Butzer and C. Markett in [18]. More precisely, does there exist a counterpart of the Poisson summation formula which somehow connects the generalized Fourier coefficients of an orthogonal series, on the one hand, with an orthogonal transform on the other? There are indications that such a formula may exist, but none has been explicitly found as yet!

3.7 Sampling Theorems in More General Function Spaces

In Section 2.2 we discussed uniform sampling for general classes of band-limited functions, namely $B_\sigma^p (1 \leq p \leq \infty)$. Then in Section 3.1 we discussed non-uniform sampling of functions that are band-limited to intervals symmetric about the origin and showed in Section 3.3 how to extend this to higher dimensions. We have also pointed out that no general sampling series expansion was known for functions band-limited to a general compact region Ω in $\mathfrak{R}^N, (N \geq 1)$ until very recently.

An earlier attempt to derive such a series expansion was made by I. Kluvánek [48], who derived a general form of the WSK sampling theorem in the setting of harmonic analysis on locally compact abelian groups. By a suitable restriction to $\mathfrak{R}^N$, we can obtain Theorem 3.6 as a special case of Kluvánek's result. Nevertheless, Kluvánek's result does not seem to easily yield any explicit sampling series expansion for a function that is band-limited to a general region in $\mathfrak{R}^N$. More recent results have been obtained by H. Feichtinger [22], H. Feichtinger and K. Gröchenig [23-25], and K. Gröchenig [30]. Their approach to sampling theorems appears to be new and promising, but a bit abstract. They have also extended their results to locally compact abelian groups using harmonic analysis techniques. This connection between the WSK sampling theorem and harmonic analysis on locally compact abelian groups is not surprising since the Poisson summation formula is known to play a fundamental role in both areas; see [88] and [89].

In the last four decades, not only new results in sampling theory have been obtained, but also new ways to view and interpret them have emerged. Connections between sampling theory and other branches of mathematical analysis, such as *Functional Analysis, Special Functions* and *Boundary-Value Problems*, have been well established.

To the best of my knowledge, F. Beutler [7] was the first to use Hilbert space concepts to prove sampling theorems. Later, he [6] used some Banach space techniques with some results of Levinson to establish sufficient conditions for error-free recovery of certain signals based upon completion properties of sets of complex exponentials. The first reference to the notion of reproducing kernels in a Hilbert space in sampling theory was in the work of Yao [115], who used it to derive various sampling theorems. Higgins [36], then, utilized some of Beutler and Yao's ideas to derive an explicit sampling series expansion for band-limited signals analogous to the WSK sampling series, but by using irregularly spaced sample points; see also [35 and 37] for some related results.

Very recently, Z. Nashed and G. Walter [66], using the notion of reproducing kernels in a Hilbert space together with some of Higgins' ideas, have obtained sampling theorems for functions in a subspace of the Sobolev space H^{-1}. In another direction, J. Benedetto [3], J. Benedetto and W. Heller [4] used the theory of frames, which has emerged in the last few years as a powerful and popular tool in mathematical analysis because of its connection with Wavelets, to obtain general sampling series that involve regular as well as irregular sample points. The notion of frames is more general than the notion of basis in a Hilbert space. We shall discuss some of these recent generalizations in Chapter 10.

3.8 Error Analysis

In this section, we shall briefly discuss several types of errors that may arise in the practical implementation of sampling theorems and influence the accuracy of the signal reconstruction. A comprehensive treatment of these types of errors can be found in [16, 40, 42, 57].

i) The truncation error $T_N f$: this is the error that results when only a finite number N of samples are used instead of the infinitely many samples needed for the signal reconstruction. For the WSK sampling theorem, $T_N f$ may take the form

$$(T_N f)(t) = f(t) - \sum_{k=-N}^{N} f(t_k)\frac{\sin\sigma(t-t_k)}{\sigma(t-t_k)} = \sum_{|k|>N} f(t_k)\frac{\sin\sigma(t-t_k)}{\sigma(t-t_k)}.$$

ii) The aliasing error $R_\sigma f$: this results if the band-limitedness condition is violated. If f is not band-limited or if it is band-limited to a larger band than $[-\sigma,\sigma]$, then the series (2.1.2) constructed by using the samples $\{f(t_k)\}$ will not converge to f. The difference between f and the series (2.1.2) is the aliasing error

$$(R_\sigma f)(t) = f(t) - \sum_{k=-\infty}^{\infty} f\left(\frac{k\pi}{\sigma}\right)\frac{\sin(\sigma t - k\pi)}{(\sigma t - k\pi)};$$

See Section 3.5.A.

iii) The amplitude error $A_\varepsilon f$: it arises if the exact sampled values $f(t_k)$ are not accurately known, but only approximations thereof, say $\tilde{f}(t_k)$, differing from $f(t_k)$ by not more than ε, are known,

$$(A_\varepsilon f)(t) = \sum_{k=-\infty}^{\infty} \left[f(t_k) - \tilde{f}(t_k)\right] \frac{\sin \sigma(t - t_k)}{\sigma(t - t_k)}.$$

Round-off errors may be considered as a special case of the amplitude error.

iv) The time-jitter error $J_\varepsilon f$: it is caused by sampling at instants $\bar{t}_k = t_k + \gamma_k$, which differ from the Nyquist sampling instants t_k by γ_k with $|\gamma_k| \le \delta$,

$$(J_\varepsilon f)(t) = \sum_{k=-\infty}^{\infty} \left[f(t_k) - f\left(\bar{t}_k\right)\right] \frac{\sin \sigma(t - t_k)}{\sigma(t - t_k)}.$$

All four types of errors can be combined in the form

$$(Ef)(t) = f(t) - \sum_{k=-N}^{N} \tilde{f}(t_k + \gamma_k) \frac{\sin(\sigma t - k\pi)}{(\sigma t - k\pi)}. \tag{3.8.1}$$

v) The information loss error If: it arises if some of the sampled data $\{f(t_k)\}$ or fractions thereof are missing

$$(If)(t) = \sum_{k=-\infty}^{\infty} \alpha_k f(t_k) \frac{\sin \sigma(t - t_k)}{\sigma(t - t_k)},$$

where $\alpha_k = 1$ for the values of k for which $f(t_k)$ is missing and $\alpha_k = 0$ otherwise. If a fraction of $f(t_k)$ is only missing, then $0 \le \alpha_k \le 1$. This is the case, for example, in digital recording when a digital channel is defective.

A significant portion of research efforts has been dedicated to error analysis, in particular, to deriving conditions under which these different types of errors can be minimized. We begin by giving some of the main results on the truncation error, and then discuss important theorems concerning the other types of errors.

3.8.A Truncation Errors

Truncation errors, which occur naturally in applications, have been studied rather extensively in engineering literature. The truncation error $(T_N f)(t)$ or $T_n(t)$, for short, can be controlled by imposing some extra conditions on f besides being band-limited.

According to A. Jerri [42], B. Tsybakov and V. Iakovlev [107] were the first to give a reasonable estimate for the truncation error. Their estimate can be written as

$$|T_N(t)| \le \frac{\sqrt{2}}{\pi} E \,\left|\sin\left(\frac{\pi t}{\Delta t}\right)\right| \sqrt{\frac{T\,\Delta t}{(T^2 - t^2)}}, \quad T > 0$$

where $-T \le t \le T$, $0 < \Delta t < (1/\sigma)$, and E is the total energy of the signal which is given by

$$E = \int_{-\sigma}^{\sigma} |F(\omega)|^2 \, d\omega .$$

H. Helms and J. Thomas [32] considered a more general type of truncation errors, where the number of terms taken from the WSK sampling series depends on t, namely,

$$T_N(t) = f(t) - \sum_{k=K(t)-N}^{K(t)+N} f(t_k) \frac{\sin \sigma(t - t_k)}{\sigma(t - t_k)}, \tag{3.8.2}$$

where N is a fixed integer and $K(t)$ is an integer satisfying

$$(\sigma/\pi)t - 1/2 \le K(t) \le (\sigma/\pi)t + 1/2 .$$

They proved several results concerning this type of truncation error, the first of which can be summarized in the following theorem.

THEOREM 3.17 (Helms and Thomas [32])

Let $f(t)$ be a band-limited signal with no frequencies greater than rw cps $(w = \sigma/2\pi)$, $0 < r < 1$. If $M = \max_{-\infty < t < \infty} |f(t)|$, then the truncation error $T_N(t)$ given by (3.8.2) satisfies

$$|T_N(t)| \le \frac{4M}{\pi^2 N(1 - r)} = \frac{4M}{\pi^2 N q}, \quad -\infty < t < \infty, \tag{3.8.3}$$

with $q = 1 - r$.

The number, $q = 1 - r$, which is called the guard-band, is a measurement of the difference between the assumed band-width of the signal $[-\sigma, \sigma]$ and the actual one.

As an example of the estimate (3.8.3), suppose that $N = 50$, $w = 1000$ cps and the highest frequency of f is 800 cps, then $T_N(t)$ is bounded by $(0.041)M$. It is also possible to give an upper bound to the truncation error for the case where the truncation interval is asymmetrically placed with respect to the time t at which the WSK series is being evaluated. If the truncation error is defined as

$$T_{N_1,N_2}(t) = f(t) - \sum_{k=K(t)-N_1}^{K(t)+N_2} f(t_k) \frac{\sin\sigma(t-t_k)}{\sigma(t-t_k)},$$

then it can be shown [32] that

$$\left|T_{N_1,N_2}(t)\right| \le \frac{2M}{\pi^2 q}\left[\frac{1}{N_1}+\frac{1}{N_2}\right], \quad -\infty < t < \infty. \tag{3.8.4}$$

Using contour integration and complex-variable methods, K. Yao and J. Thomas [117] were able to improve the estimate (3.8.4) to

$$\left|T_{N_1,N_2}(t)\right| \le \frac{M\,|\sin\sigma t|}{2\pi\cos(r\pi/2)}\left[\frac{1}{N_1}+\frac{1}{N_2}\right]. \tag{3.8.5}$$

They also showed that this truncation error can be estimated even when an arbitrary number of samples are missing between $K(t)-N_1$ and $K(t)+N_2$. For example, let samples from $K(t)+N_3$ to $K(t)+N_4$ be missing, where $0 < N_1 < N_3 \le N_4 < N_2$, then

$$\left|T_{N_1,N_2}(t)\right| \le \frac{M\,|\sin\sigma t|}{2\pi\cos(r\pi/2)}\left[\frac{1}{N_1}+\frac{1}{N_2}+\frac{1}{N_3-1}+\frac{1}{N_4}\right].$$

In addition, they have obtained estimates for the truncation error in the case where a band-limited signal f is reconstructed from the samples of $f, f^{(1)}, \ldots, f^{(m-1)}$ at $t = (km\pi)/\sigma$; see (3.6.10). In the case $m = 2$, their estimate may be put in the form

$$\left|T_{N_1,N_2}(t)\right| \le \frac{M\,|\sin^2(\sigma t/2)|}{\pi^2(\sin r\pi)/r\pi}\left[\frac{1}{N_1}+\frac{1}{N_2}\right].$$

By using real-variable methods only, J. Brown [12] was able to obtain a slightly better estimate than (3.8.5), especially when the guard-band q is small (r near 1). He considered the truncation error

$$T_{N_1,N_2}(t) = f(t) - \sum_{k=-N_1}^{N_2} f(n) \frac{\sin\pi(t-n)}{\pi(t-n)}, \tag{3.8.6}$$

where in this case $\sigma = \pi$, and obtained the following theorem.

THEOREM 3.18 (Brown [12])

Let

$$f(t) = \frac{1}{2\pi} \int_{-\pi r}^{\pi r} F(\omega) e^{i\omega t} dt \,,$$

with

$$E = \frac{1}{2\pi} \int_{-\pi r}^{\pi r} |F(\omega)|^2 d\omega < \infty \,.$$

Then for $T_{N_1,N_2}(t)$ given by (3.8.6), we have

$$|T_{N_1,N_2}(t)| \le \frac{2\sqrt{2}}{\pi^{3/2}} |\sin \pi t| \sqrt{E \tan(r\pi/2)} \left[\frac{1}{N_1} + \frac{1}{N_2}\right], \quad |t| \le 1/2 \,. \tag{3.8.7}$$

But if

$$\int_{-\pi r}^{\pi r} |F(\omega)| \, d\omega < \infty \,,$$

with

$$M = \max_{-\infty < t < \infty} |f(t)| \,,$$

then

$$|T_{N_1,N_2}(t)| \le \frac{2M}{\pi} |\sin \pi t| \, \{c_0 + \sqrt{r\pi/18} \, (\tan(r\pi/2))^{3/2}\} \left[\frac{1}{N_1} + \frac{1}{N_2}\right], \quad |t| \le 1/2 \,,$$

where

$$c_0 = \frac{1}{r\pi} \ln\left[\frac{1 + \sin(r\pi/2)}{1 - \sin(r\pi/2)}\right].$$

F. Beutler [5], on the other hand, obtained similar bounds for the truncation error in the absence of a guard-band; he required only that the Fourier transform F of the signal f be of bounded variation near the endpoints of the band. In fact, he derived his results for a more general class of signals namely, continuous signals given by Fourier-Stieljes transforms.

THEOREM 3.19 (Beutler [5])

Let $f(t)$ be a continuous function given by

$$f(t) = \frac{1}{2\pi} \int_{-\pi^+}^{\pi^-} e^{i\omega t} dF(\omega)$$

where F has finite total variation $2\pi V_F$ and $V_\delta < \infty$ for some $\delta > 0$. Then

$$|T_N(t)| \le \frac{1}{\pi N}\left[M\left(\sin\frac{\pi t}{2}\right)^2 + |\sin \pi t| \left(\csc\frac{\delta}{2}\right)\left\{V_F + \frac{3}{2}\delta V_\delta + \left(1 + \frac{3}{4}\delta + \frac{3}{8}\delta^2\right)M_F(\delta)\right\}\right],$$

where V_δ is the sum of the total variation of

$$(1/2\pi u)F(u-\pi) \quad \text{and} \quad (1/2\pi u)[F(\pi^-) - F(\pi - u)]$$

over the interval $(0,\delta)$, $M_F(\delta) = (1/2\pi)[V_F(-\pi, \delta - \pi) + V_F(\pi - \delta, \pi)]$, and $V_F(a,b)$ is the total variation of F over the interval (a,b).

If further restrictions are imposed on the signal, better estimates can be obtained for the truncation error. Some such restrictions were considered by D. Jagerman [38]. First, he obtained the following upper bound for the truncation error

$$T_N(t) = f(t) = \sum_{k=-N}^{N} f(t_k)\frac{\sin \sigma(t-t_k)}{\sigma(t-t_k)}, \quad |t| < (N\pi)/\sigma.$$

THEOREM 3.20 (Jagerman [38])

Let f be a σ-band-limited function and let

$$K_N = \left\{\frac{\pi}{\sigma}\sum_{k>N} |f(t_k)|^2\right\}^{1/2} \quad \text{and} \quad L_N = \left\{\frac{\pi}{\sigma}\sum_{k<-N} |f(t_k)|^2\right\}^{1/2},$$

where $N \ge 1$. Then

$$|T_N(t)| \le \frac{1}{\pi}|\sin \sigma t| \left\{\frac{K_N}{\sqrt{(N\pi/\sigma) - t}} + \frac{L_N}{\sqrt{(N\pi/\sigma) + t}}\right\},$$

where $|t| \le (N\pi)/\sigma$.

For $|t| \le \pi/(2\sigma)$,

$$|T_N(t)| \le \frac{|\sin \sigma t|}{\pi\sqrt{\pi/\sigma}} \frac{K_N + L_N}{\sqrt{N - 1/2}}.$$

With further restrictions on f, Jagerman obtained better upper bounds for $T_N(t)$.

THEOREM 3.21 (Jagerman [38])

Let f be a σ-band-limited function such that $t^k f(t) \in L^2(-\infty,\infty)$, for some integer $k > 0, N \geq 1$. Let

$$\tilde{E}_k = \left\{ \int_{-\infty}^{\infty} t^{2k} |f(t)|^2 dt \right\}^{1/2} .$$

Then, for $|t| < N\pi/\sigma$

$$|T_N(t)| \leq \frac{|\sin \sigma t|\,\tilde{E}_k}{\pi(\pi/\sigma)^k \sqrt{1-4^{-k}}} \left[\frac{1}{\sqrt{(N\pi/\sigma)-t}} + \frac{1}{\sqrt{(N\pi/\sigma)+t}} \right] \frac{1}{(N+1)^k}$$

But if $|t| < \pi/(2\sigma)$, then

$$|T_N(t)| \leq \frac{2}{\pi} \frac{|\sin \sigma t|\,\tilde{E}_k}{\sqrt{1-4^{-k}}(\pi/\sigma)^{k+1/2}} \frac{1}{(N+1)^k \sqrt{N-1/2}} .$$

The following is another interesting result of Jagerman on error bounds; however, the type of error considered does not fall under any classifications of the aforementioned truncation errors. First, recall that in view of the Poisson summation formula (2.2.3), it follows that if f is integrable, band-limited to $[-\sigma, \sigma]$, and $\sum_{-\infty}^{\infty} |f(t_k)| < \infty$, then

$$\int_{-\infty}^{\infty} f(t)\,dt = \frac{\pi}{\sigma} \sum_{k=-\infty}^{\infty} f(t_k) .$$

For the truncation error

$$T_N = \int_{-\infty}^{\infty} f(t)\,dt - \frac{\pi}{\sigma} \sum_{k=-N}^{N} f(t_k) ,$$

Jagerman obtained the following upper bound assuming that f satisfies the hypothesis of Theorem 3.17

$$|T_N| \leq \frac{2}{1-2^{1/2-k}} \frac{\tilde{E}_k}{(\pi/\sigma)^{k-1/2}(N+1)^{k-1/2}} .$$

In a different direction, L. Campbell [19] established an upper bound for the truncation error of the sampling series in Theorem 3.10, which is a sampling series of a generalized function with compact support.

THEOREM 3.22 (Campbell [19])

Let f, F, S, q and σ be as in Theorem 3.10, and let

$$(T_N f)(t) = f(t) - \sum_{k=-N}^{N} f(t_k) \frac{\sin \sigma(t-t_k)}{\sigma(t-t_k)} S(q\sigma[t-t_k]) .$$

Furthermore, let r be an integer, b be a positive number such that $|f(t)| \le b\,|t|^r$ for $N\pi/\sigma < |t|$, and let m be an integer $> r$. Set

$$C_m = 4(9)^{m-1}[(m-1)!]^2 (2m)!(m)^{-1/2}\pi^{-3/2} bK ,$$

where

$$K^{-1} = \int_{-1}^{1} \exp(x^2-1)^{-1} dx .$$

Then

$$|T_N f(t)| \le \frac{C_m\,|\sin \sigma t\,|}{q^m(N\pi - |\sigma t\,|)^m} \left(\frac{N\pi}{\sigma}\right)^r ,$$

where $|t| < N\pi/\sigma$.

Three remarks concerning this theorem should be made. First, when $r = 0$ and $m = 1$, the error bound is similar to the bounds given by Helms and Thomas. Second, since F is a generalized function with compact support, the existence of b and r is guaranteed since it is known that [27, p. 131] the Fourier transform of such a generalized function is an entire function f with polynomial growth as $|t| \to \infty$ on the real-axis, i.e., there exist b and r such that $|f(t)| \le b\,|t|^r$ for large t. Third, although the series in Theorem 3.10 converges for all r, it is necessary to know r before the truncation error can be estimated.

Truncation error estimates for multidimensional band limited signals have also been obtained. One such estimate was obtained also by Yao and Thomas in [117].

THEOREM 3.23 (Yao and Thomas [117])

Let $f(t_1, \ldots, t_m)$ be a function band-limited to $[-\sigma_1, \sigma_1] \times \ldots \times [-\sigma_m, \sigma_m]$, i.e.,

$$f(t_1, \ldots, t_m) = \frac{1}{(2\pi)^m} \int_{-\sigma_1}^{\sigma_1} \cdots \int_{-\sigma_m}^{\sigma_m} F(\omega_1, \ldots, \omega_m) e^{i(t_1\omega_1 + \ldots + t_m\omega_m)} d\omega_1 \ldots d\omega_m ,$$

with $\int_{-\sigma_1}^{\sigma_1} \ldots \int_{-\sigma_m}^{\sigma_m} |F(\omega_1, \ldots, \omega_m)|^2 d\omega_1 \ldots d\omega_m < \infty$.

Assume that f has no frequencies higher than $(r_1\sigma_1/2\pi), \ldots, (r_m\sigma_m/2\pi)$ cps and $|f(t_1,\ldots,t_m)| \le M$. Then for

$$T(t_1,\ldots,t_m) = f(t_1,\ldots,t_m) - \sum_{k_1=N_1^1}^{N_2^1} \cdots \sum_{k_m=N_1^m}^{N_2^m} f(t_{1,k_1},\ldots,t_{m,k_m}) \frac{\sin\sigma_1(t_1 - t_{1,k_1})}{\sigma_1(t_1 - t_{1,k_1})} \cdots \frac{\sin\sigma_m(t_m - t_{m,k_m})}{\sigma_m(t_m - t_{m,k_m})},$$

we have the estimate

$$|T(t_1,\ldots,t_m)| \le \frac{M\,|\sin\sigma_1 t_1|\ldots|\sin\sigma_m t_m|}{(2\pi)^m[\cos(\pi r_1/2)]\ldots[\cos(\pi r_m/2)]}\left[\frac{1}{N_1^1}+\frac{1}{N_2^1}\right]\cdots\left[\frac{1}{N_1^m}+\frac{1}{N_2^m}\right].$$

The next type of error we shall discuss is the aliasing error.

3.8.B Aliasing Errors

We have already seen that under the hypothesis of Theorem 3.12 an upper bound for the aliasing error may be given by

$$\|R_\sigma f\|_\infty \le \sqrt{2/\pi} \int_{|\omega|>\sigma} |F(\omega)|\,d\omega. \tag{3.8.8}$$

A similar result, yet under slightly different hypotheses, was first announced by Weiss in [112].

THEOREM 3.24 (Weiss [112])

Let $F(\omega) = \overline{F}(-\omega)$ be absolutely integrable, of bounded variation over $(-\infty,\infty)$ and $2F(\omega) = F(\omega+0) + F(\omega-0)$. Then

$$\|R_\sigma f\|_\infty \le 2\sqrt{\frac{2}{\pi}} \int_\sigma^\infty |F(\omega)|\,d\omega, \tag{3.8.9}$$

and the upper bound can be attained.

Later, Brown [13] showed that the absolute integrability of F alone was sufficient to obtain (3.8.9). He has also pointed out that the constant $2\sqrt{2/\pi}$ is best possible in the sense that it is actually attained by some functions such as

$$f(t) = \sqrt{\frac{2}{\pi}}\,\frac{\sin 2\sigma\left(t - \frac{\pi}{2\sigma}\right)}{\left(t - \frac{\pi}{2\sigma}\right)},$$

whose Fourier transform is $F(\omega) = e^{-i\pi\omega/(2\sigma)}\chi_{(-2\sigma,2\sigma)}(\omega)$. To verify this, just observe that $f(t_k) = 0$ for all k and $\max_{-\infty<t<\infty} |f(t)| = \sqrt{2/\pi}\,(2\sigma)$.

In the same paper [13], Brown obtained an upper bound for the aliasing error of band-pass signals; see Section 3.5.B. This may require an explanation: suppose that the Fourier transform F of f does not vanish outside the band-pass I_{BP}

$$I_{BP} = [-\omega_0 - \sigma, -\omega_0 + \sigma] \cup [\omega_0 - \sigma, \omega_0 + \sigma]; \quad 0 < \sigma < \omega_0$$

and that the series in (3.5.4) is constructed as if f were band-pass, then the series in (3.5.4) would not, in general, converge to f. The difference between f and that series is called the aliasing error for the band-pass signal f and will be denoted by $R_{BP}f$.

THEOREM 3.25 (Brown [13])

Let F be absolutely integrable and $|F(\omega)|$ be an even function of ω. Then for

$$f(t) = \frac{1}{\sqrt{2\pi}} \int_{-\infty}^{\infty} F(\omega) e^{-it\omega} d\omega,$$

we have

$$|R_{BP}f| \le \sqrt{\frac{2}{\pi}} \int_{\omega \notin BP} |F(\omega)| \, d\omega,$$

and the constant $\sqrt{2/\pi}$ is best possible.

3.8.C Amplitude Errors

The amplitude error is usually dealt with by stochastic methods. However, in some cases such as in round-off errors or quantization, i.e., the sampled values are replaced by the nearest discrete values, deterministic methods can be employed. In digital recording of a signal f, for example, if the stored numbers are $\tilde{f}(t_k)$, then the round-off errors are given by

$$\varepsilon_k = f(t_k) - \tilde{f}(t_k).$$

The recovered signal

$$f_r(t) = \sum_{k=-\infty}^{\infty} \tilde{f}(t_k) \frac{\sin \sigma(t - t_k)}{\sigma(t - t_k)}$$

differs from the original signal by

$$(A_\varepsilon f)(t) = \sum_{k=-\infty}^{\infty} \varepsilon_k \frac{\sin \sigma(t - t_k)}{\sigma(t - t_k)}.$$

It is worth noting that, even if $|\varepsilon_k| \le \varepsilon$ for all k, $(A_\varepsilon f)$ may exceed all bounds for some values of t. To this end, suppose

$$\varepsilon_k = \begin{cases} \dfrac{(-1)^{k+1}k\varepsilon}{|k|} & , \quad k \neq 0 \\ 0 & , \quad k = 0 \end{cases} ,$$

then

$$(A_\varepsilon f)(\pi/2\sigma) = \infty .$$

If f, is band-limited to $[-\sigma, \sigma]$, then so is $A_\varepsilon f$, and in view of (3.6.3)

$$|A_\varepsilon f(t)| \le \sqrt{\frac{\sigma E_\varepsilon}{\pi}},$$

where E_ε is the total energy of $A_\varepsilon f$. A sharper estimate for the amplitude error is given in

THEOREM 3.26 (Butzer et al. [16])

If $f \in B_\sigma^\infty$ satisfies $|f(t)| \le M_f |t|^{-\gamma}$ ($|t| \ge 1$) for some constant M_f and $0 < \gamma \le 1$, then

$$\|A_\varepsilon f\|_\infty \le \frac{4}{\gamma}(\sqrt{3}\,\varepsilon + \sqrt{2}M_f e^{1/4})\,\varepsilon \ln(1/\varepsilon),$$

for $\sigma \ge \pi$, $\varepsilon \le \min\{\pi/\sigma, 1/\sqrt{e}\}$.

3.8.D Time-Jitter Errors

The time-jitter error is similar to the amplitude error in its treatment; it is usually treated by stochastic methods, but in some cases deterministic techniques can be used. Ideally, one would reconstruct a signal f from its samples $\{f(t_k)\}$; however, in real problems the samples are usually of the form $\{f(t_k - \gamma_k)\}$, where the γ_k's are the deviations from the sampling times t_k, and the problem is to determine f from the actual samples $\{f(t_k - \gamma_k)\}$.

Papoulis [76] suggested the following approach when the γ_k's are known. First, let

$$\theta(\tau) = \sum_{k=-\infty}^{\infty} \gamma_k \frac{\sin\sigma(\tau - t_k)}{(\tau - t_k)},$$

which is band-limited by σ (but not necessarily in the classical sense) and satisfies $\theta(t_k) = \gamma_k$. Second, let

$$t = \tau - \theta(\tau) \tag{3.8.10}$$

and assume that this function has a single-valued inverse $\tau = \eta(t)$. Hence, for $\tau = t_k$, we have $t = t_k - \theta(t_k) = t_k - \gamma_k$. Thus, the nonlinear transformation (3.8.10) transforms the points $t_k - \gamma_k$ of the t-axis into the points t_k of the τ-axis. Now set

$$g(\tau) = f(\tau - \theta(\tau)) = f(t) .$$

Since $f(t_k - \gamma_k)$ are given, $g(t_k)$ are known; therefore, g can be reconstructed via

$$g(\tau) = \sum_{k=-\infty}^{\infty} g(t_k) \frac{\sin \sigma(t - t_k)}{\sigma(t - t_k)} , \tag{3.8.11}$$

provided that it is band-limited to $[-\sigma, \sigma]$. Under this assumption, f can be reconstructed as well via

$$f(t) = g(\eta(t)) = \sum_{k=-\infty}^{\infty} f(t_k - \gamma_k) \frac{\sin \sigma[\eta(t) - t_k]}{\sigma[\eta(t) - t_k]} .$$

If g is not band-limited to $[-\sigma, \sigma]$, an aliasing error will arise in (3.8.11) whose magnitude can be estimated as in Section 3.8.B.

A result similar to Theorem 3.26 for the time-jitter errors can be obtained.

THEOREM 3.27 (Butzer et al. [16])

Let f be a C^1-function satisfying the same hypothesis as in Theorem 3.26. Then

$$\|J_\varepsilon f\|_\infty \le \frac{4}{\gamma}(\sqrt{5}\, e \|f^{(1)}\|_\infty + \sqrt{8}\, M_f e^{1/4})\, \varepsilon \ln(1/\varepsilon) ,$$

where $0 < \varepsilon \le \min\{\pi/\sigma, 1/\sqrt{e}\}$, $\sigma \ge \pi$.

3.8.E Combined Errors

In practice, more than one of the aforementioned errors can occur. In the general case when the first four main types of errors occur together, an estimation of the combined error defined in (3.8.1) is evidently very useful. The following important result gives an upper bound for this error; it is the error caused by approximating a not-necessarily band-limited function by a truncated sampling series with quantized sampled values taken at jittered time instants.

THEOREM 3.28 (Butzer et al. [16])

Let f be C^1-function satisfying $|f(t)| \le M_f |t|^{-\gamma}$ for some constant M_f and $0 < \gamma \le 1$. Then for any $W (W = \sigma/\pi)$ such that $\ln W \ge 2$

$$\|Ef\|_\infty \le K(f, \gamma, p_1, p_2) \frac{\ln W}{W},$$

where

$$K(f,\gamma,p_1,p_2) = \left(1+\frac{1}{\gamma}\right)\left\{\sqrt{5}\,e\left[\left(\frac{14}{\pi}+p_2+\frac{7}{3\sqrt{5}\,\pi}\right)\|f^{(1)}\|_\infty + p_1\right] + 6e(M_f + \|f\|_\infty)\right\},$$

$$N = [W^{1+1/\gamma}+1], \quad \varepsilon = p_1/W, \quad \delta = p_2/W,$$

$$|f(\bar{t}_k) - \tilde{f}(\bar{t}_k)| \le \varepsilon, \quad \text{and} \quad |\gamma_k| \le \delta \quad \text{for all} \quad k.$$

3.9 Closing Remark

Having completed our journey through the various generalizations and extensions of the WSK sampling theorem, we must stress that our coverage is by no means extensive. Because of space limitation, we could not include many others. Furthermore, there are other important results in sampling theory that are not considered either generalizations or extensions of the WSK sampling theorem, such as results showing the connection and even the equivalence between the WSK sampling theorem and other fundamental results in mathematical analysis. To mention a few: the equivalence between the WSK sampling theorem or a generalization thereof, on the one hand, and the Poisson summation formula of Fourier analysis, the Cauchy integral formula in complex function theory, the Euler-Maclaurian summation formula of *numerical analysis*, and the Reimann-zeta function on the other hand.

While we were preparing the final version of this book, we came across an interesting article by D. Klusch (*J. Comp. Appl. Math.*, 44 (1992), 261-273) in which he has shown that the WSK sampling theorem as well as a generalization thereof for time-limited signals, can be deduced from five important results in different areas of mathematics, such as the *theory of integral transforms*, *analytic number theory*, and the *theory of special functions of mathematical physics*. First, he shows that the following are equivalent:

i) The Riemann functional equation for the zeta-function.
ii) The "modular relation" of the Jacobi theta-function.
iii) The Nielsen-Doetsch summation formula for the Bessel function of the first kind.
iv) A general summation formula for the Hankel integral transform.
v) The partial fraction expansion of the generalized periodic "Hilbert kernel."

Then, he shows that the summation formula for the Hankel integral transform implies a Poisson-type summation formula, which in turn implies the WSK sampling theorem and its generalization for time-limited signals.

In Chapters 4 through 8 we shall show another connection between the WSK sampling theorem and two other branches of mathematical analysis, namely, the *Theory of Boundary-Value Problems* and *Special Functions*. We shall show that a large number of boundary-value problems generate their own sampling theorems in a natural way, where the sampling points are the eigenvalues of the problems. We shall then exhibit the strength of sampling theory by applying some of these recent results to obtain new summation formulae in one and several variables involving special functions.

Chapters 9 and 10 will be devoted to the discussion of some novel approaches to sampling theory and to new relationships with very recent advances in mathematical analysis, such as the theory of frames, wavelets and multiresolution analysis.

References

1. I. Bar-David, An implicit sampling theorem for bounded band-limited functions, *Information and Control*, 24 (1974), 36-44.
2. M. J. Bastiaans, A generalized sampling theorem with application to computer-generated transparencies, *J. Opt. Soc. Amer.*, 68 (1978), 1658-1665.
3. J. Benedetto, irregular sampling and frames, *Wavelet-A Tutorial*, Ed. C. Chui, Academic Press (1991), 1-63.
4. J. Benedetto and W. Heller, Irregular sampling and the theory of frames, I, *Note di Matematica*, Vol. X, Suppl. 1 (1990), 103-125.
5. F. J. Beutler, On the truncation error of the cardinal sampling expansion, *IEEE Trans. Inform. Theory*, 22, 5 (1976), 568-573.
6. ________, Error free recovery of signals from irregularly spaced samples, *SIAM Rev.*, 8 (1966), 328-335.
7. ________, Sampling theorems and bases in a Hilbert space, *Information and Control*, 4 (1961), 97-117.
8. V. Blazek, Optical information processing by the Fabry-Perot resonator, *Optical Quantum Electronics*, 8 (1976), 237-240.
9. ________, Sampling theorem and the number of degrees of freedom of an image, *Optics Comm.*, 11 (1974), 144-147.
10. F. E. Bond and C. R. Chan, On sampling the zeros of bandwidth limited signals, *IRE Trans. Inform. Theory*, IT 4 (1958), 110-113.
11. J. L. Brown, Jr., First order sampling of bandpass signals - a new approach, *IEEE Trans. Inform. Theory*, IT-26 (1980), 613-615.
12. ________, Bounds for truncation error in sampling expansions of band limited signals, *IRE Trans. Inform. Theory*, 15, 4 (1969), 440-444.
13. ________, On the error in reconstructing a nonband-limited function by means of the band pass sampling theorem, *J. Math. Anal. Appls.*, 18 (1967), 75-84.
14. P. L. Butzer, A survey of the Whittaker-Shannon sampling theorem and some of its extensions, *J. Math. Res. Exposition*, 3 (1983), 185-212.
15. P. L. Butzer and R. L. Stens, Sampling theory for not necessarily band-limited functions: A historical overview, *SIAM Rev.*, 34, 1 (1992), 40-53.
16. P. L. Butzer, W. Splettstösser and R. L. Stens, The sampling theorem and linear prediction in signal analysis, *Jber. d. Dt. Math. Verein.*, 90 (1988), 1-70.
17. P. L. Butzer and G. Hinsen, Two-dimensional nonuniform sampling expansions - An iterative approach, *Arbeitsbericht, Lehrstuhl A für Mathematik*, Aachen Technical University, Aachen-Germany (1988).

18. P. L. Butzer and C. Markett, The Poisson summation formula for orthogonal systems. In: *Anniversary Volume on Approximation Theory and Functional Analysis*, P. L. Butzer, R. L. Stens and S. Z. Nagy, Eds., Birkhäuser Verlag, Basel (1984), 595-601.
19. L. L. Campbell, Sampling theorem for the Fourier transform of a distribution with bounded support, *SIAM J. Appl. Math.*, 16 (1968), 626-636.
20. _______, A comparison of the sampling theorems of Kramer and Whittaker, *J. SIAM*, 12 (1964), 117-130.
21. S. R. Curtis and A. V. Oppenheim, Reconstruction of multidimensional signals from zero crossings, *J. Optical Soc. Amer.*, 69 (1987), 221-231.
22. H. G. Feichtinger, Discretization of convolution and reconstruction of band-limited functions from irregular sampling, Progress in Approximation Theory, *Special Issue of the J. Approx. Theory*, (1991), 333-345.
23. H. G. Feichtinger and K. Gröchenig, Reconstruction of band-limited functions from irregular sampling values, manuscript (1992).
24. _______, Irregular sampling theorems and series expansions of band-limited functions, *J. Math. Anal. Appls.*, 167 (1992), 530-556.
25. _______, Multidimensional irregular sampling of band-limited functions in L^p-spaces, *Multivariate Approx. Theory*, IV (Oberwolfach), *Internat. Ser. Numer. Math.*, 90, Birkhäuser, Basel (1989), 135-142.
26. L. J. Fogel, A note on the sampling theorem, *IRE Trans. Inform. Theory*, 1 (1955), 47-48.
27. I. M. Gelfand and G. E. Shilov, *Generalized Functions*, Vol. I, II, Academic Press, New York (1968).
28. S. Goldman, *Information Theory*, Prentice Hall, New York (1953).
29. F. Gori and G. Guattari, Holographic restoration of non-uniformly sampled band-limited functions, *Optics Comm.*, 3 (1971), 147-149.
30. K. Gröchenig, A new approach to irregular sampling of band-limited functions, Recent advances in Fourier analysis and its applications, *NATO Adv. Sci. Inst. Ser. C: Math. Phys. Sci.*, Kluwer Acad. Publ., Dordrecht (1990), 251-260.
31. A. H. Haddad, K. Yao and J. B. Thomas, General methods for the derivation of sampling theorems, *IEEE Trans. Inform. Theory*, IT 13 (1967), 227-230.
32. H. D. Helms and J. B. Thomas, Truncation error of sampling theorem expansion, *Proc. IRE*, 50 (1962), 179-184.
33. R. W. Hielmming, An introduction to radio astronomy very large array, National Radio Astronomy Observatory, United States of America (1982).

34. J. R. Higgins, Five short stories about the cardinal series, *Bull. Amer. Math. Soc.*, 12 (1985), 45-89.
35. _______, *Completeness and Basis Properties of Sets of Special Functions*, Cambridge Univ. Press, Cambridge, England (1977).
36. _______, A sampling theorem for irregularly spaced sample points, *IEEE Trans. Inform. Theory*, IT 22 (1976), 621-622.
37. _______, An interpolation series associated with the Bessel-Hankel transform, *J. London Math. Soc.* 5 (1972), 707-714.
38. D. Jagerman, Bounds for truncation error of the sampling expansion, *SIAM J. Appl. Math.*, 14, 4 (1966), 714-723.
39. D. L. Jagerman and L. J. Fogel, Some general aspects of the sampling theorem, *IEEE Trans. Inform. Theory* 2 (1956), 139-156.
40. A. J. Jerri, *Integral and Discrete Transforms with Applications and Error Analysis*, Mercel Dekker Publ., New York (1992).
41. _______, A note on sampling expansion for a transform with parabolic cylindrical kernel, *Inform. Sci.*, 26 (1982), 155-158.
42. _______, The Shannon sampling theorem—its various extensions and applications: a tutorial review, *Proc. IEEE* 65 (1977), 1565-1596.
43. _______, Sampling expansion for Laguerre-L_v^α transforms, *J. Res. Nat. Bur. Standards* sec. B (80), (B) 3 (1976), 415-418.
44. _______, On the application of some interpolating functions in physics, *J. Res. Nat. Bur. Standards* sec. B (80), (B) 3 (1969), 241-245.
45. _______, On the equivalence of Kramer's and Shannon's sampling theorems, *IEEE Trans. Inform. Theory*, IT-15 (1969), 497-499.
46. A. J. Jerri and D. W. Kreisler, Sampling expansions with derivatives for finite Hankel and other transforms, *SIAM J. Math. Anal.*, 6 (1975), 262-267.
47. S. C. Kak, Sampling theorem in Walsh-Fourier analysis, *Electronics Lett.* 6 (1970), 447-448.
48. I. Kluvánek, Sampling theorem in abstract harmonic analysis, *Nat.-Fyz. Casopis Sloven. Akad. Vied.* 15 (1965), 43-48.
49. A. Kohlenberg, Exact interpolation of band-limited functions, *J. Appl. Phys.* 24 (1953), 1432-1436.
50. H. P. Kramer, A generalized sampling theorem, *J. Math. Phys.*, 63 (1957), 68-72.
51. N. Levinson, *Gap and Density Theorems*, Amer. Math. Soc. Colloq. Publs., Ser., Vol. 26, Amer. Math. Soc., Providence, RI (1940).
52. J. C. R. Licklider and I. Pollack, Effects of differentiation, integration and peak clipping upon the intelligibility of speech, *J. Acoust. Soc. Amer.* 20 (1958), 42-51.
53. D. A. Linden, A discussion of sampling theorems, *Proc. IRE* 47 (1959), 1219-1226.

54. D. A. Linden and N. M. Abramson, A generalization of the sampling theorem, *Inform. Control*, 3 (1960), 26-31; Errata, Ibid., 4 (1961), 95-96.
55. JR. B. F. Logan, Information in zero crossing of bandpass signals., *Bell Syst. Tech J.*, 56 (1977), 487-510.
56. R. J. Marks II, editor, *Advanced Topics on Shannon Sampling and Interpolation Theory*, Springer-Verlag, New York (1992).
57. ________, *Introduction to Shannon Sampling and Interpolation Theory*, Springer-Verlag, New York (1991).
58. F. Marvasti, *A Unified Approach to Zero Crossings and Nonuniform Sampling of Single and Multidimensional Signals and Systems*, 1st ed., Oak Park, IL (1987).
59. _______, Zero crossings, bandwidth compression, and restoration of nonlinearly distorted band-limited signals, *J. Optical Soc. Amer.*, 3 (1986), 651-654.
60. F. C. Mehta, A general sampling expansion, *Inform. Sci.*, 16 (1978), 41-46.
61. ________, Sampling expansion for band-limited signals through some special functions, *J. Cybernetics* (1975), 61-68.
62. R. Mersereau, The processing of hexagonally sampled two dimensional signals, *Proc. IEEE*, 67 (1979), 930-949.
63. R. Mersereau and T. Speake, The processing of periodically sampled multidimensional signals, *IEEE Trans. Acoust. Speech Signal Process.* ASSP-31 (1983), 188-194.
64. W. Montgomery, K-order sampling of N-dimensional band-limited functions, *Int. J. Contr.*, 1 (1965), 7-12.
65. D. Mugler and W. Splettstösser, Reconstruction of two dimensional signals from irregularly spaced samples, *Proc. 6 Aachner Symp. für Signaltheorie Informatik Fachberichte* 153, Springer-Verlag, New York (1987), 41-44.
66. Z. Nashed and G. Walter, General sampling theorems for functions in reproducing kernel Hilbert spaces, *Math-Control Signal Systems*, 4 (1991), 363-390.
67. S. M. Nikol'skii, *Approximation of Functions of Several Variables and Imbedding Theorems*, Springer-Verlag, New York (1975).
68. A. V. Oppenheim, S. R. Curtis and J. S. Lim, Signal reconstruction from Fourier transform sign information, *IEEE Trans. Acoust. Speech Signal Processing*, ASSP 33 (1985), 643-657.
69. R. Paley and N. Wiener, *Fourier Transforms in the Complex Domain*, Amer. Math. Soc. Colloq. Publs., Ser., Vol 19, Amer. Math. Soc., Providence, RI (1934).

70. M. R. Palmer, J. J. Clark and P. D. Lawrence, A transformation method for the reconstruction of functions from nonuniformly spaced samples, *IEEE Trans. Acoust. Speech Signal Processing*, ASSP 33 (1985), 1151-1165.
71. S. X. Pan and A. C. Kak, A computational study of reconstruction algorithms for differaction tomography: interpolation versus filtered backprojection, *IEEE Trans. Acoust. Speech Signal Processing*, ASSP 31 (1983), 1262-1275.
72. A. Papoulis, *Signal Analysis*, McGraw-Hill, New York (1977).
73. _______, *Systems and Transforms with Applications in Optics*, McGraw-Hill, New York (1968).
74. _______, Limits on band-limited signals, *Proc. IEEE*, 55, 10 (1967), 1677-1686.
75. _______, Truncated sampling expansions, *IEEE Trans. Automat. Contr.*, 12 (1967), 604-605.
76. _______, Error analysis in sampling theory, *Proc. IEEE*, 54, 7 (1966), 947-955.
77. _______, *The Fourier Integral and its Application*, McGraw-Hill, New York (1962).
78. E. Parzen, A simple proof and some extensions of sampling theorems, Tech. Rep., 7, Stanford Univ., Stanford, California (1956).
79. D. P. Petersen, Sampling of space-time stochastic processes with application to information and decision systems, D.E.S. dissertation, Rensselaer Poltechnic Institute, Troy, New York (1963).
80. D. P. Petersen and D. Middleton, Reconstruction of multidimensional stochastic fields from discrete measurements of amplitude and gradient, *Inform. Contr.*, 1 (1974), 445-476.
81. _______, Sampling and reconstruction of wave number-limited function in N-dimensional Euclidean space, *Inform. Control*, 5 (1962), 279-323.
82. E. Pfaffelhuber, Sampling series for band-limited generalized functions, *IEEE Trans. Inform. Theory*, IT 17 (1971), 650-654.
83. F. Pichler, Walsh functions—introduction to the theory, Signal Processing (Proc. NATO Advanced Study Institute for Signal Processing, J. W. R. Griffiths et al., eds.), Academic Press, New York (1973), 23-41.
84. R. T. Prosser, A multidimensional sampling theorem, *J. Math. Anal. Appl.*, 16 (1966), 574-584.
85. M. Rawn, On nonuniform sampling expansions using entire interpolating functions and on the stability of Bessel-type sampling expansions, *IEEE Trans. Inform. Theory*, 35 (1989), 549-557.
86. F. Reza, *An Introduction to Information Theory*, McGraw-Hill, New York (1961).

87. D. Rotem and Y. Y. Zeevi, Image reconstruction from zero crossings, *IEEE Trans. Acoust. Speech Signal Processing*, ASSP 34 (1986), 1269-1277.
88. W. Schempp, Radar ambiguity functions, nilpotent harmonic analysis, and holomorphic theta series, *Special Functions: Group Theoretical Aspects and Applications*, R. Askey et al., Eds., Reidel Pub. Co., Hingham, MA (1984).
89. _______, Gruppentheoretische Aspekte der Signalubertragung und der kardinal Interpolationssplines, 1. *Math. Methods Appl. Sci.* 5 (1983), 195-215.
90. G. Schwierz, W. Härer and K. Wiesent, Sampling and discretization problems in X-ray-CT, *Mathematical Aspects of Computerized Tomography*, Springer-Verlag, Berlin (1981).
91. K. Seip, An irregular sampling theorem for functions bandlimited in a generalized sense, *SIAM J. Appl. Math.*, 47, 5 (1987), 1112-1116.
92. A. Sekey, A computer simulation study of real-zero interpolation, *IEEE Trans. Audio Electroacoust.*, 18 (1970), 43-54.
93. S. D. Selvaratnam, Shannon-Whittaker sampling theorems, Ph.D. thesis, University of Wisconsin-Milwaukee (1987).
94. B. D. Sharma and F. C. Mehta, Generalized band-pass sampling theorem, *Math. Balkanica* 6 (1976), 204-217.
95. I. N. Sneddon, *The Use of Integral Transforms*, McGraw-Hill, New York (1972).
96. W. Splettstösser, Sampling approximation of continuous functions with multidimensional domain, *IEEE Trans. Inform. Theory*, IT-28 (1982), 809-814.
97. _______, Error estimates for sampling approximation of non-band-limited functions, *Math. Meth. in the Appl. Sci.* 1 (1979), 127-137.
98. _______, On generalized sampling sums based on convolution integrals, *Arch. Elek. Obertr.*, 32 (1978), 267-275.
99. H. Stark, Polar sampling theorems of use in optics, *Proc. SPIE Int. Soc. Opt. Eng.*, 358 (1982), 24-30.
100. _______, Sampling theorems in polar coordinates, *J. Opt. Soc. Amer.*, 69 (1979), 1519-1525.
101. H. Stark and C. S. Sarna, Image reconstruction using polar sampling theorems, *Appl. Optics*, 18 (1979), 2086-2088.
102. R. L. Stens, A unified approach to sampling theorems for derivatives and Hilbert transforms, *Signal Process*, 5 (1983), 139-151.
103. M. Theis, Über eine interpolations formel von de la Vallée Poussin, *Math. Z.*, 3 (1919), 93-113.

104. E. C. Titchmarsh, *Eigenfunction Expansions Associated with Second-Order Differential Equations*, Vol. 1, 2nd. ed., Clarendon Press, Oxford (1962).
105. _______, The zeros of certain integral functions, *Proc. London Math. Soc.*, 25 (1926), 283-302.
106. H. Triebel, *Theory of Function Spaces*, Akad. Verlagsges., Leipzig (1983).
107. B. S. Tsybakov and V. P. Iakovlev, On the accuracy of reconstructing a function with a finite number of terms of Kotel'nikov series, *Radio Eng. Electron. (Phys.)*, 4, 3 (1959), 274.
108. S. Ullman, D. Marr and T. Poggio, Bandpass channels, zero crossings, and early visual information processing, *J. Optical Soc. Amer.*, 69 (1979), 914-916.
109. Ch. J. De la Vallée Poussin, Sur la convergence des formules d'interpolation entre ordonnée équidistantes, *Bull. Cl. Sci. Acad. Ray. Belg.*, 4 (1908), 319-410.
110. H. B. Voelcker, Toward a Unified theory of modulation - part II: zero manipulation, *Proc. IEEE*, 54 (1966), 735-755.
111. J. L. Walsh, A closed set of normal orthogonal functions, *Amer. J. Math.*, 55 (1923), 5-24.
112. P. Weiss, An estimation of the error arising from misapplication of the sampling theorem, *Amer. Math. Soc. Notices*, 10 (1963), 351.
113. _______, Sampling theorems associated with Sturm-Liouville systems, *Bull. Amer. Math. Soc.*, 63 (1957), 242.
114. J. M. Whittaker, *Interpolatory Function Theory*, Cambridge University Press, Cambridge, England (1935).
115. K. Yao, Applications of reproducing kernel Hilbert space of band-limited signal models, *Inform. Control*, 11 (1967), 429-444.
116. K. Yao and J. Thomas, On some stability and interpolatory properties of nonuniform sampling expansions, *IEEE Trans. Circuit Theory*, CT-14 (1967), 404-407.
117. _______, On truncation error for sampling representations of band-limited signals, *IEEE Trans. Aero. Electron. Syst.*, Vol. ASE 2 (1966), 640-646.
118. J. L. Yen, On nonuniform sampling of bandwidth-limited signals, *IRE Trans. Circuit Theory* CT, 3 (1956), 251-257.
119. R. M. Young, *An Introduction to Nonharmonic Fourier Series*, Academic Press, New York (1980).
120. A. Zayed, On Kramer sampling theorem associated with general Sturm-Liouville problems and Lagrange interpolation, *SIAM J. Appl. Math.*, 51 (1991), 575-604.
121. _______, Sampling expansion for the continuous Bessel transform, *J. Appl. Anal.*, 27 (1988), 47-64.

122. A. Zayed, G. Hinsen and P. Butzer, On Lagrange interpolation and Kramer-type sampling theorems associated with Sturm-Liouville problems, *SIAM J. Appl. Math.*, 50 (1990), 893-909.
123. A. H. Zemanian, *Generalized Integral Transformations*, Interscience Publ., New York, 1968.

4

SAMPLING THEOREMS ASSOCIATED WITH STURM-LIOUVILLE BOUNDARY-VALUE PROBLEMS

4.0 Introduction

In this chapter, we will continue the investigation we started in Section 3.2 of the Weiss-Kramer sampling theorem and will provide answers to some of the questions raised therein.

Let us recall from Section 3.2 that:

i) Although Kramer's assumptions may appear to be quite general, in practice, they are not. For, his main assumption that the eigenfunctions of the self-adjoint boundary-value problem (3.2.4) and (3.2.5) are all generated by one single function does not generally hold if the order of the differential operator (3.2.4) (cf. (3.2.6)) is larger than 2, or if the boundary conditions (3.2.5) are of mixed types.

ii) Kramer gave an example to illustrate the possibility of extending his sampling theorem to singular self-adjoint boundary-value problems; however, he never explicitly discussed that extension. In fact, to the best of my knowledge, there has not been any formal extension of Kramer's sampling theorem to singular boundary-value problems; some examples have been given in [15-18, 23, and 25], but no general theory seems to exist in the literature. Among these examples is the case where the interval $(a,b)=(0,\infty)$ and $\{K(x,t_k)\}_{k=0}^{\infty}$ are the Laguerre polynomials. A. Jerri [17] derived a Kramer-type sampling expansion for this case, but his results were not very satisfactory since they did not generalize in a natural way the results in the finite interval case; somewhat better results have been obtained in [25]. In the case where $(a,b)=(-\infty,\infty)$, an attempt was made by Mehta [23] to derive a Kramer-type sampling expansion using the Hermite polynomials as

$\{K(x, t_k)\}_{k=0}^{\infty}$; however, his result was in error and was later corrected by Jerri in [15].

iii) L. Campbell [2] has compared the Whittaker-Shannon-Kotel'nikov (WSK) sampling theorem with the Weiss-Kramer sampling theorem associated with *regular* Sturm-Liouville boundary-value problems and concluded that the latter provided no improvement over the former. That is any function that can be expanded in a Weiss-Kramer sampling series can also be expanded in a WSK sampling series. He has shown, further, that his conclusion is also valid for two singular Sturm-Liouville boundary-value problems, namely, the Legendre and the Bessel case. The Bessel case can, in fact, be traced back to the work of J. M. Whittaker [33, p. 71].

First, we shall begin by giving a deeper insight into Campbell's result by showing that both the Weiss-Kramer and the WSK sampling series are, in fact, special cases of the Lagrange-type interpolation series. Already shown in Section 3.1 was the fact that the WSK sampling series is a special case of the Lagrange-type interpolation series; therefore, we only need to show that the Weiss-Kramer sampling series is a special case of the Lagrange-type interpolation series. Second, we shall extend the Weiss-Kramer sampling theorem to *singular* Sturm-Liouville boundary-value problems and show, among other things, that even for these singular problems if a Weiss-Kramer sampling theorem exists, then the associated sampling series is nothing more than a Lagrange-type interpolation series.

To demonstrate the strength of our new techniques, we shall apply them to some concrete examples to obtain not only several known results, but also some new ones. Finally, included at the end of this chapter is an example showing that the discreteness of the spectrum of a singular Sturm-Liouville boundary-value problem is, surprisingly, not sufficient for the existence of a Weiss-Kramer sampling theorem.

Sampling theorems associated with boundary-value problems other than the Sturm-Liouville ones will be discussed in Chapters 5 and 6, wherein the order of the differential operator will not be restricted to 2 and the boundary conditions will be of more general types than those considered in the Sturm-Liouville problems.

4.1 The Regular Case*

In this section we shall show that, when the kernel in the Weiss-Kramer sampling theorem arises from a regular Sturm-Liouville boundary-value

---

*Some material in Sections 1 through 4 is based on the author's article that appeared in *SIAM Journal of Applied Mathematics*, Vol. 51, 1991, pp. 575-604 and is reprinted with permission from SIAM.

problem, the associated sampling series (3.2.2) is nothing more than a Lagrange-type interpolation series.

THEOREM 4.1

Consider the following regular Sturm-Liouville problem:

$$y'' - q(x)y = -\lambda y = -t^2 y\,, \quad -\infty < a \le x \le b < \infty\,, \tag{4.1.1}$$

$$y(a)\cos\alpha + y'(a)\sin\alpha = 0\,, \tag{4.1.2}$$

$$y(b)\cos\beta + y'(b)\sin\beta = 0\,, \tag{4.1.3}$$

where $q(x)$ is continuous on (a,b) and tends to finite limits as $x \to a^+$ and $x \to b^-$. Let $\phi(x,\lambda)$ and $\chi(x,\lambda)$ be the solutions of (4.1.1) such that

$$\phi(a,\lambda) = \sin\alpha\,, \quad \phi'(a,\lambda) = -\cos\alpha\,, \tag{4.1.4}$$

$$\chi(b,\lambda) = \sin\beta\,, \quad \chi'(b,\lambda) = -\cos\beta\,. \tag{4.1.5}$$

Let $F(x) \in L^2(a,b)$,

$$f(\lambda) = \int_a^b F(x)\phi(x,\lambda)\,dx \tag{4.1.6}$$

and

$$f^*(\lambda) = \int_a^b F(x)\chi(x,\lambda)\,dx\,. \tag{4.1.7}$$

Then, $f(\lambda)$ and $f^*(\lambda)$ are entire functions of order 1/2 and type η with $0 \le \eta \le b-a$ that admit the following sampling representations:

$$f(\lambda) = \sum_{n=0}^{\infty} f(\lambda_n)\,\frac{G(\lambda)}{(\lambda-\lambda_n)G'(\lambda_n)}\,, \tag{4.1.8}$$

$$f^*(\lambda) = \sum_{n=0}^{\infty} f^*(\lambda_n)\,\frac{G(\lambda)}{(\lambda-\lambda_n)G'(\lambda_n)}\,, \tag{4.1.9}$$

where $\{\lambda_n\}_{n=0}^{\infty}$ are the eigenvalues of the problem (4.1.1) to (4.1.3) and $G(\lambda)$ is the Wronskian $W(\phi,\chi) = \phi(x,\lambda)\chi'(x,\lambda) - \phi'(x,\lambda)\chi(x,\lambda)$ of the two functions ϕ and χ which, without loss of generality, may be written in the form

$$G(\lambda) = \begin{cases} \prod_{n=0}^{\infty}\left(1-\dfrac{\lambda}{\lambda_n}\right) & \text{if none of the eigenvalues is zero}, \\ \lambda\prod_{n=1}^{\infty}\left(1-\dfrac{\lambda}{\lambda_n}\right) & \text{if one of the eigenvalues say } \lambda_0 = 0. \end{cases} \tag{4.1.10}$$

The two series (4.1.8) and (4.1.9) converge uniformly on any compact subsets of the complex plane and they are related via the relation $f^*(\lambda_n) = k_n f(\lambda_n)$ for all n, where $k_n = \phi(x,\lambda_n)/\chi(x,\lambda_n)$ which is neither zero nor infinity.

Proof. First, we recall the following facts from [27, pp. 7-11, 19]: the Wronskian $W(\phi,\chi)$ is independent of x; in fact, it is an entire function of order 1/2 in λ whose zeros are all real, simple, and located exactly at the eigenvalues $\{\lambda_n\}_{n=0}^{\infty}$. Since $\lambda_n = 0(n^2)$ as $n \to \infty$ (more precisely $\lambda_n \sim n^2\pi^2/(b-a)^2$ as $n \to \infty$; [27, p. 19]) the product in (4.1.10) converges and defines an entire function of order 1/2 [27, p. 251], which will be denoted temporarily by $\hat{G}(\lambda)$. By Hadamard's factorization theorem for entire functions (cf. Theorem 1.1)

$$G(\lambda) = h(\lambda)\hat{G}(\lambda),$$

where $h(\lambda)$ is an entire function of order zero with no zeros. Thus,

$$\frac{G(\lambda)}{G'(\lambda_n)} = \frac{h(\lambda)\hat{G}(\lambda)}{h(\lambda_n)\hat{G}'(\lambda_n)}$$

and (4.1.6), (4.1.8) ((4.1.7), (4.1.9)) remain valid for the function $f(\lambda)/h(\lambda)$ ($f^*(\lambda)/h(\lambda)$). Therefore, without loss of generality, we may assume that $G(\lambda) = \hat{G}(\lambda)$.

It is evident that $\phi(x,\lambda_n)$ is an eigenfunction corresponding to the eigenvalue λ_n. Hence, we adopt the notation

$$\psi_n(x) = \phi(x,\lambda_n).$$

Since both $F(x)$ and $\phi(x,\lambda)$ are in $L^2(a,b)$, it follows that

$$F(x) = \sum_{n=0}^{\infty} \hat{F}(n) \frac{\psi_n(x)}{\|\psi_n\|^2}, \tag{4.1.11}$$

$$\phi(x,\lambda) = \sum_{n=0}^{\infty} \frac{\langle\phi,\psi_n\rangle}{\|\psi_n\|^2} \psi_n(x), \tag{4.1.12}$$

where

$$\hat{F}(n) = \int_a^b F(x)\psi_n(x)\,dx, \tag{4.1.13}$$

and

$$\langle\phi,\psi_n\rangle = \int_a^b \phi(x,\lambda)\psi_n(x)\,dx. \tag{4.1.14}$$

The two series in (4.1.11) and (4.1.12) converge in the sense of $L^2(a,b)$. From (4.1.6), (4.1.11), (4.1.12), and Parseval's equality we obtain

$$f(\lambda) = \sum_{n=0}^{\infty} f(\lambda_n) \frac{\langle\phi,\psi_n\rangle}{\|\psi_n\|^2}, \tag{4.1.15}$$

since

$$f(\lambda_n) = \int_a^b F(x)\phi(x,\lambda_n)\,dx = \int_a^b F(x)\psi_n(x)\,dx = \hat{F}(n). \tag{4.1.16}$$

From the relation

$$\int_a^b U(x)(LV(x))\,dx - \int_a^b V(x)(LU(x))\,dx = W_{x=b}(U,V) - W_{x=a}(U,V), \tag{4.1.17}$$

where

$$L = \frac{d^2}{dx^2} - q(x) \tag{4.1.18}$$

we have for $U(x) = \phi(x,\lambda)$ and $V(x) = \phi(x,\lambda')$

$$(\lambda - \lambda') \int_a^b \phi(x,\lambda)\phi(x,\lambda')\,dx = \phi(b,\lambda)\phi'(b,\lambda') - \phi'(b,\lambda)\phi(b,\lambda'). \tag{4.1.19}$$

Note that $W_{x=a}(\phi(x,\lambda), \phi(x,\lambda')) = 0$ because $\phi(x,\lambda)$ and $\phi(x,\lambda')$ satisfy the same boundary-condition (4.1.4) at $x = a$. Since the Wronskian $W(\phi,\chi)$ of ϕ and χ is independent of x, we may evaluate it at $x = b$ to obtain

$$\begin{aligned} G(\lambda) = W_{x=b}(\phi,\chi) &= \phi(b,\lambda)\chi'(b,\lambda) - \phi'(b,\lambda)\chi(b,\lambda) \\ &= -\cos\beta\,\phi(b,\lambda) - \sin\beta\,\phi'(b,\lambda). \end{aligned} \tag{4.1.20}$$

If $\sin\beta \neq 0$, then

$$G(\lambda)\phi(b,\lambda') - G(\lambda')\phi(b,\lambda) = \sin\beta[\phi'(b,\lambda')\phi(b,\lambda) - \phi'(b,\lambda)\phi(b,\lambda')],$$

which, when combined with (4.1.19), yields

$$(\lambda - \lambda')\sin\beta \int_a^b \phi(x,\lambda)\phi(x,\lambda')\,dx = G(\lambda)\phi(b,\lambda') - G(\lambda')\phi(b,\lambda). \tag{4.1.21}$$

By taking the limit in (4.1.21) as $\lambda' \to \lambda_n$, we obtain

$$\langle \phi, \psi_n \rangle = \int_a^b \phi(x,\lambda)\psi_n(x)\,dx = \frac{G(\lambda)\psi_n(b)}{(\lambda - \lambda_n)\sin\beta}, \tag{4.1.22}$$

and once more upon taking the limit in (4.1.22) as $\lambda \to \lambda_n$, we obtain

$$\|\psi_n\|^2 = \int_a^b |\psi_n(x)|^2\,dx = \frac{G'(\lambda_n)\psi_n(b)}{\sin\beta}. \tag{4.1.23}$$

By substituting (4.1.22) and (4.1.23) into (4.1.15) we obtain (4.1.8). Note that $\psi_n(b) \neq 0$. For, suppose that $\psi_n(b) = 0$; then since $\psi_n(x)$ is an eigenfunction, it satisfies (4.1.3). Thus, $\psi_n'(b)\sin\beta = 0$, but since $\sin\beta \neq 0$, then $\psi_n'(b) = 0$. Hence $\psi_n(x) \equiv 0$, which is a contradiction.

Similarly, if $\sin\beta = 0$, we have

$$(\lambda - \lambda')\cos\beta \int_a^b \phi(x,\lambda)\phi(x,\lambda')\,dx = G(\lambda')\phi'(b,\lambda) - G(\lambda)\phi'(b,\lambda'), \tag{4.1.24}$$

which upon taking the limit as $\lambda' \to \lambda_n$ yields

$$\langle \phi, \psi_n \rangle = \int_a^b \phi(x,\lambda)\psi_n(x)\,dx = \frac{-G(\lambda)\psi_n'(b)}{(\lambda - \lambda_n)\cos\beta}. \tag{4.1.25}$$

This in turn, as in (4.1.23), yields

$$\|\psi_n\|^2 = \int_a^b |\psi_n(x)|^2\,dx = \frac{-G'(\lambda_n)\psi_n'(b)}{\cos\beta}. \tag{4.1.26}$$

Upon substituting (4.1.25) and (4.1.26) into (4.1.15), we obtain (4.1.8).

That $f(\lambda)$ is an entire function of order 1/2 and type η with $0 \le \eta \le b - a$ follows from the relation

$$|f(\lambda)| \le \|F\|_2 \max_{a \le x \le b} |\phi(x,\lambda)| \sqrt{b-a}$$

and the fact that $\phi(x,\lambda)$ has these properties [27, p. 10].

Similar analysis can easily be applied to $\chi(x,\lambda)$ to prove the result for $f^*(\lambda)$. The relation between the two series (4.1.8) and (4.1.9) follows immediately in view of the fact that at an eigenvalue λ_n, $\chi(x,\lambda_n) = k_n\,\phi(x,\lambda_n)$ [27, p. 8] where $0 \ne k_n \ne \infty$.

It remains to show the uniform convergence of the series (4.1.8); the proof for (4.1.9) is similar. Let

$$S_N(x) = \sum_{n=0}^{N} \hat{F}(n)\frac{\psi_n(x)}{\|\psi_n\|^2}.$$

Hence, by the Cauchy-Schwarz inequality, (4.1.15), (4.1.22) and (4.1.23), we have for $\sin\beta \ne 0$,

$$\left|f(\lambda) - \sum_{n=0}^{N} f(\lambda_n)\frac{G(\lambda)}{(\lambda-\lambda_n)G'(\lambda n)}\right|^2 = \left|f(\lambda) - \sum_{n=0}^{N}\hat{F}(n)\frac{\langle\phi,\psi_n\rangle}{\|\psi_n\|^2}\right|^2$$

$$= \left|\int_a^b \phi(x,\lambda)\{F(x) - S_N(x)\}dx\right|^2 \le \left(\int_a^b |\phi(x,\lambda)|^2 dx\right)\|F - S_N\|^2 \le C(K)\,\|F - S_N\|^2,$$

but the last term goes to zero as $N \to \infty$, where

$$C(K) = \max_{\lambda \in K}\left(\int_a^b |\phi(x,\lambda)|^2\,dx\right) < \infty,$$

is a constant that depends only on the compact set K ([24], p. 229). The proof when $\sin\beta = 0$ is the same, except one uses (4.1.25) and (4.1.26) instead of (4.1.22) and (4.1.23). ∎

The main result in [38] can now be stated as a corollary.

Corollary 4.1.1. Let $\tilde{G}(t) = G(\lambda)$, $\tilde{\phi}(x,t) = \phi(x,\lambda)$, $\tilde{\chi}(x,t) = \chi(x,\lambda)$ where $\lambda = t^2$, $\lambda_n = t_{\pm n}^2$ with $t_{-n} = -t_n$, $n = 0, 1, 2, \ldots$. Let $F(x) \in L^2(a,b)$ and

$$\tilde{f}(t) = \int_a^b F(x)\tilde{\phi}(x,t)\,dx\,,$$

$$\tilde{f}^*(t) = \int_a^b F(x)\tilde{\chi}(x,t)\,dx\,.$$

Then $\tilde{f}(t)$ and $\tilde{f}^*(t)$ are entire functions of exponential type (of order one) with type η, $0 \le \eta \le b-a$ that admit the following sampling expansions:

i) If none of the eigenvalues is zero, then

$$\left.\begin{matrix}\tilde{f}(t)\\ \tilde{f}^*(t)\end{matrix}\right\} = \sum_{n=0}^{\infty}\left\{\begin{matrix}\tilde{f}(t_n)\\ \tilde{f}^*(t_n)\end{matrix}\right\}\frac{\tilde{G}(t)(2t_n)}{(t^2-t_n^2)\tilde{G}'(t_n)}\,. \tag{4.1.27}$$

ii) If one of the eigenvalues, say $\lambda_0 = 0$, then

$$\left.\begin{matrix}\tilde{f}(t)\\ \tilde{f}^*(t)\end{matrix}\right\} = \left\{\begin{matrix}\tilde{f}(0)\\ \tilde{f}^*(0)\end{matrix}\right\}\frac{\tilde{G}(t)}{t^2} + \sum_{n=1}^{\infty}\left\{\begin{matrix}\tilde{f}(t_n)\\ \tilde{f}^*(t_n)\end{matrix}\right\}\frac{\tilde{G}(t)(2t_n)}{(t^2-t_n^2)\tilde{G}'(t_n)}\,. \tag{4.1.28}$$

Proof. Clearly, $\tilde{f}(t) = f(\lambda) = f(t^2)$, $\tilde{f}^*(t) = f^*(\lambda) = f^*(t^2)$. We only prove the result for $\tilde{f}(t)$ since the proof for $\tilde{f}^*(t)$ is identical:

i) By changing λ to t^2 in (4.1.8), and noting that

$$\frac{d\tilde{G}(t)}{dt} = 2tG'(\lambda)\,,$$

we have

$$\tilde{G}'(t_n) = 2t_nG'(\lambda_n)\,.$$

Equation (4.1.8) now yields (4.1.27) provided that $t_n \neq 0$ for any $n = 0, 1, 2, \ldots$.

ii) If $\lambda_0 = 0$, the first term in (4.1.8) becomes

$$f(0)\frac{G(\lambda)}{\lambda G'(0)} = \tilde{f}(0)\frac{\tilde{G}(t)}{t^2},$$

since $G'(0) = 1$ as can easily be seen from (4.1.10). The remaining terms in (4.1.8) are treated as in case (i) to give (4.1.28). ∎

Corollary 4.1.2. Under the same assumptions as in Corollary 4.1.1, we have

$$\tilde{f}(t) = \sum_{n=-\infty}^{\infty} \tilde{f}(t_n)\frac{G(t)}{G'(t_n)(t-t_n)}$$

if none of the eigenvalues is zero and

$$\tilde{f}(t) = \tilde{f}(0)\frac{G(t)}{t^2} + \sum_{\substack{n=-\infty \\ n\neq 0}}^{\infty} \tilde{f}(t_n)\frac{G(t)}{G'(t_n)(t-t_n)},$$

if one of the eigenvalues, say $\lambda_0 = 0$. Similar expansions hold for $\tilde{f}^*(t)$.

Proof. This follows immediately upon observing that $f(t), G(t)$ are even functions in t and that $G'(t)$ is an odd function in t, hence $G'(t_{-n}) = G'(-t_n) = -G'(t_n)$. ∎

Corollary 4.1.3. Let $\psi(x,\lambda) = c_1\phi(x,\lambda) + c_2\chi(x,\lambda)$ and $F(x) \in L^2(a,b)$. Then, if

$$f(\lambda) = \int_a^b F(x)\psi(x,\lambda)\,dx,$$

then

$$f(\lambda) = \sum_{n=0}^{\infty} f(\lambda_n)\frac{G(\lambda)}{(\lambda-\lambda_n)G'(\lambda_n)}.$$

4.2 The Singular Case on a Halfline

The aim of this section is two-fold:

i) to extend the Weiss-Kramer sampling theorem to a singular Sturm-Liouville boundary-value problem, namely a singular problem on a halfline,

ii) to show that even for this singular case, if a Weiss-Kramer sampling series exists, it is nothing more than a Lagrange-type interpolation series.

More precisely, we shall extend the results of Theorem 4.1 to the case where either the function $q(x)$ in (4.1.1) has a singularity at one of the endpoints or the interval (a,b) extends to infinity in one direction and $q(x)$ is continuous, in particular, at the finite endpoint. Since the analysis in these two cases is similar (cf. [3, 27)], we will restrict ourselves, without loss of generality, to the case where $(a,b)=(0,\infty)$ and $q(x)$ is continuous on $[0,\infty)$.

Consider the following singular Sturm-Liouville boundary-value problem:

$$y'' - q(x)y = -\lambda y\,, \quad 0 \le x < \infty\,, \tag{4.2.1}$$

$$y(0)\cos\alpha + y'(0)\sin\alpha = 0\,, \tag{4.2.2}$$

where $q(x)$ is continuous on $[0,\infty)$.

Let $\phi(x)=\phi(x,\lambda)$, $\theta(x)=\theta(x,\lambda)$ be the solutions of (4.2.1) such that

$$\phi(0)=\sin\alpha\,, \quad \phi'(0) = -\cos\alpha\,,$$

$$\theta(0)=\cos\alpha\,, \quad \theta'(0)=\sin\alpha\,.$$

Then, it is known [22, Chap. 2] or [27, Chap. 2, 3] that there exists a complex valued function $m(\lambda)$ such that for every nonreal λ, (4.2.1) has a solution

$$\psi(x,\lambda)=\theta(x,\lambda)+m(\lambda)\phi(x,\lambda) \tag{4.2.3}$$

belonging to $L^2(0,\infty)$. In the limit point case $m(\lambda)$ is unique, while in the limit circle case there are uncountably many such functions. Moreover, in the limit circle case $\phi(x,\lambda)$ is in $L^2(0,\infty)$ and hence every solution of (4.2.1) is in $L^2(0,\infty)$. The function $m(\lambda)$ is analytic in the upper and lower halfplanes and if it has poles on the real axis, they are all simple. The functions denoted by $m(\lambda)$ in the upper and lower halfplanes are not necessarily analytic continuations of each other. But throughout this section we assume that they form a single analytic function whose only singularities are poles on the nonnegative real axis. Therefore, throughout this section $m(\lambda)$ will denote a meromorphic function that is real-valued on the real axis and whose singularities are simple poles on the real axis. We will denote the poles of $m(\lambda)$ by $\{\lambda_n\}_{n=0}^{\infty}$ and their corresponding residues by $\{r_n\}_{n=0}^{\infty}$. For $m(\lambda)$ to satisfy these assumptions, certain conditions must be imposed on $q(x)$ (see [27, Chaps. V, VII]). The following facts are known [27, p. 28].

The λ_n's are the eigenvalues of (4.2.1), (4.2.2) and they form a monotone increasing sequence of real numbers whose only limit point is ∞. The eigenfunction $\phi_n(x)$ corresponding to the eigenvalue λ_n is given by

$$\phi_n(x) = \phi(x, \lambda_n), \tag{4.2.4}$$

where

$$\lim_{\lambda \to \lambda_n} (\lambda - \lambda_n) \psi(x, \lambda) = r_n \phi(x, \lambda_n). \tag{4.2.5}$$

Moreover, for any nonreal λ and λ',

$$\int_0^\infty \psi(x, \lambda) \psi(x, \lambda') \, dx = \frac{m(\lambda) - m(\lambda')}{\lambda' - \lambda}. \tag{4.2.6}$$

By multiplying (4.2.6) by $(\lambda' - \lambda_n)$, then taking the limit as $\lambda' \to \lambda_n$ and employing (4.2.5), we obtain

$$\int_0^\infty \psi(x, \lambda) \phi_n(x) \, dx = \frac{1}{\lambda - \lambda_n}. \tag{4.2.7}$$

Applying the same argument once more to (4.2.7) yields

$$\int_0^\infty |\phi_n(x)|^2 \, dx = \frac{1}{r_n}. \tag{4.2.8}$$

We also have [27, pp. 38, 41]; [22, p. 126]

$$\sum_{n=0}^\infty \left| \frac{\sqrt{r_n}\, \phi_n(x)}{\lambda - \lambda_n} \right|^2 < K \tag{4.2.9}$$

uniformly for x on any compact subset of $\Re$ and $\lambda \neq \lambda_n$ for any n. As a special case of (4.2.9) we have

$$\sum_{n=0}^\infty \frac{r_n}{|\lambda - \lambda_n|^2} < \infty \tag{4.2.10}$$

for any $\lambda \neq \lambda_n$, $n = 0, 1, 2, \ldots$.

It should be noted that one of the main differences between the singular and regular cases is the distribution of the eigenvalues. Unlike the regular case, the eigenvalues $\{\lambda_n\}_{n=0}^{\infty}$ in the singular case are not necessarily of order $0(n^2)$ as $n \to \infty$. In fact, in problem (4.2.1) and (4.2.2) if $q(x) = x^k$, $k > 0$, then

$$\lambda_n = \left\{ \frac{2\sqrt{\pi}\, k\Gamma(3/2 + 1/k)}{\Gamma(1/k)} n + 0(1) \right\}^{2k/(k+2)} \quad \text{as } n \to \infty \quad \text{(cf. [27], p. 144)},$$

for which the canonical product $G(t)$ given by (4.1.10) may not exist.

Now we state and prove our main theorem of this section, which is a generalization of Theorem 4.1.

THEOREM 4.2

Consider the singular Sturm-Liouville problem

$$y'' - q(x)y = -\lambda y, \quad 0 \le x < \infty,$$

$$y(0)\cos\alpha + y'(0)\sin\alpha = 0,$$

where $q(x)$ is continuous on $[0,\infty)$. Assume that $m(\lambda)$ is a meromorphic function that is real-valued on the real axis and whose only singularities are simple poles $\{\lambda_n\}_{n=0}^{\infty}$ on the nonnegative real axis; λ_0 will be reserved for the eigenvalue zero.

Let p be the smallest integer for which the series $\sum_{n=1}^{\infty} 1/(\lambda_n)^{p+1}$ converges.

i) If none of the λ_n's is zero, set

$$G(\lambda) = \begin{cases} \prod_{n=0}^{\infty}\left(1 - \frac{\lambda}{\lambda_n}\right) \exp\left[\left(\frac{\lambda}{\lambda_n}\right) + \frac{1}{2}\left(\frac{\lambda}{\lambda_n}\right)^2 + \ldots + \frac{1}{p}\left(\frac{\lambda}{\lambda_n}\right)^p\right] & \text{if } p = 1, 2, \ldots, \\ \prod_{n=0}^{\infty}\left(1 - \frac{\lambda}{\lambda_n}\right) & \text{if } p = 0. \end{cases} \tag{4.2.11}$$

ii) If one of the λ_n's is zero, say $\lambda_0 = 0$, set

$$G(\lambda) = \begin{cases} \lambda \prod_{n=0}^{\infty}\left(1 - \frac{\lambda}{\lambda_n}\right) \exp\left[\left(\frac{\lambda}{\lambda_n}\right) + \frac{1}{2}\left(\frac{\lambda}{\lambda_n}\right)^2 + \ldots + \frac{1}{p}\left(\frac{\lambda}{\lambda_n}\right)^p\right] & \text{if } p = 1, 2, \ldots, \\ \lambda \prod_{n=0}^{\infty}\left(1 - \frac{\lambda}{\lambda_n}\right) & \text{if } p = 0. \end{cases} \tag{4.2.12}$$

Let $\Phi(x,\lambda) = G(\lambda)\psi(x,\lambda)$, $F(x) \in L^2(0,\infty)$ and

$$f(\lambda) = \int_0^\infty F(x)\,\Phi(x,\lambda)\,dx\,. \tag{4.2.13}$$

Then $f(\lambda)$ is an entire function that admits the following sampling representation:

$$f(\lambda) = \sum_{n=0}^{\infty} f(\lambda_n)\frac{G(\lambda)}{(\lambda-\lambda_n)G'(\lambda_n)}, \tag{4.2.14}$$

where the series converges uniformly on any compact subset of the complex λ-plane.

Proof. It is easy to see that the products in (4.2.11) and (4.2.12) are well defined and that $G(\lambda)$ is an entire function of order ρ with $p \le \rho \le p+1$ [21, Thm. 7, p. 16] (cf. [28]), where ρ is the convergence exponent of the sequence $\{\lambda_n\}_{n=1}^{\infty}$ (cf. [21, p. 9]). Since the poles of $m(\lambda)$ are exactly at the zeros of $G(\lambda)$ and both $\theta(x,\lambda)$, $\phi(x,\lambda)$ are entire functions in λ, it follows that $\Phi(x,\lambda)$ is also an entire function in λ.

By applying the Cauchy-Schwarz inequality to (4.2.13), we obtain

$$|f(\lambda)|^2 \le \|F\|_2^2\,|G(\lambda)|^2\left(\int_0^\infty |\psi(x,\lambda)|^2\,dx\right). \tag{4.2.15}$$

Hence, in the limit circle case, $f(\lambda)$ exists for all λ since $\psi(x,\lambda)$ is in $L^2(0,\infty)$ for all λ. However, in the limit point case this is only true for $\operatorname{Im}\lambda \ne 0$. To show that, even in this case, $f(\lambda)$ exists for all λ, we recall from (4.2.5) and the definition of $\Phi(x,\lambda)$ that

$$\Phi(x,\lambda_n) = \lim_{\lambda\to\lambda_n}\Phi(x,\lambda) = G'(\lambda_n)\,r_n\,\phi_n(x)\,. \tag{4.2.16}$$

Since $F(x)$ and $\psi(x,\lambda)$ are in $L^2(0,\infty)$, provided that $\operatorname{Im}\lambda \ne 0$, we have

$$F(x) = \sum_{n=0}^{\infty}\frac{\hat F(n)}{\|\phi_n\|^2}\,\phi_n(x)\,, \tag{4.2.17}$$

where

$$\hat F(n) = \int_0^\infty F(x)\,\phi_n(x)\,dx\,, \quad \sum_{n=0}^{\infty}\frac{|\hat F(n)|^2}{\|\phi_n\|^2} < \infty \tag{4.2.18}$$

and by (4.2.7) (cf. [27])

$$\psi(x,\lambda) = \sum_{n=0}^{\infty} \frac{1}{(\lambda-\lambda_n)} \frac{\phi_n(x)}{\|\phi_n\|^2}, \quad \text{Im } \lambda \neq 0, \tag{4.2.19}$$

where the series in (4.2.17) and (4.2.19) converge in the sense of $L^2(0,\infty)$.

By Parseval's equality, we obtain from (4.2.13), (4.2.17), and (4.2.19)

$$f(\lambda) = \sum_{n=0}^{\infty} \frac{\hat{F}(n)G(\lambda)}{(\lambda-\lambda_n)\|\phi_n\|^2}. \tag{4.2.20}$$

Hence,

$$|f(\lambda)|^2 \le |G(\lambda)|^2 \left(\sum_{n=0}^{\infty} \frac{|\hat{F}(n)|^2}{\|\phi_n\|^2} \right) \left(\sum_{n=0}^{\infty} \frac{r_n}{|\lambda-\lambda_n|^2} \right), \tag{4.2.21}$$

and in view of (4.2.10) and (4.2.18) $f(\lambda)$ exists for all λ except possibly for $\lambda = \lambda_n$, $n = 0, 1, 2, \ldots$. But this restriction is easily removed upon observing that

$$f(\lambda_n) = \lim_{\lambda \to \lambda_n} f(\lambda) = G'(\lambda_n) r_n \int_0^{\infty} F(x)\phi_n(x)\,dx = G'(\lambda_n)\, r_n \hat{F}(n). \tag{4.2.22}$$

By combining (4.2.20) and (4.2.22) we obtain (4.2.14) since $r_n\|\phi_n\|^2 = 1$ (see (4.2.8)). To show that the series in (4.2.14) converges uniformly on any compact subset K of the complex λ-plane, we recall from (4.2.10) that the series $\sum_{n=0}^{\infty} r_n / |\lambda-\lambda_n|^2$ is uniformly bounded on any compact subset of the complex λ-plane not containing any of the points λ_n. But since the λ_n's are the zeros of $G(\lambda)$, it is easy to see that the series $\sum_{n=0}^{\infty} r_n |G(\lambda)|^2 / |\lambda-\lambda_n|^2$ is uniformly bounded on any compact subset K of the λ-plane, i.e., there exists a positive constant C_K such that

$$\sum_{n=0}^{\infty} \frac{r_n |G(\lambda)|^2}{|\lambda-\lambda_n|^2} \le C_K \quad \text{for all} \quad \lambda \in K.$$

From (4.2.14) and the Cauchy-Schwarz inequality we have

$$\left| f(\lambda) - \sum_{n=0}^{N} f(\lambda_n) \frac{G(\lambda)}{(\lambda-\lambda_n)G'(\lambda_n)} \right| = \left| \sum_{n=N+1}^{\infty} \frac{G(\lambda)\hat{F}(n)}{(\lambda-\lambda_n)\|\phi_n\|^2} \right|$$

$$\le \left(\sum_{n=N+1}^{\infty} \frac{|\hat{F}(n)|^2}{\|\phi_n\|^2} \right)^{1/2} \left(\sum_{n=N+1}^{\infty} \frac{|G(\lambda)|^2}{|\lambda-\lambda_n|^2 \|\phi_n\|^2} \right)^{1/2}$$

$$\le \sqrt{C_K} \left(\sum_{n=N+1}^{\infty} \frac{|\hat{F}(n)|}{\|\phi_n\|^2} \right)^{1/2} \to 0$$

as $N \to \infty$ by (4.2.18).

To show that $f(\lambda)$ is an entire function we remark that from the uniform convergence of (4.2.14), it follows that $f(\lambda)$ is analytic on any compact subset K of the complex λ-plane, i.e., $f(\lambda)$ is entire. ∎

Corollary 4.2.1. Let $\tilde{G}(t) = G(\lambda)$ and $\tilde{\Phi}(x,t) = \Phi(x,\lambda)$ where $\lambda = t^2$, $\lambda_n = t_n^2$ and $t_{-n} = -t_n$, $n = 0, 1, 2, \dots$. Let $F(x) \in L^2(0,\infty)$ and

$$f(t) = \int_0^\infty F(x)\, \tilde{\Phi}(x,t)\, dx \,.$$

Then, $f(t)$ is an entire function that admits the following sampling representations:

$$f(t) = \begin{cases} \displaystyle\sum_{n=0}^{\infty} f(t_n) \frac{\tilde{G}(t)(2t_n)}{\tilde{G}'(t_n)(t^2-t_n^2)} & \text{if } t_n \ne 0 \text{ for all } n\,, \\ \displaystyle f(0)\frac{\tilde{G}(t)}{t^2} + \sum_{n=1}^{\infty} f(t_n) \frac{\tilde{G}(t)(2t_n)}{\tilde{G}'(t_n)(t^2-t_n^2)} & \text{if one of the } t_n\text{'s say } t_0 = 0\,. \end{cases}$$

Proof. The proof is similar to the proof of Corollary 4.1.1. ∎

Corollary 4.2.2. Under the same assumptions as in Corollary 4.2.1 we have

$$f(t) = \begin{cases} \displaystyle\sum_{n=-\infty}^{\infty} f(t_n) \frac{\tilde{G}(t)}{\tilde{G}'(t_n)(t-t_n)} & \text{if } t_n \ne 0 \text{ for all } n\,, \\ \displaystyle f(0)\frac{\tilde{G}(t)}{t^2} + \sum_{\substack{n=-\infty \\ n\ne 0}}^{\infty} f(t_n) \frac{\tilde{G}(t)}{\tilde{G}'(t_n)(t-t_n)} & \text{if one of the } t_n\text{'s say } t_0 = 0\,. \end{cases}$$

Proof. The proof is the same as the one given in Corollary 4.1.2. ∎

It is evident from the proof of Theorem 4.2 (cf. (4.2.13), (4.2.14), (4.2.20)) that Kramer's sampling theorem holds for the interval $[0,\infty)$ by choosing $K(x,\lambda) = \Phi(x,\lambda)$, the set $E = \{\lambda_n\}_{n=0}^{\infty}$, and hence

$$S_n^*(\lambda) = G(\lambda)/\{(\lambda-\lambda_n)\|\phi_n\|^2\}.$$

In the limit circle case, we may use $\phi(x,\lambda)$ in Theorem 4.2 instead of $\Phi(x,\lambda)$ since in this case $\phi(x,\lambda)\in L^2(0,\infty)$ and $\phi(x,\lambda_n) = \phi_n(x)$.

4.3 The Singular Case on the Whole Line

Now we consider the case where either the function $q(x)$ has singularities at both endpoints of the interval (a,b) or the interval (a,b) extends to infinity in both directions, or it extends to infinity in one direction and $q(x)$ has a singularity at the other finite endpoint. Since the analysis in all these cases is similar we shall, without loss of generality, confine our attention to the case where (a,b) extends to infinity in both directions and $q(x)$ is continuous.

Consider

$$y'' - q(x)y = -\lambda y\,, \quad -\infty < x < \infty\,, \tag{4.3.1}$$

where $q(x)$ is continuous on $(-\infty,\infty)$.

Let $\phi(x) = \phi(x,\lambda)$ and $\theta(x) = \theta(x,\lambda)$ be the solutions of (4.3.1) such that

$$\phi(0) = 0\,, \quad \phi'(0) = -1\,, \tag{4.3.2}$$

$$\theta(0) = 1\,, \quad \theta'(0) = 0\,. \tag{4.3.3}$$

By the results mentioned at the beginning of the previous section, there exist functions $m_1(\lambda)$ and $m_2(\lambda)$, analytic in the upper and lower halfplanes such that

$$\psi_1(x,\lambda) = \theta(x,\lambda) + m_1(\lambda)\,\phi(x,\lambda) \tag{4.3.4}$$

is in $L^2(0,\infty)$ and

$$\psi_2(x,\lambda) = \theta(x,\lambda) + m_2(\lambda)\,\phi(x,\lambda) \tag{4.3.5}$$

is in $L^2(-\infty,0)$.

It is easy to see that the Wronskian of ψ_1, ψ_2

$$W(\psi_1,\psi_2) = m_1(\lambda) - m_2(\lambda) \tag{4.3.6}$$

and

$$\int_0^\infty \psi_1(x,\lambda)\,\psi_1(x,\lambda')\,dx = \frac{m_1(\lambda)-m_1(\lambda')}{\lambda'-\lambda}, \tag{4.3.7}$$

$$\int_{-\infty}^0 \psi_2(x,\lambda)\,\psi_2(x,\lambda')\,dx = \frac{m_2(\lambda)-m_2(\lambda')}{\lambda-\lambda'}. \tag{4.3.8}$$

Throughout this section we shall assume that both $m_1(\lambda)$ and $m_2(\lambda)$ are meromorphic functions satisfying the same conditions as $m(\lambda)$ given in Section 4.2. A sufficient condition for this to hold, for example, is that $\lim_{x\to\pm\infty} q(x) = \infty$ (cf. [27, p. 127]). It is known [27, p. 43] that the eigenvalues $\{\lambda_n\}_{n=0}^\infty$ are the points where:

i) $m_1(\lambda)$, $m_2(\lambda)$ have poles of the form $m_1(\lambda) \sim a_1/(\lambda-\lambda_n)$,
$m_2(\lambda) \sim a_2/(\lambda-\lambda_n)$, $a_1 \neq a_2$, or

ii) $m_1(\lambda_n) = m_2(\lambda_n) \neq 0$, or

iii) $m_1(\lambda) \sim a_1(\lambda-\lambda_n)$, $m_2(\lambda) \sim a_2(\lambda-\lambda_n)$, $a_1 \neq a_2$.

Now we are able to state our main theorem for this section.

THEOREM 4.3

Consider the differential equation

$$y'' - q(x)y = -\lambda y, \quad -\infty < x < \infty,$$

where $q(x)$ is an even continuous function on $(-\infty,\infty)$. Let $m_i(\lambda)$, $\psi_i(x,\lambda)$, $i = 1,2$ have the same meaning as above. Denote the zeros and the poles of $m_1(\lambda)$ by $\{\lambda_{2n}\}_{n=0}^\infty$ and $\{\lambda_{2n+1}\}_{n=0}^\infty$, respectively, and the corresponding residues by $\{r_{2n+1}\}_{n=0}^\infty$. Let $G(\lambda)$ be the canonical product of $\{\lambda_{2n+1}\}_{n=0}^\infty$ as defined in Theorem 4.2. If $F(x) \in L^2(-\infty,\infty)$ and

$$f(\lambda) = \int_0^\infty F(x)\,\Phi_1(x,\lambda)\,dx + \int_{-\infty}^0 F(x)\,\Phi_2(x,\lambda)\,dx, \tag{4.3.9}$$

where

$$\Phi_i(x,\lambda) = G(\lambda)\,\psi_i(x,\lambda), \quad i = 1,2, \tag{4.3.10}$$

then

$$f(\lambda) = \sum_{n=0}^\infty \frac{f(\lambda_{2n})G(\lambda)m_1(\lambda)}{G(\lambda_{2n})(\lambda-\lambda_{2n})m_1'(\lambda_{2n})} + \sum_{n=0}^\infty \frac{f(\lambda_{2n+1})G(\lambda)}{(\lambda-\lambda_{2n+1})G'(\lambda_{2n+1})}, \tag{4.3.11}$$

where the two series converge uniformly on compact subsets of the complex λ-plane.

Proof. Since $q(x)$ is even, then $\phi(x,\lambda)$ is an odd function in x, $\theta(x,\lambda)$ is an even function in x, and hence $\psi_1(-x,\lambda) = \psi_2(x,\lambda)$. Therefore, $m_1(\lambda) = -m_2(\lambda)$, which implies that

$$\begin{aligned} \psi_1(x,\lambda) &= \theta(x,\lambda) + m_1(\lambda)\,\phi(x,\lambda)\,, \\ \psi_2(x,\lambda) &= \theta(x,\lambda) - m_1(\lambda)\,\phi(x,\lambda)\,, \end{aligned} \tag{4.3.12}$$

$$\Phi_1(x,\lambda) = \Phi_2(-x,\lambda)\,, \tag{4.3.13}$$

and that $\{\lambda_n\}_{n=0}^{2}$ are the eigenvalues [27, p. 43].

Moreover,

$$\begin{aligned} \Phi_1(x,\lambda_{2n}) &= \lim_{\lambda\to\lambda_{2n}} G(\lambda)\,\psi_1(x,\lambda) = G(\lambda_{2n})\theta(x,\lambda_{2n})\,, \\ \Phi_1(x,\lambda_{2n+1}) &= \lim_{\lambda\to\lambda_{2n+1}} G(\lambda)\,\psi_1(x,\lambda) = G'(\lambda_{2n+1})\,r_{2n+1}\phi(x,\lambda_{2n+1})\,, \end{aligned} \tag{4.3.14}$$

where $\theta(x,\lambda_{2n})$ and $\phi(x,\lambda_{2n+1})$ are the eigenfunctions corresponding to the eigenvalues λ_{2n}, λ_{2n+1}, $n = 0,1,2,3,\ldots$, respectively. Set

$$\psi_{2n}(x) = \theta(x,\lambda_{2n})\,, \quad \psi_{2n+1}(x) = \phi(x,\lambda_{2n+1}) \tag{4.3.15}$$

and note that the eigenfunctions $\psi_n(x)$ are either even or odd depending on whether n is even or odd. We have

$$\begin{aligned} \Phi_1(x,\lambda) &= \sum_{n=0}^{\infty} \frac{\langle \Phi_1,\psi_n\rangle}{\|\psi_n\|_1^2}\,\psi_n(x)\,, \\ \Phi_2(x,\lambda) &= \sum_{n=0}^{\infty} \frac{\langle \Phi_2,\psi_n\rangle}{\|\psi_n\|_2^2}\,\psi_n(x)\,, \end{aligned} \tag{4.3.16}$$

where

$$\begin{aligned} \langle \Phi_1,\psi_n\rangle &= \int_0^{\infty} \Phi_1(x,\lambda)\,\psi_n(x)\,dx\,, \\ \langle \Phi_2,\psi_n\rangle &= \int_{-\infty}^{0} \Phi_2(x,\lambda)\,\psi_n(x)\,dx\,, \end{aligned} \tag{4.3.17}$$

$$\begin{aligned} \|\psi_n\|_1^2 &= \int_0^{\infty} |\psi_n(x)|^2\,dx\,, \\ \|\psi_n\|_2^2 &= \int_{-\infty}^{0} |\psi_n(x)|^2\,dx\,, \\ \|\psi_n\|^2 &= \int_{-\infty}^{\infty} |\psi_n(x)|^2\,dx\,, \end{aligned} \tag{4.3.18}$$

and the two series in (4.3.16) converge in the sense of $L^2(0,\infty)$ and $L^2(-\infty,0)$, respectively.

From (4.3.9) and (4.3.16) we obtain

$$f(\lambda) = \sum_{n=0}^{\infty} \frac{\langle \Phi_1, \psi_n\rangle}{\|\psi_n\|_1^2} \int_0^{\infty} F(x)\psi_n(x)\,dx + \sum_{n=0}^{\infty} \frac{\langle \Phi_2, \psi_2\rangle}{\|\psi_n\|_2^2} \int_{-\infty}^{0} F(x)\psi_n(x)\,dx\,.$$

But since

$$\begin{aligned} \langle \Phi_1, \psi_n\rangle &= \langle \Phi_2, \psi_n\rangle \qquad \text{if } n \text{ is even}\,, \\ \langle \Phi_1, \psi_n\rangle &= -\langle \Phi_2, \psi_n\rangle \qquad \text{if } n \text{ is odd} \end{aligned} \tag{4.3.19}$$

(cf. (4.3.12), (4.3.17)), we have

$$\begin{aligned} f(\lambda) = {} & 2\sum_{n=0}^{\infty} \frac{\langle \Phi_1, \psi_{2n}\rangle}{\|\psi_{2n}\|^2} \left[\int_{-\infty}^{\infty} F(x)\psi_{2n}(x)\,dx \right] \\ & +2\sum_{n=0}^{\infty} \frac{\langle \Phi_1, \psi_{2n+1}\rangle}{\|\psi_{2n+1}\|^2} \left[\int_0^{\infty} F(x)\psi_{2n+1}(x)\,dx - \int_{-\infty}^{0} F(x)\psi_{2n+1}(x)\,dx \right] \end{aligned} \tag{4.3.20}$$

since $\|\psi_n\|_1^2 = \|\psi_n\|_2^2 = \frac{1}{2}\|\psi_n\|^2$.

In view of (4.3.9), (4.3.13), and (4.3.14) we have

$$\begin{aligned} f(\lambda_{2n}) &= \int_0^{\infty} F(x)G(\lambda_{2n})\psi_{2n}(x)\,dx + \int_{-\infty}^{0} F(x)G(\lambda_{2n})\psi_{2n}(x)\,dx \\ &= G(\lambda_{2n}) \int_{-\infty}^{\infty} F(x)\psi_{2n}(x)\,dx\,, \end{aligned} \tag{4.3.21}$$

and

$$\begin{aligned} f(\lambda_{2n+1}) = {} & \int_0^{\infty} F(x)G'(\lambda_{2n+1})r_{2n+1}\psi_{2n+1}(x)\,dx \\ & - \int_{-\infty}^{0} F(x)G'(\lambda_{2n+1})r_{2n+1}\psi_{2n+1}(x)\,dx \\ = {} & G'(\lambda_{2n+1})r_{2n+1}\left[\int_0^{\infty} F(x)\psi_{2n+1}(x)\,dx - \int_{-\infty}^{0} F(x)\psi_{2n+1}(x)\,dx \right]. \end{aligned} \tag{4.3.22}$$

Thus, by employing (4.3.21) and (4.3.22), equation (4.3.20) becomes

$$\begin{aligned} f(\lambda) = 2 \sum_{n=0}^{\infty} \frac{\langle \Phi_1, \psi_{2n} \rangle}{\| \psi_{2n} \|^2} \frac{f(\lambda_{2n})}{G(\lambda_{2n})} \\ + 2 \sum_{n=0}^{\infty} \frac{\langle \Phi_1, \psi_{2n+1} \rangle}{\| \psi_{2n+1} \|^2} \frac{f(\lambda_{2n+1})}{G'(\lambda_{2n+1}) r_{2n+1}} . \end{aligned} \tag{4.3.23}$$

By multiplying (4.3.7) by $G(\lambda)G(\lambda')$, taking the limits as $\lambda' \to \lambda_n$, then as $\lambda \to \lambda_n$, and using (4.3.14), (4.3.15), we obtain

$$\langle \Phi_1, \psi_{2n} \rangle = \int_0^{\infty} \Phi_1(x, \lambda) \psi_{2n}(x)\, dx = \frac{G(\lambda) m_1(\lambda)}{\lambda_{2n} - \lambda} . \tag{4.3.24}$$

$$\| \psi_{2n} \|_1^2 = -m_1'(\lambda_{2n}) , \tag{4.3.25}$$

$$\langle \Phi_1, \psi_{2n+1} \rangle = \int_0^{\infty} \Phi_1(x, \lambda) \psi_{2n+1}(x)\, dx = \frac{G(\lambda)}{\lambda - \lambda_{2n+1}} , \tag{4.3.26}$$

$$\| \psi_{2n+1} \|_1^2 = \frac{1}{r_{2n+1}} . \tag{4.3.27}$$

Upon substituting (4.3.24) through (4.3.27) into (4.3.23), we obtain (4.3.11). The convergence of the series in (4.3.11) can be treated as in Theorem 4.2. ▮

Corollary 4.3.1. Suppose that $m_1(\lambda)$ has no zeros, i.e., all the eigenfunctions are odd. Let

$$f(\lambda) = \int_0^{\infty} F(x) \phi_1(x, \lambda)\, dx + \int_{-\infty}^{0} F(x) \phi_2(x, \lambda)\, dx$$

for some $F(x) \in L^2(-\infty, \infty)$; then

$$f(\lambda) = \sum_{n=0}^{\infty} f(\lambda_n) \frac{G(\lambda)}{(\lambda - \lambda_n) G'(\lambda_n)} ,$$

where $\{\lambda_n\}_{n=0}^{\infty}$ are the eigenvalues.

Corollary 4.3.2. Suppose that $m_1(\lambda)$ has no poles, i.e., all the eigenfunctions are even. Let

$$f(\lambda) = \int_0^\infty F(x)\psi_1(x,\lambda)\,dx + \int_{-\infty}^0 F(x)\psi_2(x,\lambda)\,dx$$

for some $F(x) \in L^2(-\infty,\infty)$; then

$$f(\lambda) = \sum_{n=0}^\infty f(\lambda_n)\frac{m_1(\lambda)}{(\lambda-\lambda_n)m_1'(\lambda_n)},$$

where $\{\lambda_n\}_{n=0}^\infty$ are the eigenvalues.

THEOREM 4.4

In addition to the assumptions of Theorem 4.3, let us further assume that the limit circle case holds at both $\pm\infty$. Then, there exists a solution $\phi_0(x,\lambda)$ of (4.3.1) with the property that $\phi_0(x,\lambda_n)$ is the eigenfunction corresponding to the eigenvalue λ_n, and moreover, if

$$f(\lambda) = \int_{-\infty}^\infty F(x)\phi_0(x,\lambda)\,dx \quad \text{for some} \quad F(x) \in L^2(-\infty,\infty), \tag{4.3.28}$$

then

$$f(\lambda) = \sum_{n=0}^\infty F(\lambda_n)\frac{G(\lambda)}{(\lambda-\lambda_n)G'(\lambda_n)}, \tag{4.3.29}$$

where $G(\lambda)$ is the canonical product of $\{\lambda_n\}_{n=0}^\infty$ as given by (4.2.11) or (4.2.12). $G(\lambda)$ is an entire function of order ≤ 2.

Proof. Since $m_1(\lambda)$ is a meromorphic function, it can be written as $m_1(\lambda) = A(\lambda)/B(\lambda)$ where $A(\lambda)$ and $B(\lambda)$ are entire functions. The eigenvalues are at the noncommon zeros of $A(\lambda)$ and $B(\lambda)$. Let us denote the zeros of $A(\lambda)$ by $\{\lambda_{2n}\}_{n=0}^\infty$ and those of $B(\lambda)$ by $\{\lambda_{2n+1}\}_{n=0}^\infty$ and assume without loss of generality that both $A(\lambda)$ and $B(\lambda)$ are the canonical products of their zeros. Define $\phi_0(x,\lambda)$ by

$$\phi_0(x,\lambda) = B(\lambda)\psi_1(x,\lambda) = B(\lambda)\theta(x,\lambda) + A(\lambda)\phi(x,\lambda), \tag{4.3.30}$$

and the normalized eigenfunctions $\{\psi_n(x)\}_{n=0}^{\infty}$ by

$$\psi_{2n}(x) = B(\lambda_{2n})\theta(x,\lambda_{2n}) \quad \text{and} \quad \psi_{2n+1}(x) = A(\lambda_{2n+1})\phi(x,\lambda_{2n+1}). \tag{4.3.31}$$

Thus,

$$\phi_0(x,\lambda_n) = \psi_n(x). \tag{4.3.32}$$

Also note that the eigenfunctions are even if n is even and odd if n is odd. It is easy to see that the Wronskian $W(\phi_0(x,\lambda),\phi_0(-x,\lambda)) = 2AB$, $W(\phi,\theta) = 2AB \neq 0$, hence $\phi_0(x,\lambda)$ and $\phi_0(-x,\lambda)$ are two linearly independent solutions of (4.3.1). Because in the limit circle case all solutions of (4.3.1) are in $L^2(-\infty,\infty)$, it is evident that $f(\lambda)$ is well defined. We have

$$F(x) = \sum_{n=0}^{\infty} \hat{F}(n) \frac{\psi_n(x)}{\|\psi_n\|^2}, \text{ where } \hat{F}(n) = \int_{-\infty}^{\infty} F(x)\psi_n(x)\,dx, \tag{4.3.33}$$

and

$$\phi_0(x,\lambda) = \sum_{n=0}^{\infty} \frac{\langle\phi_0,\psi_n\rangle}{\|\psi_n\|^2}\psi_n(x), \text{ where } \langle\phi_0,\psi_n\rangle = \int_{-\infty}^{\infty} \phi_0(x,\lambda)\psi_n(x)\,dx. \tag{4.3.34}$$

The series in (4.3.33) and (4.3.34) converge in the sense of $L^2(-\infty,\infty)$.

From (4.3.28), (4.3.32), and (4.3.33), we have

$$f(\lambda_n) = \int_{-\infty}^{\infty} F(x)\phi_0(x,\lambda_n)\,dx = \int_{-\infty}^{\infty} F(x)\psi_n(x)\,dx = \hat{F}(n). \tag{4.3.35}$$

By using (4.3.28), (4.3.33) through (4.3.35), and Parseval's equality, we obtain

$$f(\lambda) = \sum_{n=0}^{\infty} f(\lambda_n) \frac{\langle\phi_0,\psi_n\rangle}{\|\psi_n\|^2}. \tag{4.3.36}$$

As in (4.1.19), we can easily verify that

$$(\lambda-\lambda') \int_{-a}^{a} \phi_0(x,\lambda)\phi_0(-x,\lambda')\,dx = 2W_{x=a}(\phi_0(x,\lambda),\phi_0(-x,\lambda')),$$

which upon taking the limit as $a \to \infty$ yields

$$(\lambda-\lambda')\int_{-\infty}^{\infty}\phi_0(x,\lambda)\phi_0(-x,\lambda')\,dx = 2W_{x=\infty}(\phi_0(x,\lambda),\phi_0(-x,\lambda')) \tag{4.3.37}$$
$$= G(\lambda,\lambda').$$

The limit exists since $\phi_0(\pm x,\lambda)$ is in $L^2(-\infty,\infty)$. By taking the limit in (4.3.37) as $\lambda' \to \lambda_n$, we obtain

$$\int_{-\infty}^{\infty}\phi_0(x,\lambda)\psi_n(x)\,dx = \frac{(-1)^n G(\lambda,\lambda_n)}{(\lambda-\lambda_n)} = \frac{(-1)^n}{(\lambda-\lambda_n)}G_n(\lambda), \tag{4.3.38}$$

where $G_n(\lambda) = G(\lambda,\lambda_n)$. In deriving (4.3.38), we have used the fact that

$$\phi_0(-x,\lambda_n) = (-1)^n\phi_0(x,\lambda_n) = (-1)^n\psi_n(x).$$

By taking the limit in (4.3.38) as $\lambda \to \lambda_n$, we obtain

$$\|\psi_n\|^2 = \int_{-\infty}^{\infty}|\psi_n(x)|^2\,dx = (-1)^n G_n'(\lambda_n). \tag{4.3.39}$$

By combining (4.3.36), (4.3.38), and (4.3.39) we obtain

$$f(\lambda) = \sum_{n=0}^{\infty} f(\lambda_n)\,\frac{G_n(\lambda)}{(\lambda-\lambda_n)G_n'(\lambda_n)}. \tag{4.3.40}$$

From (4.3.30), (4.3.37), and (4.3.38) it is easy to see that $G_n(\lambda)$ is an entire function whose zeros are located exactly at $\{\lambda_m\}_{m=0}^{\infty}$ and that the order of $G_n(\lambda) = \max(\text{order of } A(\lambda) \text{ and } B(\lambda))$ is less than or equal to the order of $G(\lambda)$. But since $G(\lambda)$ and $G_n(\lambda)$ have exactly the same zeros, it follows from Hadamard's factorization theorem for entire functions that

$$G_n(\lambda) = C_n G(\lambda), \quad n = 0,1,2,\ldots. \tag{4.3.41}$$

Thus, upon employing (4.3.41) in (4.3.40), we obtain (4.3.29). Finally, the validity of the statement about the order of $G(\lambda)$ follows from [27, p. 251] and the fact that $\Sigma_{n=1}^{\infty} 1/|\lambda_n|^2 < \infty$ (cf. [3, p. 259]). ∎

Corollary 4.4.1. Under the same assumptions as in Theorem 4.4, if

$$f(\lambda) = \int_{-\infty}^{\infty} F(x)\phi_0(-x,\lambda)\,dx\,, \quad F(x) \in L^2(-\infty,\infty)\,,$$

then

$$f(\lambda) = \sum_{n=0}^{\infty} f(\lambda_n)\,\frac{G(\lambda)}{(\lambda-\lambda_n)G'(\lambda_n)}\,.$$

From Theorem 4.4 it is evident that Kramer's sampling theorem holds for the interval $(-\infty,\infty)$ by choosing $K(x,\lambda) = \Phi_0(x,\lambda)$, the set $E = \{\lambda_n\}_{n=0}^{\infty}$, and hence

$$S_n^*(\lambda) = \frac{G(\lambda)}{(\lambda-\lambda_n)G'(\lambda_n)}\,.$$

Theorem 4.4 remains valid in the limit point case provided that further restrictions are imposed. For example, since $\psi_1(x,\lambda)$ is not, in general, in $L^2(-\infty,0)$ we may further assume that $F(x)$ is a nice function on $(-\infty,0)$ so that the integral in (4.3.28) is absolutely convergent. It is also essential that the integral in (4.3.34) exists.

4.4 Examples

In this section we give several examples covering the regular and the singular cases on both the half and the whole line. Some of the examples, in particular Examples D1 and D2, are believed to be new. Example C2, which deals with the sampling expansion of the finite continuous Jacobi transform, has recently been obtained by Koornwinder and Walter [19]. However, the main advantage of our method over theirs is that instead of using ad hoc techniques pertaining to the Jacobi polynomials to derive the sampling expansion of the finite continuous Jacobi transform, we use the general procedure described in the previous sections and conclude their results as a special case.

4.4.A The Regular Case

Example A1. The finite cosine transform. Consider the regular Sturm-Liouville problem

$$y'' = -\lambda y\,, \quad 0 \le x \le \pi\,,$$

$$y'(0) = 0 = y'(\pi)\,.$$

It is easy to see that in the notation of Theorem 4.1

$$\phi(x,\lambda) = \cos\sqrt{\lambda}x \quad \text{and} \quad \chi(x,\lambda) = \cos\sqrt{\lambda}(\pi - x).$$

The eigenvalues are $\lambda_n = n^2$, $n = 0, 1, 2, \ldots$, hence

$$G(\lambda) = \lambda \prod_{n=1}^{\infty}\left(1 - \frac{\lambda}{n^2}\right) = \frac{\sqrt{\lambda}\sin\pi\sqrt{\lambda}}{\pi}$$

and

$$G'(\lambda_n) = \begin{cases} \frac{1}{2}(-1)^n & \text{if} \quad n \neq 0, \\ 1 & \text{if} \quad n = 0. \end{cases}$$

Therefore, by Theorem 4.1 we have that if

$$f(\lambda) = \int_0^\pi F(x)\cos\sqrt{\lambda}x\, dx\,,$$

$$f^*(\lambda) = \int_0^\pi F(x)\cos\sqrt{\lambda}(\pi - x)\, dx\,,$$

for some $F(x) \in L^2(0,\pi)$, then

$$\begin{Bmatrix} f(\lambda) \\ f^*(\lambda) \end{Bmatrix} = \begin{Bmatrix} f(0) \\ f^*(0) \end{Bmatrix} \frac{\sin\pi\sqrt{\lambda}}{\pi\sqrt{\lambda}} + 2\sum_{n=1}^{\infty} \begin{Bmatrix} f(n^2) \\ f^*(n^2) \end{Bmatrix} \frac{\sqrt{\lambda}\sin\pi(\sqrt{\lambda} - n)}{\pi(\lambda - n^2)}. \tag{4.4.1}$$

Upon replacing λ by t^2 in (4.4.1) we obtain Corollary 2 of [38, sec. 4] for $f(\lambda)$. The change of variable, $z = \pi - x$, in the integral defining f^* shows that f^* and f are related, hence, so are their sampling expansions.

4.4.B The Singular Case on a Halfline

Example **B1.** The finite Hankel transform ($\nu \geq 1$). Consider the following singular Sturm-Liouville problem:

$$y'' - \frac{\nu^2 - \frac{1}{4}}{x^2} y = -\lambda y\,, \quad 0 < x \leq b < \infty\,, \quad \nu \geq 1\,,$$

$$y(b) = 0\,.$$

This is equivalent to a singular Sturm-Liouville problem on a halfline since the function $q(x) = (\nu^2 - 1/4)/x^2$ has a singularity at one of the endpoints.

It is known that [27, p. 81] for this problem the limit point case holds and in the notation of Section 2 we have

$$\phi(x,\lambda)=\frac{1}{2}\pi\sqrt{xb}\,\{J_\nu(x\sqrt{\lambda})Y_\nu(b\sqrt{\lambda})-Y_\nu(x\sqrt{\lambda})J_\nu(b\sqrt{\lambda})\}\,,$$

$$\theta(x,\lambda)=\frac{1}{2}\pi\sqrt{xb\lambda}\{J_\nu(x\sqrt{\lambda})Y_\nu'(b\sqrt{\lambda})-Y_\nu(x\sqrt{\lambda})J_\nu'(b\sqrt{\lambda})\}+\frac{\phi(x,\lambda)}{2b}\,,$$

$$m(\lambda)=-\sqrt{\lambda}\,\frac{J_\nu'(b\sqrt{\lambda})}{J_\nu(b\sqrt{\lambda})}-\frac{1}{2b}\,,$$

where $J_\nu(z)$ and $Y_\nu(z)$ are the Bessel functions of order ν of the first and second kinds, respectively. Hence, the eigenvalues λ_n are the zeros of $J_\nu(b\sqrt{\lambda})$, and consequently the function $G(\lambda)$ of Theorem 4.2 is given by

$$G(\lambda)=\prod_{n=1}^{\infty}\left(1-\frac{\lambda b^2}{\alpha_{\nu,n}^2}\right),$$

where $\alpha_{\nu,n}$ is the nth positive zero of $J_\nu(z)$; $n=1,2,\ldots$. With some easy calculations we can show that

$$J_\nu(b\sqrt{\lambda})\,\psi(x,\lambda)=J_\nu(b\sqrt{\lambda})\,\{\theta(x,\lambda)+m(\lambda)\phi(x,\lambda)\}=\sqrt{x/b}\,J_\nu(x\sqrt{\lambda})\,.$$

But in view of the relation [32, p. 498]

$$J_\nu(b\sqrt{\lambda})=\frac{(b\sqrt{\lambda})^\nu}{2^\nu\Gamma(\nu+1)}G(\lambda)\,,$$

we have, using the notation of Theorem 4.2,

$$\Phi(x,\lambda)=G(\lambda)\,\psi(x,\lambda)=\frac{2^\nu\Gamma(\nu+1)}{(b\sqrt{\lambda})^\nu}\sqrt{\frac{x}{b}}\,J_\nu(x\sqrt{\lambda})\,.$$

Therefore, Theorem 4.2 now takes on the following form. If

$$f(\lambda)=c_\nu\int_0^b F(x)\sqrt{x}\,\lambda^{-\nu/2}J_\nu(x\sqrt{\lambda})\,dx\,,\quad c_\nu=\frac{2^\nu\Gamma(\nu+1)}{b^{\nu+1/2}} \tag{4.4.2}$$

for some $F(x)\in L^2(0,b)$, then

$$f(\lambda)=\sum_{n=1}^{\infty}f(\lambda_n)\frac{2\lambda_n^{(\nu+1)/2}\lambda^{-\nu/2}J_\nu(b\sqrt{\lambda})}{(\lambda-\lambda_n)\,b\,J_\nu'(b\sqrt{\lambda_n})}\,,$$

which reduces to formula (5.4) of [38] upon putting $\lambda = t^2$, $\lambda_n = t_n^2$, and $b = 1$ (see also [2, 13, 14, 20 and 33]).

Formula (5.9) of [38] can also be reproduced in the same fashion. The integral in (4.4.2) is known as the finite Hankel transform of $F(x)$.

Example **B2.** $(0 < \nu < 1, \nu \neq 1/2)$. In the above example if $\nu = 1/2$, $q(x)$ becomes zero, and we have a regular Sturm-Liouville problem that can be treated as in section 1 (see also [38, Ex. 2, Cor. 3, sec. 5]).

Let us assume that $0 < \nu < 1$, $\nu \neq 1/2$. For this problem the limit circle case holds [27, p. 82] and the function $m(\lambda)$ can be given by

$$m(\lambda) = -\sqrt{\lambda}\,\frac{c(\sqrt{\lambda})^{-\nu}J_\nu'(b\sqrt{\lambda}) - (\sqrt{\lambda})^\nu J_{-\nu}'(b\sqrt{\lambda})}{c(\sqrt{\lambda})^{-\nu}J_\nu(b\sqrt{\lambda}) - (\sqrt{\lambda})^\nu J_{-\nu}(b\sqrt{\lambda})} - \frac{1}{2b},$$

where c is an arbitrary constant. The eigenvalues λ_n are the zeros of $cJ_\nu(b\sqrt{\lambda}) - \lambda^\nu J_{-\nu}(b\sqrt{\lambda})$. By employing the notation and the argument given in Theorem 4.4 and its following remarks, we can choose

$$\phi_0(x,\lambda) = \phi(x,\lambda) = -\frac{\pi\sqrt{xb}}{2\sin\nu\pi}\{J_\nu(x\sqrt{\lambda})J_{-\nu}(b\sqrt{\lambda}) - J_{-\nu}(x\sqrt{\lambda})J_\nu(b\sqrt{\lambda})\}$$

so that

$$\phi_0(x,\lambda_n) = \frac{\pi\sqrt{xb}}{2\sin\nu\pi}J_\nu(b\sqrt{\lambda_n})\{c\lambda_n^{-\nu}J_\nu(x\sqrt{\lambda_n}) - J_{-\nu}(x\sqrt{\lambda_n})\}$$

is the eigenfunction corresponding to the eigenvalue λ_n. The presence of the arbitrary constant c in the eigenfunctions is due to the fact that in the limit circle case all the solutions of the differential equation are in $L^2(a,b)$.

Therefore, by Theorem 4.2 we have that if

$$f(\lambda) = \int_0^b F(x)\,\phi_0(x,\lambda)\,dx\,, \quad F(x) \in L^2(0,b)\,,$$

then

$$f(\lambda) = \sum_{n=1}^{\infty} f(\lambda_n)\frac{G(\lambda)}{(\lambda-\lambda_n)G'(\lambda_n)},$$

where $G(\lambda)$ is the canonical product of the zeros of the function $cJ_\nu(b\sqrt{\lambda}) - \lambda^\nu J_{-\nu}(b\sqrt{\lambda})$.

Example **B3.** ($\nu = 0$). If $\nu = 0$, the limit circle case still holds. The function $m(\lambda)$ is given by [27, p. 84]

$$m(\lambda) = -\sqrt{\lambda}\,\frac{cJ_0'(b\sqrt{\lambda}) - Y_0'(b\sqrt{\lambda}) + (2/\pi)J_0'(b\sqrt{\lambda})\log\sqrt{\lambda}}{cJ_0(b\sqrt{\lambda}) - Y_0(b\sqrt{\lambda}) + (2/\pi)J_0(b\sqrt{\lambda})\log\sqrt{\lambda}} - \frac{1}{2b},$$

where c is an arbitrary constant. Hence, the eigenvalues λ_n are the solutions of the equation

$$cJ_0(b\sqrt{\lambda}) - Y_0(b\sqrt{\lambda}) + \frac{2}{\pi}J_0(b\sqrt{\lambda})\log\sqrt{\lambda} = 0,$$

$$\phi_0(x,\lambda) = \sqrt{x}\left\{cJ_0(x\sqrt{\lambda}) - Y_0(x\sqrt{\lambda}) + \frac{2}{\pi}J_0(x\sqrt{\lambda})\log\sqrt{\lambda}\right\}$$

and the eigenfunctions are

$$\psi_n(x) = \sqrt{x}\left\{cJ_0(x\sqrt{\lambda_n}) - Y_0(x\sqrt{\lambda_n}) + \frac{2}{\pi}J_0(x\sqrt{\lambda_n})\log\sqrt{\lambda_n}\right\}, \quad n = 1,2,3,\ldots.$$

Therefore, by Theorem 4.2, we have that if

$$f(\lambda) = \int_0^b F(x)\,\phi_0(x,\lambda)\,dx, \quad F(x) \in L^2(0,b),$$

then

$$f(\lambda) = \sum_{n=1}^{\infty} f(\lambda_n)\frac{G(\lambda)}{(\lambda - \lambda_n)G'(\lambda_n)},$$

where $G(\lambda)$ is the canonical product of the zeros of the function

$$cJ_0(b\sqrt{\lambda}) - Y_0(b\sqrt{\lambda}) + \frac{2}{\pi}J_0(b\sqrt{\lambda})\log\sqrt{\lambda}.$$

As a special case of Examples B2 and B3 if we let $c \to \infty$, then $f(\lambda)$ becomes the finite Hankel transform of $F(x)$ and the sampling series representation of $f(\lambda)$ in this case becomes similar to the one given in Example B1.

4.4.C. The Singular Case on the Whole Line

Example C1. The continuous Legendre transform. Consider

$$y'' - \left(\frac{1}{4}\sec^2 x\right)y = -\lambda y, \quad -\frac{\pi}{2} < x < \frac{\pi}{2}.$$

This is equivalent to a singular Sturm-Liouville problem on the whole line since the function $q(x) = (1/4)\sec^2 x$ has singularities at both finite endpoints. Since $q(x)$ is even, Theorem 4.3 applies. It is known that for this problem the limit circles case holds [27, p. 78], hence we can apply Theorem 4.4 as well. The function $m_1(\lambda)$ of Theorem 4.4 that renders the ordinary Legendre expansion is given by

$$m_1(\lambda) = \frac{\phi_0'(0,\lambda)}{\phi_0(0,\lambda)},$$

where

$$\phi_0(-x,\lambda) = 2\sqrt{2}\pi\sqrt{\cos x}\,P_{\sqrt{\lambda}-1/2}(\sin x),$$

and

$$P_t(z) = {}_2F_1\left(-t, t+1; 1; \frac{1-z}{2}\right), \quad t \in \mathbf{C}.$$

The function $P_t(z)$ is known as the Legendre function of order t (see [1], [26]). It is not difficult to see that the eigenvalues λ_n and eigenfunctions $\psi_n(x)$ are given by $\lambda_n = (n+1/2)^2$, $n = 0, 1, 2, \ldots$ and $\psi_n(x) = 2\sqrt{2}\pi\sqrt{\cos x}\,P_n(\sin x)$, where $P_n(z)$ is the Legendre polynomial of degree n. Hence,

$$G(\lambda) = \prod_{n=0}^{\infty}\left(1 - \frac{\lambda}{\left(n+\frac{1}{2}\right)^2}\right) = \cos\pi\sqrt{\lambda}$$

and

$$G'(\lambda_n) = \frac{(-1)^{n+1}\pi}{(2n+1)}.$$

Therefore, Theorem 4.4 now takes on the form: if

$$f(\lambda) = 2\sqrt{2}\pi \int_{-\pi/2}^{\pi/2} F(x)\sqrt{\cos x}\,P_{\sqrt{\lambda}-1/2}(\sin x)\,dx \tag{4.4.3}$$

for some $F(x) \in L^2(-\pi/2, \pi/2)$, then

$$\begin{aligned} f(\lambda) &= \sum_{n=0}^{\infty} f\left(\left(n+\frac{1}{2}\right)^2\right) \frac{(2n+1)\cos\pi\sqrt{\lambda}}{\left[\lambda - \left(n+\frac{1}{2}\right)^2\right](-1)^{n+1}\pi} \\ &= \sum_{n=0}^{\infty} f\left(\left(n+\frac{1}{2}\right)^2\right) \frac{(2n+1)\sin\pi\left(\sqrt{\lambda} - \left(n+\frac{1}{2}\right)\right)}{\pi\left[\lambda - \left(n+\frac{1}{2}\right)^2\right]}, \end{aligned} \tag{4.4.4}$$

which, upon setting $\lambda = t^2$, yields formula (5.16) in [38].

The integral in (4.4.3) is known as the continuous Legendre transform of $F(x)$ and has been studied by many people (see [1, 29, 31]). Formula (4.4.4) was obtained earlier by Campbell [2] and Butzer, Stens, and Wehrens [1] by using different techniques.

The sampling expansion of $f^*(\lambda)$, which is given by

$$f^*(\lambda) = 2\sqrt{2}\,\pi \int_{-\pi/2}^{\pi/2} F(x)\sqrt{\cos x}\, P_{\sqrt{\lambda}-1/2}(-\sin x)\, dx\,,$$

yields no new information since the change of variable, $z = -x$, in the above integral shows that f^* and f are related; hence so are their sampling expansions.

Example C2. The finite continuous Jacobi transform. Consider

$$y'' - \left[\frac{\alpha^2 - \frac{1}{4}}{4(\sin x/2)^2} + \frac{\beta^2 - \frac{1}{4}}{4(\cos x/2)^2}\right] y = -\lambda y\,, \quad 0 < x < \pi\,, \quad \alpha, \beta > -1\,.$$

This is equivalent to a singular Sturm-Liouville problem on the whole line since

$$q(x) = \frac{\alpha^2 - \frac{1}{4}}{4(\sin x/2)^2} + \frac{\beta^2 - \frac{1}{4}}{4(\cos x/2)^2}$$

has singularities at both finite endpoints. Since $q(x)$ is even, Theorem 4.3 applies. It is easy to see that the limit circle case holds if $-1 < \alpha, \beta < 1$; hence we can use Theorem 4.4. However, if $\alpha, \beta \geq 1$, the limit point case prevails but the spectrum is still discrete since $\lim_{x \to 0^+} q(x) = \infty = \lim_{x \to \pi^-} q(x)$. It is known that the eigenvalues are $\lambda_n = (n+\gamma)^2$, $n = 0, 1, 2, \dots$, where $2\gamma = \alpha + \beta + 1$ and the corresponding normalized eigenfunctions $\psi_n(x)$ are

$$\psi_n(x) = \frac{\Gamma(\alpha+1)\Gamma(n+1)}{\Gamma(n+\alpha+1)} \left(\sin\frac{x}{2}\right)^{\alpha+1/2} \left(\cos\frac{x}{2}\right)^{\beta+1/2} P_n^{(\alpha,\beta)}(\cos x)\,,$$

where $P_n^{(\alpha,\beta)}(z)$ is the Jacobi polynomial of degree n [26].

The eigenfunctions can also be written in terms of the hypergeometric function as

$$\psi_n(x) = \left(\sin\frac{x}{2}\right)^{\alpha+1/2} \left(\cos\frac{x}{2}\right)^{\beta+1/2} {}_2F_1\!\left(-n, n+2\gamma; \alpha+1; \sin^2\frac{x}{2}\right).$$

By using the notation of Theorems 4.3 and 4.4, we may write

$$\phi_0(x,\lambda) = \left(\sin\frac{x}{2}\right)^{\alpha+1/2}\left(\cos\frac{x}{2}\right)^{\beta+1/2} R^{(\alpha,\beta)}_{\sqrt{\lambda}-\gamma}(\cos x),$$

where

$$R^{(\alpha,\beta)}_t(z) = {}_2F_1\left(-t, t+2\gamma; \alpha+1; \frac{1-z}{2}\right),$$

which is known as the Jacobi function of order t (see [5, 30]). The entire function $G(\lambda)$ is given by

$$G(\lambda) = \begin{cases} \prod\limits_{n=0}^{\infty}\left(1 - \dfrac{\lambda}{(n+\lambda)^2}\right) & \text{if } \gamma \neq 0, \\ \lambda\prod\limits_{n=1}^{\infty}\left(1 - \dfrac{\lambda}{n^2}\right) & \text{if } \gamma = 0, \end{cases} \tag{4.4.5}$$

which in view of formulae (1.3) and (1.4) in [6, p. 5], can be written as

$$G(\lambda) = \begin{cases} \dfrac{\Gamma^2(\gamma)}{\Gamma(\gamma+\sqrt{\lambda})\Gamma(\gamma-\sqrt{\lambda})} & \text{if } \gamma \neq 0, \\ \sqrt{\lambda}\,\dfrac{\sin\sqrt{\lambda}\,\pi}{\pi} & \text{if } \gamma = 0. \end{cases} \tag{4.4.6}$$

With some easy calculations using the fact

$$\begin{aligned} \lim_{x\to n}\frac{\Gamma'(-x)}{\Gamma^2(-x)} &= \lim_{x\to n}\frac{\Gamma(1+x)\Gamma'(-x)}{\Gamma(-x)\Gamma(1+x)\Gamma(-x)} \\ &= \lim_{x\to n}\frac{\Gamma(1+x)\sin\pi x\,\Gamma'(-x)}{(-\pi)\Gamma(-x)} \\ &= \Gamma(1+n)(-1)^{n+1}, \end{aligned} \tag{4.4.7}$$

we obtain

$$G'(\lambda_n) = \begin{cases} \dfrac{(-1)^{n+1}\Gamma^2(\gamma)\Gamma(1+n)}{2(\gamma+n)\Gamma(2\gamma+n)} & \text{if } \gamma \neq 0, \\ \dfrac{1}{2}(-1)^n \text{ for } n = 1,2,\ldots \text{ and } 1 \text{ for } n = 0 & \text{if } \gamma = 0. \end{cases}$$

Therefore, if

$$f(\lambda)=\int_0^\pi F(x)\left(\sin\frac{x}{2}\right)^{\alpha+1/2}\left(\cos\frac{x}{2}\right)^{\beta+1/2}R^{(\alpha,\beta)}_{\sqrt{\lambda}-\gamma}(\cos x)\,dx\,, \qquad (4.4.8)$$

where the integral converges absolutely, then

$$f(\lambda)=\begin{cases}\displaystyle\sum_{n=0}^{\infty} f((n+\gamma)^2)\cdot\frac{(-1)^{n+1}2(n+\gamma)\Gamma(n+2\gamma)}{\Gamma(\gamma+\sqrt{\lambda})\Gamma(\gamma-\sqrt{\lambda})[\lambda-(n+\gamma)^2]\Gamma(1+n)} & \text{if } \gamma\neq 0, \qquad (4.4.9)\\[2ex] \displaystyle f(0)\,\frac{\sin\pi\sqrt{\lambda}}{\pi\sqrt{\lambda}}+\sum_{n=1}^{\infty} f(n^2)\,\frac{2\sqrt{\lambda}\sin\pi(\sqrt{\lambda}-n)}{\pi[\lambda-n^2]} & \text{if } \gamma=0. \qquad (4.4.10)\end{cases}$$

Except for the first term in (4.4.9), i.e., for $n=0$, each term in (4.4.9) reduces to the corresponding term in (4.4.10) when $\gamma\to 0$.

In the limit circle case the integral in (4.4.8) converges absolutely provided that $F(x)\in L^2(0,\pi)$, while in the limit point case we need to impose more stringent conditions on $F(x)$; such conditions can be found in [19 and 30]. Moreover, we can easily see that in the limit point case the integral in (4.3.34) also exists.

The integral in (4.4.8) is known as the finite continuous Jacobi transform and has recently been studied by several authors (see [4], 5, 12, 19, 30, 34-37]).

Formulae (4.4.9) and (4.4.10) were obtained earlier by Walter and Zayed [30] when 2γ is a positive integer, and more recently by Koornwinder and Walter [19] for a general γ (see also [18] for some special cases).

4.4.D. One-Sided Cardinal Series Associated with Linear Forms of Eigenvalues

Now let us make the following observation. If, for a Sturm-Liouville boundary-value problem, the eigenvalues λ_n are of the form $\lambda_n=an+b, n=0,1,2,\dots,a\neq 0$, then the associated sampling series is essentially a one-sided WSK sampling series. For, in this case,

$$G(\lambda)=\begin{cases}\displaystyle\prod_{n=0}^{\infty}\left(1-\frac{\lambda}{an+b}\right)e^{\lambda/(an+b)}=\frac{\exp\left(-\frac{\lambda}{a}\psi\left(\frac{b}{a}\right)\right)\Gamma\left(\frac{b}{a}\right)}{\Gamma\left(\frac{b}{a}-\frac{\lambda}{a}\right)} & \text{if } b\neq 0,\\[2ex] \displaystyle\lambda\prod_{n=1}^{\infty}\left(1-\frac{\lambda}{an}\right)e^{\lambda/(an)}=\frac{-a\exp\left(\frac{\gamma}{a}\lambda\right)}{\Gamma\left(-\frac{\lambda}{a}\right)} & \text{if } b=0,\end{cases}$$

where γ is the Euler constant. Hence,

$$G'(\lambda_n) = \begin{cases} a^{-1}\Gamma\left(\frac{b}{a}\right)(-1)^{n+1}\Gamma(1+n)\exp\left(-\left(n+\frac{b}{a}\right)\psi\left(\frac{b}{a}\right)\right) & \text{if } b \neq 0, \\ e^{\gamma n}\Gamma(1+n)(-1)^n & \text{if } b = 0, \end{cases}$$

and the sampling series for $f(\lambda)$ takes on the form

$$f(\lambda) = \begin{cases} \sum_{n=0}^{\infty} f(an+b)\,\dfrac{a(-1)^{n+1}\exp\left(\psi\left(\frac{b}{a}\right)\left(n+\frac{b}{a}-\frac{\lambda}{a}\right)\right)}{\Gamma\left(\frac{b}{a}-\frac{\lambda}{a}\right)\Gamma(1+n)[\lambda-an-b]} & \text{if } b \neq 0, \\ \sum_{n=0}^{\infty} f(an)\,\dfrac{(-1)^{n+1}a\exp\left(\gamma\left(\frac{\lambda}{a}-n\right)\right)}{\Gamma\left(-\frac{\lambda}{a}\right)\Gamma(1+n)[\lambda-an]} & \text{if } b = 0, \end{cases}$$

or

$$f(a\lambda+b)e^{\lambda\psi(b/a)} = \sum_{n=0}^{\infty} f(an+b)\,\frac{(-1)^{n+1}e^{n\psi(b/a)}}{\Gamma(-\lambda)\Gamma(1+n)[\lambda-n]} \quad \text{if } b \neq 0,$$

$$f(a\lambda)e^{-\gamma\lambda} = \sum_{n=0}^{\infty} f(an)\,\frac{(-1)^{n+1}e^{-\gamma n}}{\Gamma(-\lambda)\Gamma(1+n)[\lambda-n]} \quad \text{if } b = 0,$$

or

$$\tilde{f}(\lambda) = \sum_{n=0}^{\infty} \tilde{f}(n)\,\frac{\sin\pi(\lambda-n)}{\pi(\lambda-n)},$$

where

$$\tilde{f}(\lambda) = \begin{cases} \dfrac{f(a\lambda+b)e^{\lambda\psi(b/a)}}{\Gamma(1+\lambda)} & \text{if } b \neq 0, \\ \dfrac{f(a\lambda)e^{-\gamma\lambda}}{\Gamma(1+\lambda)} & \text{if } b = 0. \end{cases}$$

This observation will be displayed in the next three examples which, of course, arise from singular Sturm-Liouville boundary value problems, since in the regular case $\lambda_n \sim n^2$ as $n \to \infty$. Similarly, the following observation holds for regular and some singular (e.g., in the limit circle case) Sturm-Liouville problems: if the eigenvalues λ_n are quadratic in n, i.e., they are of the form $\lambda_n = an^2 + b$, $a \neq 0$, then the associated sampling series is essentially the same as the one given in Example C2. The proof is similar to the one previously given and will be left to the reader.

Example D1. The continuous Laguerre transform with a squared argument. Consider

$$y'' - \left(x^2 + \frac{\alpha^2 - \frac{1}{4}}{x^2}\right) y = -\lambda y, \quad 0 < x < \infty, \quad \alpha > -1. \tag{4.4.11}$$

This is equivalent to a singular Sturm-Liouville problem on the whole line since the interval (a, b) extends to infinity in one direction and the function $q(x) = x^2 + (\alpha^2 - 1/4)/x^2$ has a singularity at the other finite endpoint except when $\alpha = \pm 1/2$. For this problem the limit point case holds, but since $q(x)$ tends monotonically to infinity as $x \to \infty$, the spectrum is discrete. In fact, the eigenvalues λ_n and eigenfunctions $\psi_n(x)$ are given by $\lambda_n = 4n + 2\alpha + 2$ and $\psi_n(x) = x^{\alpha+1/2} e^{-x^2/2} L_n^\alpha(x^2)$, where $L_n^\alpha(z)$ is the generalized Laguerre polynomial of degree n (cf. [26, 27]). Two independent solutions of (4.4.11) are

$$\phi_0(x, \lambda) = x^{-1/2} W_{\lambda/4, \alpha/2}(x^2) \quad \text{and} \quad \psi_0(x, \lambda) = x^{-1/2} M_{\lambda/4, \alpha/2}(x^2), \tag{4.4.12}$$

where $W_{k,\mu}(x)$ and $M_{k,\mu}(x)$ are the Whittaker functions that can be related to the confluent hypergeometric functions $\Phi(a, c; x)$ and $\Psi(a, c; x)$ via

$$M_{k,\mu}(x) = e^{-x/2} x^{c/2} \Phi(a, c; x), \quad W_{k,\mu}(x) = e^{-x/2} x^{c/2} \Psi(a, c; x) \tag{4.4.13}$$

with $k = -a + c/2$, $\mu = c/2 - 1/2$; $\Phi(a, c; x)$ and $\Psi(a, c; x)$ are solutions of the differential equation $xy'' + (c - x)y' - ay = 0$ (cf. [6, p. 252]. By using the asymptotic expansions of $\Phi(a, c; x)$, $\Psi(a, c; x)$ (cf. [6, p. 278]) as $x \to 0$ and $x \to \infty$, we can show from (4.4.12), (4.4.13) that

$$\phi_0(x, \lambda) = O\left(x^{\lambda/2 - 1/2} e^{-x^2/2}\right), \quad \psi_0(x, \lambda) = O(x^{-\lambda/2 - 1/2}) \quad \text{as} \quad x \to \infty$$

and that $\phi_0(x, \lambda) \in L^2(0, \delta)$ for any $\delta > 0$ provided that $-1 < \alpha < 1$, while $\psi_0(x, \lambda) \in L^2(0, \delta)$ for all α. Therefore, $\phi_0(x, \lambda) \in L^2(0, \infty)$ for $-1 < \alpha < 1$. Moreover, both $\phi_0(x, \lambda)$ and $\psi_0(x, \lambda)$ reduce to a multiple of $\psi_n(x)$ when $\lambda = \lambda_n$.

The canonical product of the λ_n's as given by Theorem 4.2 now takes on the form

$$G(\lambda) = \prod_{n=0}^{\infty} \left(1 - \frac{\lambda}{4n + 2\alpha + 2}\right) \exp\left(\frac{\lambda}{4n + 2\alpha + 2}\right).$$

With the aid of Mellin's formula [6, p. 6],

$$e^{y\psi(x)} (\Gamma(x)/\Gamma(x + y)) = \prod_{n=0}^{\infty} [1 + y/(x + n)] e^{-y/(x+n)}.$$

where

$$\psi(z) = \Gamma'(z)/\Gamma(z),$$

$G(\lambda)$ can be simplified to

$$G(\lambda) = \frac{\exp\left(-\frac{\lambda}{4}\psi\left(\frac{\alpha}{2}+\frac{1}{2}\right)\right)\Gamma\left(\frac{\alpha}{2}+\frac{1}{2}\right)}{\Gamma\left(\frac{\alpha}{2}+\frac{1}{2}-\frac{\lambda}{4}\right)}.$$

With some easy calculations using (4.4.7), we obtain

$$G'(\lambda_n) = \frac{(-1)^{n+1}}{4}\Gamma\left(\frac{\alpha+1}{2}\right)\Gamma(1+n)\exp\left(-\left(n+\frac{\alpha+1}{2}\right)\psi\left(\frac{\alpha+1}{2}\right)\right).$$

Therefore, if

$$f(\lambda) = \int_0^\infty F(x)x^{-1/2}W_{\lambda/4,\alpha/2}(x^2)\,dx, \tag{4.4.14}$$

where the integral converges absolutely, then

$$f(\lambda) = \sum_{n=0}^\infty f(4n+2\alpha+2)\,\frac{4(-1)^{n+1}\exp\left[\frac{1}{4}\psi\left(\frac{\alpha+1}{2}\right)[(4n+2\alpha+2)-\lambda]\right]}{\Gamma\left(\frac{\alpha+1}{2}-\frac{\lambda}{4}\right)\Gamma(n+1)[\lambda-(4n+2\alpha+2)]}$$

or

$$f(4\lambda+2\alpha+2) = \sum_{n=0}^\infty f(4n+2\alpha+2)\,\frac{(-1)^{n+1}\exp\{(n-\lambda)\psi((\alpha+1)/2)\}}{\Gamma(-\lambda)\Gamma(n+1)[\lambda-n]}$$

or

$$\tilde{f}(\lambda) = \sum_{n=0}^\infty \tilde{f}(n)\,\frac{\sin\pi(\lambda-n)}{\pi(\lambda-n)}, \tag{4.4.15}$$

where

$$\tilde{f}(\lambda) = \frac{f(4\lambda+2\alpha+2)}{\Gamma(1+\lambda)}\exp\left(\lambda\psi\left(\frac{\alpha+2}{2}\right)\right).$$

Here we have used the fact that $\Gamma(-\lambda)\Gamma(1+\lambda) = -\pi\csc(\pi\lambda)$.

A sufficient condition for the integral in (4.4.14) to converge absolutely is that $F(x)\in L^2(0,\infty)$ if $-1<\alpha<1$ and $F(x)\in L^2(\delta,\infty)$, $\delta>0$, $F(x)=O(x^\eta)$ as $x\to 0$ if $\alpha>1$, where $\eta>\alpha-3/2$; here we have used the fact that $\phi_0(x,\lambda)=O(x^{1/2-\alpha})$ as $x\to 0$ when $\alpha>0$. Finally, we remark that the integral

(4.3.34) is easily seen to converge absolutely for $\phi_0(x,\lambda)$ and $\psi_n(x)$ for all $\alpha > -1$.

Although we have derived our results using the differential operator $L = d^2/dx^2 - q(x)$, we can easily see [3, Chap. 9] that the same results could have been obtained if we had used the operator $d/dx(p(x)d/dx) - q(x)$ instead, where $p(x)$ and $q(x)$ are real valued continuous functions with $p(x) > 0$.

The next example utilizes this last operator.

Example **D2**. The continuous Laguerre transform. Consider

$$\frac{d}{dx}\left(x\frac{dy}{dx}\right) - \left(\frac{x}{4} + \frac{\alpha^2}{4x}\right)y = -\lambda y\,, \quad 0 < x < \infty\,, \quad \alpha > -1\,. \tag{4.4.16}$$

This is equivalent to a singular Sturm-Liouville problem on the whole line since $q(x) = x/4 + \alpha^2/4x$ has a singularity at the finite endpoint. Since $\lim_{x\to 0^+} q(x) = \infty = \lim_{x\to\infty} q(x)$, the limit point case holds. Moreover, the spectrum is discrete; in fact, the eigenvalues λ_n and eigenfunctions $\psi_n(x)$ are given by $\lambda_n = n + (\alpha+1)/2$ and $\psi_n(x) = e^{-x/2}x^{\alpha/2}L_n^{\alpha}(x)$ (Laguerre functions), $n = 0, 1, 2, \ldots$. Two independent solutions of (4.4.16) are given by

$$\begin{aligned} \phi_0(x,\lambda) &= e^{-x/2}x^{\alpha/2}\Psi\left(\frac{\alpha+1}{2} - \lambda, \alpha+1; x\right), \\ \psi_0(x,\lambda) &= e^{-x/2}x^{\alpha/2}\Phi\left(\frac{\alpha+1}{2} - \lambda, \alpha+1; x\right), \end{aligned} \tag{4.4.17}$$

where $\Psi(a,c;x)$ and $\Phi(a,c;x)$ are the confluent hypergeometric functions (see Example D1). Both $\phi_0(x,\lambda)$ and $\psi_0(x,\lambda)$ reduce to a multiple of $\psi_n(x)$ when $\lambda = \lambda_n$. As in Example D1, it is easily seen that $\phi_0(x,\lambda) \in L^2(0,\infty)$ if $-1 < \alpha < 1$, $\phi_0(x,\lambda) \in L^2(\delta,\infty)$ for all $\alpha > -1$ and any $\delta > 0$, while $\psi_0(x,\lambda) \notin L^2(0,\infty)$ for any α. Moreover,

$$G(\lambda) = \prod_{n=0}^{\infty}\left(1 - \frac{\lambda}{n + ((\alpha+1)/2)}\right)\exp\left(\frac{\lambda}{n + (\alpha + 1/2)}\right),$$

which, in view of Mellin's formula, can be reduced to

$$G(\lambda) = \frac{\exp\left(-\lambda\psi\left(\frac{\alpha+1}{2}\right)\right)\Gamma\left(\frac{\alpha+1}{2}\right)}{\Gamma\left(\frac{\alpha+1}{2} - \lambda\right)},$$

where $\psi(z) = \Gamma'(z)/\Gamma(z)$. Thus, by (4.4.7), we have

$$G'(\lambda_n) = (-1)^{n+1}\Gamma\left(\frac{\alpha+1}{2}\right)\Gamma(1+n)\exp\left\{-\left[n+\left(\frac{\alpha+1}{2}\right)\right]\psi\left(\frac{\alpha+1}{2}\right)\right\}.$$

Therefore, if

$$f(\lambda) = \int_0^\infty F(x)e^{-x/2}x^{\alpha/2}\psi\left(\frac{\alpha+1}{2} - \lambda, \alpha+1; x\right)dx\,, \qquad (4.4.18)$$

provided that the integral converges absolutely, then

$$f(\lambda) = \sum_{n=0}^\infty f\left(n+\frac{\alpha+1}{2}\right)\frac{(-1)^{n+1}\exp\left\{\psi\left(\frac{\alpha+1}{2}\right)\left[n+\frac{\alpha+1}{2}-\lambda\right]\right\}}{\Gamma\left(\frac{\alpha+1}{2}-\lambda\right)\Gamma(1+n)\left[\lambda-\left(n+\frac{\alpha+1}{2}\right)\right]}\,,$$

or

$$f\left(\lambda+\frac{\alpha+1}{2}\right) = \sum_{n=0}^\infty f\left(n+\frac{\alpha+1}{2}\right)\frac{(-1)^{n+1}\exp\{(n-\lambda)\psi((\alpha+1)/2)\}}{\Gamma(-\lambda)\Gamma(1+n)[\lambda-n]}\,,$$

or

$$\tilde{f}(\lambda) = \sum_{n=0}^\infty \tilde{f}(n)\frac{\sin\pi(\lambda-n)}{\pi(\lambda-n)}\,, \qquad (4.4.19)$$

where

$$\tilde{f}(\lambda) = \frac{f(\lambda+(\alpha+1)/2)\exp(\lambda x^{(\alpha+1)/2})}{\Gamma(1+\lambda)}\,.$$

A sufficient condition for the integral in (4.4.18) to converge absolutely is that $F(x)\in L^2(0,\infty)$ if $-1<\alpha<1$ and $F(x)\in L^2(\delta,\infty)$, $F(x) = O(x^\eta)$ as $x\to 0$ if $\alpha>1$, where $\eta>\alpha/2-1$. Again it can be easily verified that the integral (4.3.34) exists for $\phi_0(x,\lambda)$ and $\psi_n(x)$.

The integral in (4.4.18) is known as the continuous Laguerre transform (see [9]-[11], [25]).

In [17], Jerri derived a sampling expansion for the "continuous Laguerre transform," defined by

$$f(\lambda) = \frac{\Gamma(\lambda+\alpha+1)}{\Gamma(\lambda+1)\Gamma(\alpha+1)}\int_0^\infty F(x)x^\alpha e^{-x}\Phi\left(-\lambda,\alpha+1;\frac{\nu x}{\nu-1}\right)dx\,,$$

where $\alpha>-1$, $\lambda\ge 0$, $-1<\nu<1$.

However, his results are not satisfactory for the following reasons:

i) His definition of the continuous transform does not reduce in a nice way to the discrete Laguerre transform when λ is a nonnegative integer.
ii) The sampling expansion involves a double series.
iii) His results, in general, do not generalize in a natural way the results in the finite interval case.

In a more recent attempt to derive a sampling expansion for the continuous Laguerre transform, Selvaratnam [25] defined the continuous Laguerre transform of a function $F(x)$ as

$$\hat{f}(\lambda) = \int_0^\infty F(x) e^{-x/2} L_\lambda(x)\, dx \,, \tag{4.4.20}$$

where $L_\lambda(x)$ is the continuous extension of the Laguerre polynomials defined by

$$L_\lambda(x) = \sum_{n=0}^\infty L_n(x) \frac{\sin \pi(\lambda - n)}{\pi(\lambda - n)} \,,$$

in which $L_n(x)$ is the standard Laguerre polynomial of degree n. It can be easily shown that $L_\lambda(x)$ is C^∞-function that satisfies the Laguerre differential equations, i.e., $xy'' + (1-x)y' + \lambda y = 0$, which is the same differential equation satisfied by $L_n(x)$ when λ is replaced by n. From the definition of $L_\lambda(x)$, it is evident that it reduces to $L_n(x)$ when $\lambda = n$. It was shown in [25] that if

$$\hat{f}(\lambda) = \int_0^\infty F(x) e^{-x/2} L_\lambda(x)\, dx \,, \quad F(x) \in L^2(0, \infty) \,,$$

then

$$\hat{f}(\lambda) = \sum_{n=0}^\infty \hat{f}(n) \frac{\sin \pi(\lambda - n)}{\pi(\lambda - n)} \,. \tag{4.4.21}$$

This is indeed a special case of our results in Example D2 (see (4.4.19)). It is not difficult to see that the ratio between the continuous Laguerre transforms defined in (4.4.20) and in (4.4.18) (when $\alpha = 0$) is an entire function in λ.

We conclude this chapter with the following very important example that shows that the discreteness of the spectrum of a singular boundary-value problem is, surprisingly, not sufficient for the existence of a Weiss-Kramer-type sampling theorem.

4.5 Counter-example

Example. The continuous Hermite transform (the parabolic cylindrical transform). Consider

$$y'' - x^2 y = -\lambda y, \quad -\infty < x < \infty. \tag{4.5.1}$$

Since $q(x) = x^2$ tends monotonically to infinity as $x \to \pm\infty$, the limit point case prevails at $\pm\infty$, but the spectrum is discrete. The eigenvalues λ_n and eigenfunctions are given by $\lambda_n = 2n + 1$ and $\psi_n(x) = e^{-x^2/2} H_n(x)$, $n = 0, 1, 2, \ldots$, where $H_n(x)$ is the Hermite polynomial of degree n. Equation (4.5.1) is indeed a special case of (4.4.11) when $\alpha = \pm 1/2$. Thus, two linearly independent solutions of (4.5.1) may be obtained from (4.4.12) by putting $\alpha = \pm 1/2$. Since $\psi_0(x, \lambda) = x^{-1/2} M_{\lambda/4, \pm 1/4}(x^2)$ is not in $L^2(0, \infty)$, we will only be interested in $\phi_0(x, \lambda) = x^{-1/2} W_{\lambda/4, 1/4}(x^2)$ since $W_{k,\mu}(x) = W_{k,-\mu}(x)$. It is customary to write the solution of (4.5.1) in terms of the parabolic cylindrical function $D_\lambda(x)$ defined by

$$D_\lambda(x) = 2^{1/4 + \lambda/2} x^{-1/2} W_{1/4 + \lambda/2, 1/4}\left(\frac{x^2}{2}\right).$$

Thus,

$$\phi_0(x, \lambda) = 2^{1/4 - \lambda/4} D_{\lambda/2 - 1/2}(\sqrt{2}\, x),$$

and in view of the relation

$$H_n(x) = 2^{n/2} e^{x^2/2} D_n(\sqrt{2}\, x),$$

we obtain

$$\phi_0(x, \lambda_n) = 2^{-n} \psi_n(x).$$

At this stage it may appear that all the conditions needed to have a sampling expansion for the continuous Hermite transform are satisfied. However, a careful examination of the analysis outlined in Section 3 will reveal that no sampling expansion similar to the ones given by Theorems 4.3 or 4.4 exists. For, we can easily see from the asymptotic expansion of $D_\lambda(x)$ [7, p. 1060] that $\phi_0(x, \lambda)$ is not in $L^2(-\infty, \infty)$ except when $\lambda = \lambda_n$. Thus, the integral (4.3.28) fails to exist unless we impose strong conditions on $F(x)$, e.g., $F(x)e^{x^2}$ is integrable over $(-\infty, 0)$. But even if such a condition is imposed, the integral (4.3.34) will still fail to exist. Hence, no sampling expansion as stated in the previous section can be found.

It should be pointed out that Mehta [23] had attempted to derive a sampling expansion for the continuous Hermite transform. Unfortunately, he did not pay careful attention to the convergence of the integrals involved and obtained divergent integrals and vacuous results. In a later paper [15], Jerri explained the flaws in Mehta's argument and showed that no sampling expansion for the continuous Hermite transform on the whole line is possible. However, he also pointed out that if the continuous Hermite transform is only defined on the half-line $(0,\infty)$, then its sampling expansion is possible to obtain. In fact, his results are special cases of (4.4.14) and (4.4.15) when $\alpha = \pm 1/2$. Since upon setting $\alpha = 1/2$ in (4.4.14) and (4.4.15) we obtain that if

$$f(\lambda) = 2^{(1-\lambda)/4} \int_0^\infty F(x) D_{\lambda/2-1/2}(\sqrt{2}\,x)\,dx\,, \quad F(x) \in L^2(0,\infty)\,, \tag{4.5.2}$$

then

$$f(\lambda) = \sum_{n=0}^{\infty} f(4n+3)\,\frac{4(-1)^{n+1}\exp\left\{\frac{1}{4}\psi\left(\frac{3}{4}\right)[(4n+3)-\lambda]\right\}}{\Gamma\left(\frac{3}{4}-\frac{\lambda}{4}\right)\Gamma(n+1)[\lambda-(4n+3)]}\,,$$

or

$$\tilde{f}(\lambda) = \sum_{n=0}^{\infty} \tilde{f}(n)\,\frac{\sin\pi(\lambda-n)}{\pi(\lambda-n)}\,, \tag{4.5.3}$$

which is formula (2.4a) of [15] for $f(2\lambda) = 2^{\lambda}\Gamma(1+\lambda)\tilde{f}(\lambda)$, where

$$\tilde{f}(\lambda) = \frac{f(4\lambda+3)\,e^{\lambda\psi(3/4)}}{\Gamma(1+\lambda)}\,.$$

Similarly, when $\alpha = -1/2$, equations (4.4.14) and (4.4.15) become: if

$$f(\lambda) = 2^{(1-\lambda)/4} \int_0^\infty F(x) D_{\lambda/2-1/2}(\sqrt{2}\,x)\,dx\,, \quad F(x) \in L^2(0,\infty)\,,$$

then

$$f(\lambda) = \sum_{n=0}^{\infty} f(4n+1)\,\frac{4(-1)^{n+1}\exp\left\{\frac{1}{4}\psi\left(\frac{1}{4}\right)[(4n+1)-\lambda]\right\}}{\Gamma\left(\frac{1}{4}-\frac{\lambda}{4}\right)\Gamma(n+1)[\lambda-(4n+1)]}\,,$$

which can be reduced to

$$\tilde{f}(\lambda) = \sum_{n=0}^{\infty} \tilde{f}(n) \frac{\sin \pi(\lambda - n)}{\pi(\lambda - n)}. \tag{4.5.4}$$

This is formula (2.4b) of [15] for $f(2\lambda + 1) = 2^{\lambda}\Gamma(1 + \lambda)\tilde{f}(\lambda)$, where

$$\tilde{f}(\lambda) = \frac{f(4\lambda + 1) e^{\lambda \psi(1/4)}}{\Gamma(1 + \lambda)}.$$

The integral in (4.5.2) was called the parabolic cylindrical transform by Jerri [15], and the generalized Hermite transform by Glaeske [8]. But for the sake of consistency with the notation used in the previous examples we will call it the continuous Hermite transform.

Apparently, Jerri did not realize that his sampling functions $S_n(\lambda)$ in formulae (2.4a) and (2.4b) are essentially the sampling functions appearing in the WSK sampling theorem, i.e.,

$$S_n(\lambda) = \sin \pi(\lambda - n)/\pi(\lambda - n), \quad n = 0, 1, 2, \ldots,$$

as shown in (4.5.3) and (4.5.4).

This completes our study of sampling theorems associated with Sturm-Liouville boundary-value problems. In the following chapter, we shall consider sampling theorems associated with more general types of boundary-value problems. But before we leave this chapter, we would like to conclude it with an open question.

4.6 Open Question

The one-sided cardinal series associated with linear forms of eigenvalues that was considered in Section 4.4.D has led us to ask if, by using the different techniques introduced in this chapter, we can derive a two-sided cardinal series of the form

$$f(\lambda) = \sum_{n=0}^{\infty} f(an + b) \frac{\sin \pi(\lambda - n)}{\pi(\lambda - n)} + \sum_{n=-\infty}^{-1} f(an + c) \frac{\sin \pi(\lambda - n)}{\pi(\lambda - n)},$$

with $b \neq c$, $a \neq 0$.

This question is intimately related to the question of whether there is a boundary-value problem (most likely associated with first order differential operator) whose eigenvalues are $\{an + b\}_{n=0}^{\infty} \cup \{an + c\}_{n=-1}^{-\infty}$ (cf. SIAM Rev., 34 (1992), problem 92-1).

In a private communication, Professor C. C. Grosjean (State University of Gent, Belgium), derived for some appropriate entire function f, the following two-sided sampling series

$$f(\lambda) = \frac{-a}{\Gamma\left(\frac{b-\lambda}{a}\right)\Gamma\left(\frac{\lambda-c}{a}+1\right)}\left\{\sum_{n=0}^{\infty} f(an+b)\frac{(-1)^n\,\Gamma\left(n+1+\frac{b-c}{a}\right)}{n!\,(\lambda-an-b)}\right.$$

$$\left.+\sum_{n=-\infty}^{-1} f(an+c)\frac{(-1)^n\,\Gamma\left(-n+\frac{b-c}{a}\right)}{(-n-1)!\,(\lambda-an-c)}\right\},$$

which reduces to

$$f(\lambda) = \sum_{n=-\infty}^{\infty} f(an+b)\frac{\sin\pi\left(\frac{\lambda-b}{a}-n\right)}{\pi\left(\frac{\lambda-b}{a}-n\right)}$$

when $b = c$.

His derivation is independent of the work presented in this chapter, and therefore this leaves the question of whether there exists a boundary-value problem, with the above prescribed eigenvalues, unanswered.

The derivation of the above sampling series uses properties of the Γ-function and since it is known that the Γ-function does not satisfy a differential equation of finite order, but rather satisfies a difference equation, Professor Grosjean has suggested investigating sampling theorems associated with difference equations as well. We believe that his suggestion is worthwhile and may open a new avenue for research.

References

1. P. Butzer, R. Stens, and M. Wehrens, The continuous Legendre transform, its inverse transform and applications, *Internat. J. Math. Math. Sci.*, 3 (1980), 47-67.

2. L. Campbell, A comparison of the sampling theorems of Kramer and Whittaker, *Soc. Indust. Appl. Math.*, 12 (1964), 117-130.

3. E. Coddington and N. Levinson, *Theory of Ordinary Differential Equations*, McGraw-Hill, New York (1955).

4. E. Deeba and E. Koh, The second continuous Jacobi transform, *Internat. J. Math. Math. Sci.*, 8 (1985), 345-354.

5. ______, The continuous Jacobi transform, *Internat. J. Math. Math. Sci.* 6 (1983), 145-160.

6. A. Erdelyi, W. Magnus, F. Oberhettinger, and F. Tricomi, (Bateman Manuscript Project), *Higher Transcendental Functions*, Vol. I, McGraw-Hill, New York (1943).

7. I. Gradshteyn and I. Ryzhik, *Tables of Integrals, Series and Products*, Academic Press, New York (1965).

8. H. Glaeske, Operational properties of a generalized Hermite transformation, *Aequationes Math.*, 32 (1987), 155-170.

9. ______, Jenaer Beiträge zur Theorie de Integraltransformationen, Wiss. Ztschr. Friedrich-Schiller-Univ. Jena, *Naturwiss R.*, 36 jg., H.1 (1987), 31-37.

10. ______, The Laguerre transform of some elementary functions, *Z. Anal.* 3 (1984), 237-244.

11. ______, Die Laguerre-Pinney transformation, *Aequationes Math.*, 22 (1981), 73-85.

12. H. Glaeske and T. Runst, The discrete Jacobi transform of generalized functions, *Math. Nachr.*, 132 (1987), 239-251.

13. J. Higgins, Some orthogonal and complete sets of Bessel functions associated with the vibrating plate, *Math. Proc. Cambridge Philos. Soc.*, 91 (1982), 503-515.

14. ______, An interpolation series associated with the Bessel-Hankel transform, *J. London Math. Soc.*, 5 (1972), 707-714.

15. A. Jerri, A note on sampling expansion for a transform with parabolic cylindrical kernel, *Inform. Sci.*, 26 (1982), 155-158.

16. ______, The Shannon sampling theorem—its various extensions and applications: a tutorial review, *Proc. IEEE*, 65, 11 (1977), 1565-1596.

17. ______, Sampling expansion for Laguerre-L_ν^α transforms, *J. Res. Nat. Bur. Standards* Sec. B (80), (B) 3 (1976), 415-418.

18. ______, On the application of some interpolating functions in physics, *J. Res. Nat. Bur. Standards* Sec. B (80), (B) 3 (1969), 241-245.

19. T. Koornwinder and G. Walter, The finite continuous Jacobi transform and its inverse, *J. Approx. Theory*, 60 (1990), 83-100.

20. H. Kramer, A generalized sampling theorem, *J. Math. Phys.*, 38 (1959), 68-72.

21. B. Levin, *Distribution of zeros of entire functions*, Translations of Mathematical Monographs, Vol. 5, American Mathematical Society, Providence, RI (1964).

22. B. Levitan and I. Sargsjan, *Introduction to Spectral Theory: Self-Adjoint Ordinary Differential Operators*, Translations of Mathematical Monographs, Vol. 39, American Mathematical Society, Providence, RI (1975).

23. F. Mehta, A general sampling expansion, *Inform. Sci.*, 16 (1978), 41-46.

24. W. Rudin, *Real and Complex Analysis*, 2nd ed., McGraw-Hill, New York (1974).

25. S. D. Selvaratnam, Shannon-Whittaker sampling theorems, Ph.D. thesis, University of Wisconsin, Milwaukee, WI (1987).

26. G. Szegö, *Orthogonal Polynomials*, American Mathematical Society Colloquium Publications, Vol. 23, American Mathematical Society, Providence, RI (1939).

27. E. Titchmarch, *Eigenfunction Expansions Associated with Second-Order Differential Equations*, Part I, 2nd ed., Clarendon Press, Oxford (1962).

28. ______, *The Theory of Functions*, Oxford University Press, Oxford (1978).

29. G. Walter, A finite continuous Gegenbauer transform and its inverse, *SIAM J. Appl. Math.*, 48 (1988), 680-688.

30. G. Walter and A. Zayed, The continuous (α, β)-Jacobi transform and its inverse when $\alpha + \beta + 1$ is a positive integer, *Trans. Amer. Math. Soc.*, 305 (1988), 653-664.

31. ______, On the singularities of continuous Legendre transforms, *Proc. Amer. Math. Soc.*, 97 (1986), 673-681.

32. G. Watson, *A Treatise on the Theory of Bessel Functions*, 2nd ed., Cambridge University Press, Cambridge (1962).

33. J. Whittaker, *Interpolatory Function Theory*, Cambridge University Press, Cambridge (1935).

34. A. Zayed, Sampling expansion for the continuous Bessel transform, *J. Appl. Anal.*, 27 (1988), 47-64.

35. ______, On the singularities of the continuous Jacobi transform when $\alpha + \beta = 0$, *Proc. Amer. Math. Soc.*, 101 (1987), 67-75.

36. ______, A generalized inversion formula for the continuous Jacobi transform, *Internat. J. Math. Math. Sci.*, 10 (1987), 671-692.

37. A. Zayed and E. Deeba, to appear in the *Journal of Applicable Analysis.*

38. A. Zayed, G. Hinsen, and P. Butzer, On Lagrange interpolation and Kramer-type sampling theorems associated with Sturm-Liouville problems, *SIAM J. Appl. Math.*, 50 (1990), 893-909.

5

SAMPLING THEOREMS ASSOCIATED WITH SELF-ADJOINT BOUNDARY-VALUE PROBLEMS

5.0 Introduction

In Chapter 4 we discussed sampling theorems associated with regular and singular Sturm-Liouville boundary-value problems and showed that if a Weiss-Kramer-type sampling theorem exists for such problems, then the associated sampling series is a Lagrange-type interpolation series. We also know that although almost all the known examples associated with Kramer's theorem arise from Sturm-Liouville problems, the theorem is true for any regular self-adjoint boundary-value problem whose eigenfunctions are all generated by one single function, when the eigenvalue parameter λ is replaced by the eigenvalues.

In this chapter we shall discuss, with more details than Kramer did, general conditions under which Kramer's sampling theorem will hold for regular self-adjoint boundary-value problems. We shall neither restrict the order of the differential operator to 2 nor require the boundary-conditions be of separate types. It will be shown that these two restrictions on the order of the differential operator and the type of the boundary-conditions are not, by any means, essential and can be eliminated; however, what is really essential here is the simplicity of the eigenvalues. We shall show, further, that analogous to the Sturm-Liouville case, Kramer's sampling series associated with a certain class of regular self-adjoint boundary-value problems is just a Lagrange-type interpolation series. This lends credence to Campbell's conjecture on the equivalence of the Whittaker-Shannon-Kotel'nikov (WSK) and Kramer sampling theorems when the latter arises from regular self-adjoint boundary-value problems with nth ($n > 2$) order differential operator.

5.1 Preliminaries*

In this section we introduce some of the notations and relations that will be used in the sequel; then we prove some useful lemmas and a theorem.

Consider the regular differential expression l defined by

$$l(y) = P_0(x)y^{(m)} + P_1(x)y^{(m-1)} + \ldots + P_m(x)y \tag{5.1.1}$$

in the interval $I = [a, b]$, $-\infty < a < b < \infty$, where $1/P_0, P_1, \ldots, P_m$, are assumed to be real-valued and summable on I. It is known [2] that l is self-adjoint if and only if m is even and l can be put in the form

$$l(y) = \sum_{k=0}^{n} (-1)^k (P_{n-k}(x) y^{(k)})^{(k)}, \tag{5.1.2}$$

where $m = 2n$. For this to make sense, we assume that P_k has continuous derivatives up to the order $(n-k)$ inclusive on I, $k = 0, \ldots, n$. We define the quasi-derivatives of a function y by the formulae (cf. [2])

$$y^{[k]} = \frac{d^k y}{dx^k} \quad \text{for} \quad k = 0, 1, \ldots, n-1;$$

$$y^{[n]} = P_0 \frac{d^n y}{dx^n}$$

$$y^{[n+k]} = P_k \frac{d^{n-k} y}{dx^{n-k}} - \frac{d}{dx}(y^{[n+k-1]}), \quad \text{for} \quad k = 1, 2, \ldots, n.$$

Hence, it easily follows that

$$l(y) = y^{[2n]}.$$

Let y and z be two functions for which the expression l makes sense. Then, by integration by parts, we obtain Lagrange's identity ([1, 2])

$$\int_a^b z l(y) dx - \int_a^b y l(z) dx = [y, z]_{x=b} - [y, z]_{x=a}, \tag{5.1.3}$$

where

$$[y, z] = \sum_{k=1}^{n} \{ y^{[k-1]} z^{[2n-k]} - y^{[2n-k]} z^{[k-1]} \}. \tag{5.1.4}$$

*Some material presented in Sections 5.1 and 5.2 is based on the author's article that appeared in *J. Math. Anal. Appls.*, 158 (1991), and is printed with permission from Academic Press.

For $y \in C^{(2n-1)}(I)$, let

$$U_j(y) = \sum_{k=1}^{2n} (\alpha_{j,k} y^{[k-1]}(a) + \beta_{j,k} y^{[k-1]}(b)), \quad j = 1,2,\ldots,2n\,, \tag{5.1.5}$$

and assume that $U_j (j = 1,2,\ldots,2n)$ are linearly independent linear forms in $y^{[k-1]}(a)$ and $y^{[k-1]}(b)$, $k = 1,\ldots,2n$. Now consider the boundary-value problem:

$$l(y) = \lambda y\,, \quad a \leq x \leq b \tag{5.1.6}$$

$$U_j(y) = \sum_{k=1}^{2n} (\alpha_{j,k} y^{[k-1]}(a) + \beta_{j,k} y^{[k-1]}(b)) = 0\,, \quad j = 1,2,\ldots,2n\,, \tag{5.1.7}$$

where l is given by (5.1.2).

Since this problem always has the trivial solution $y = 0$, we shall only be interested in the nontrivial ones.

It is shown [3, p. 77] that this problem is self-adjoint if and only if

$$\sum_{\nu=1}^{n} (\alpha_{j,\nu}\alpha_{k,2n-\nu+1} - \alpha_{j,2n-\nu+1}\alpha_{k,\nu}) = \sum_{\nu=1}^{n} (\beta_{j,\nu}\beta_{k,2n-\nu+1} - \beta_{j,2n-\nu+1}\beta_{k,\nu})\,. \tag{5.1.8}$$

From now on we shall only consider such a problem.

In particular, for any two solutions y, z of (5.1.6), (5.1.7), we have

$$[y,z]_a = [y,z]_b\,. \tag{5.1.9}$$

It is known (see [1 and 2]) that problem (5.1.6), (5.1.7) has a discrete set of real eigenvalues $\{\lambda_k\}_{k=0}^{\infty}$ (may be void) and these eigenvalues can have no finite limit-point. If we denote the corresponding eigenfunctions by $\{\phi_k(x)\}_{k=0}^{\infty}$, then

$$\int_a^b \phi_k(x)\phi_n(x)\,dx = 0 \quad \text{if} \quad k \neq n\,. \tag{5.1.10}$$

Let $A = (\alpha_{j,k})$, $B = (\beta_{j,k})$; $j,k = 1,2,\ldots,2n$, and

$$Y(x) = \begin{pmatrix} y^{[0]}(x) \\ y^{[1]}(x) \\ \cdot \\ \cdot \\ \cdot \\ y^{[2n-1]}(x) \end{pmatrix}.$$

Then the boundary conditions (5.1.7) can be written in the form

$$AY(a)+BY(b)=0, \tag{5.1.11}$$

and hence can also be viewed as a system of $2n$ linear homogeneous equations in $4n$ variables $y^{[k-1]}(a)$, $y^{[k-1]}(b)$, $k=1,2,\ldots,2n$. Since these boundary conditions are assumed to be linearly independent, the rank of the coefficient matrix $(A\,|\,B)$ is $2n$ and; hence, $2n$ of these variables can be expressed in terms of the others. This will be stated more explicitly in the following lemma.

Lemma 5.1.1. If A is nonsingular, then

$$y^{[k-1]}(a)=\sum_{j=1}^{2n}\gamma_{k,j}y^{[j-1]}(b), \quad k=1,2,\ldots,2n \tag{5.1.12}$$

for some constants $\gamma_{j,k}$; $j,k=1,2,\ldots,2n$ satisfying

$$\sum_{\nu=1}^{n}(\gamma_{j,\nu}\gamma_{k,2n-\nu+1}-\gamma_{j,2n-\nu+1}\gamma_{k,\nu})=\begin{cases}1 & \text{if} \quad j+k=2n+1\\ 0 & \text{if} \quad j+k\neq 2n+1.\end{cases} \tag{5.1.13}$$

Similarly, if B is nonsingular, then $y^{[k-1]}(b)$, $k=1,2,\ldots,2n$, can be expressed in terms of $y^{[j-1]}(a)$, $j=1,\ldots,2n$, as in (5.1.12) and (5.1.13). Moreover,

$$\sum_{k=1}^{n}(\gamma_{k,j}\gamma_{2n-k+1,i}-\gamma_{2n-k+1,j}\gamma_{k,i})=\begin{cases}1 & \text{if} \quad i+j=2n+1\\ 0 & \text{if} \quad i+j\neq 2n+1.\end{cases} \tag{5.1.14}$$

Proof. Equation (5.1.12) follows immediately from (5.1.11) with $\gamma_{j,k}=-\Sigma_{\nu=1}^{2n}(a^{-1})_{j,\nu}(b)_{\nu,k}$. To show (5.1.13), we first observe that if $W_j(y)=0$, $j=1,\ldots,2n$, are boundary conditions of the form (5.1.7) which are self-adjoint, i.e., they satisfy (5.1.8) and if $V_k(y)=0$, $k=1,\ldots,2n$, are boundary conditions obtained from $W_j(y)=0$ by elementary operations, then we can easily verify that $V_k(y)=0$ are also self-adjoint. Clearly, (5.1.12) can also be obtained from (5.1.7) by elementary operations, hence (5.1.12) defines $2n$ self-adjoint boundary-conditions. Equation (5.1.13) now follows from (5.1.8) by setting

$$\alpha_{j,\nu}=\delta_{j,\nu}=\begin{cases}1 & \text{if} \quad j=\nu\\ 0 & \text{if} \quad j\neq\nu\end{cases} \quad \text{and} \quad \beta_{j,\nu}=-\gamma_{j,\nu}.$$

Since (5.1.12) defines self-adjoint boundary-conditions, it follows that for any $y(x)$ and $z(x)$ satisfying (5.1.12), we have (cf. (5.1.9))

$$[y,z]_{x=a}=[y,z]_{x=b}.$$

But from (5.1.4) and (5.1.12) we have

$$\begin{aligned}
[y,z]_{x=a} &= \sum_{k=1}^{n}[y^{[k-1]}(a)z^{[2n-k]}(a) - y^{[2n-k]}(a)z^{[k-1]}(a)] \\
&= \sum_{k=1}^{n}\begin{vmatrix} y^{[k-1]}(a) & y^{[2n-k]}(a) \\ z^{[k-1]}(a) & z^{[2n-k]}(a)\end{vmatrix} \\
&= \sum_{k=1}^{n}\sum_{j=1}^{2n}\sum_{i=1}^{2n}\gamma_{k,j}\gamma_{2n-k+1,i}\begin{vmatrix} y^{[j-1]}(b) & y^{[i-1]}(b) \\ z^{[j-1]}(b) & z^{[i-1]}(b)\end{vmatrix} \\
&= \sum_{k=1}^{n}\sum_{j=1}^{2n}\sum_{i=j}^{2n}(\gamma_{k,j}\gamma_{2n-k+1,i} - \gamma_{k,i}\gamma_{2n-k+1,j}) \\
&\qquad \times\begin{vmatrix} y^{[j-1]}(b) & y^{[i-1]}(b) \\ z^{[j-1]}(b) & z^{[i-1]}(b)\end{vmatrix} \\
&= \sum_{j=1}^{2n}\sum_{i=j}^{2n}\zeta_{j,i}\begin{vmatrix} y^{[j-1]}(b) & y^{[i-1]}(b) \\ z^{[j-1]}(b) & z^{[i-1]}(b)\end{vmatrix} \\
&= \sum_{j=1}^{n}\zeta_{j,2n+1-j}\begin{vmatrix} y^{[j-1]}(b) & y^{[2n-j]}(b) \\ z^{[j-1]}(b) & z^{[2n-j]}(b)\end{vmatrix} \\
&\qquad +\left(\sum_{j=1}^{n}\sum_{\substack{i=j \\ i\neq 2n+1-j}}^{2n} + \sum_{j=n+1}^{2n}\sum_{i=j}^{2n}\right)\zeta_{j,i}\begin{vmatrix} y^{[j-1]}(b) & y^{[i-1]}(b) \\ z^{[j-1]}(b) & z^{[i-1]}(b)\end{vmatrix},
\end{aligned}$$

where

$$\zeta_{j,i} = \sum_{k=1}^{n}(\gamma_{k,j}\gamma_{2n-k+1,i} - \gamma_{k,i}\gamma_{2n-k+1,j}). \tag{5.1.15}$$

Since

$$[y,z]_{x=b} = \sum_{j=1}^{n}\begin{vmatrix} y^{[j-1]}(b) & y^{[2n-j]}(b) \\ z^{[j-1]}(b) & z^{[2n-j]}(b)\end{vmatrix}$$

then for (5.1.9) to hold we must have

$$\zeta_{j,i} = \begin{cases} 1 & \text{if} \quad i+j = 2n+1,\ j = 1,\dots,n \\ 0 & \text{otherwise}, \end{cases}$$

which is (5.1.14). ▮

Remarks.

i) If a system $V = \{V_j(y) = 0; j = 1,\ldots,2n\}$ of boundary conditions is obtained from another system $U = \{U_j(y) = 0; j = 1,\ldots,2n\}$ by elementary operations, we shall say that V is equivalent to U. If U is of the form (5.1.7), then it is easy to see that the coefficient matrix corresponding to V can be obtained from the coefficient matrix corresponding to U by elementary row operations.
ii) The coefficient matrix corresponding to (5.1.12), which is $(I_{2n} \mid -\Gamma)$, where $\Gamma = (\gamma_{j,k})$, can obviously be obtained from $(A \mid B)$ by elementary row operations.
iii) If Γ is symmetric, then (5.1.14) is a consequence of (5.1.13).

We have just seen that if A is nonsingular, then $y^{[k-1]}(a)$, $k = 1,\ldots,2n$, can be expressed in terms of $y^{[j-1]}(b)$, $j = 1,\ldots,2n$, as described in Lemma 5.1.1 and similarly if B is nonsingular, then $y^{[j-1]}(b)$ can be expressed in terms of $y^{[k-1]}(a)$. However, if both A and B are singular, then in view of the fact that the rank of the matrix $(A \mid B)$ is $2n$, it follows that $2n$ of the variables $y^{[k-1]}(a)$, $y^{[k-1]}(b)$, $k = 1,\ldots,2n$, can be expressed in terms of the remaining ones as described in Lemma 5.1.1. This case is essentially the same as in the case where A is nonsingular except that the notation is more complicated. Therefore, from now on, we shall always assume, without loss of generality, that A is nonsingular.

Consider the self-adjoint boundary-value problem (5.1.6), (5.1.7) which we restate for convenience

$$l(y) = \lambda y\,, \quad a \le x \le b\,, \tag{5.1.6}$$

$$U_j(y) = 0\,, \quad j = 1,\ldots,2n\,, \tag{5.1.7}$$

where $l(y)$ is given by (5.1.2) and $U_j(y)$ satisfies (5.1.8).

Let $y_1(x,\lambda), \ldots, y_{2n}(x,\lambda)$ be the fundamental system of solutions of (5.1.6) which satisfy the initial conditions

$$y_j^{(k-1)}(a,\lambda) = \delta_{j,k}\,; \quad j,k = 1,\ldots,2n\,.$$

Thus, $y(x,\lambda) = \Sigma_{i=1}^{2n} C_i y_i(x,\lambda)$ is the general solution of (5.1.6). For y to satisfy (5.1.7), we must have

$$\sum_{i=1}^{2n} C_i U_j(y_i) = 0\,, \quad j = 1,\ldots,2n\,,$$

which is a homogeneous system of $2n$ linear equations in the $2n$ unknowns C_i. To have a nontrivial solution, the rank of the matrix $(U_j(y_i))_{i,j=1}^{2n}$ must be less than $2n$. The determinant of this matrix is a function in λ and independent of x; hence we may set $W(\lambda) = \det(U_j(y_i))$. It is known [2, pp. 14,15] that $W(\lambda)$ is an entire function in λ which is not identically zero and whose zeros are the eigenvalues of problem (5.1.6), (5.1.7). It is also known that if $\tilde{\lambda}$ is a simple zero of $W(\lambda)$, then the rank of the matrix $(U_j(y_i))$ is $2n-1$.

Definition 5.1.1. The boundary-value problem (5.1.6) and (5.1.7) is said to be one dimensional if there exists a $(2n-1)\times(2n-1)$ submatrix $A(\lambda)$ of $W(\lambda) = (U_i(y_j))$; $i,j = 1,2,\ldots,2n$, with the property that $\det(A(\lambda)) \neq 0$.

If the boundary-value problem (5.1.6), (5.1.7) is one dimensional, then all its eigenvalues are simple. For, if λ^* is an eigenvalue with multiplicity r, $r>1$, then $W(\lambda^*)$ has rank $2n-r<2n-1$; hence $\det(A)=0$ for any $(2n-1)\times(2n-1)$ submatrix A of $W(\lambda)$, which is a contradiction.

Lemma 5.1.2. Let the boundary-value problem (5.1.6) and (5.1.7) be one dimensional. Then, one can choose $2n-1$ boundary conditions from (5.1.7) so that (5.1.6), together with these $2n-1$ boundary conditions, has a unique solution (up to a multiplicative constant) which is an entire function in λ.

Proof. To simplify the notation, we may assume, without loss of generality, that

$$A(\lambda) = \begin{pmatrix} U_1(y_1) & \cdots & U_1(y_{2n-1}) \\ \cdot & \cdot & \cdot \\ \cdot & \cdot & \cdot \\ \cdot & \cdot & \cdot \\ U_{2n-1}(y_1) & \cdots & U_{2n-1}(y_{2n-1}) \end{pmatrix}.$$

This can always be done by reordering the rows and columns of $W(\lambda)$, i.e., by re-enumerating the boundary conditions (5.1.7) and the fundamental solutions $y_j(x,\lambda)$, $j = 1,2,\ldots,2n$.

Now consider the boundary-value problem

$$l(y) = \lambda y\,, \quad a \le x \le b$$

$$U_i(y) = 0\,, \quad i = 1,2,\ldots,2n-1\,.$$

Let $y_j(x,\lambda)$, $j = 1,2,\ldots,2n$, be the fundamental system of solutions of (5.1.6) described above. Then, for $y(x,\lambda) = \Sigma_{j=1}^{2n} C_j y_j(x,\lambda)$ to be a solution of this problem, we must have

$$\sum_{j=1}^{2n} C_j U_i(y_j) = 0\,; \quad i = 1,2,\ldots,2n-1\,. \tag{5.1.16}$$

This is a system of $2n-1$ homogeneous linear equations in the $2n$ unknowns C_j. The coefficient matrix of this system is

$$V(\lambda) = \begin{pmatrix} U_1(y_1) & \cdots & U_1(y_{2n}) \\ \cdot & \cdot & \cdot \\ \cdot & \cdot & \cdot \\ \cdot & \cdot & \cdot \\ U_{2n-1}(y_1) & \cdots & U_{2n-1}(y_{2n}) \end{pmatrix},$$

which has rank equal to $2n-1$ since it contains a $(2n-1)\times(2n-1)$ submatrix $A(\lambda)$ with the property that $\det(A(\lambda)) \neq 0$. Therefore, (5.1.16) has a unique solution (up to a multiplicative constant), which is easily seen to be

$$C_j = -C_{2n}\,\frac{\det(A_j(\lambda))}{\det(A(\lambda))}, \quad j = 1,2,\ldots,2n-1,$$

where $A_j(\lambda)$ is the matrix obtained by replacing the jth column of $A(\lambda)$ by

$$\begin{pmatrix} U_1(y_{2n}) \\ \cdot \\ \cdot \\ \cdot \\ U_{2n-1}(y_{2n}) \end{pmatrix}.$$

Since each $y_j(x,\lambda)$ is an entire function in λ [2, p. 13], then so is $\det(A(\lambda))$. And in view of the fact that $\det(A(\lambda)) \neq 0$, it follows that C_j, $j = 1,2,\ldots,2n-1$, is also an entire function in λ; hence, so is $y(x,\lambda)$. ∎

Hereafter, we assume that problem (5.1.6), (5.1.7) is one dimensional. Now consider the problem

$$l(y) = \lambda y\,, \quad a \le x \le b\,, \tag{5.1.6}$$

$$U_j(y) = 0\,, \quad j = 1,\ldots,2n-1\,, \tag{5.1.17}$$

and let $\phi(x,\lambda)$ be the unique solution of (5.1.6) and (5.1.17) determined by Lemma 5.1.2. Let us define $\omega(\lambda)$ by

$$\omega(\lambda) = U_{2n}(\phi) = \sum_{k=1}^{2n} (\alpha_{2n,k}\phi^{[k-1]}(a,\lambda) + \beta_{2n,k}\phi^{[k-1]}(b,\lambda))\,. \tag{5.1.18}$$

Clearly, $\omega(\lambda)$ is an entire function in λ since both $\phi^{[k-1]}(a,\lambda)$ and $\phi^{[k-1]}(b,\lambda)$ are, and $\tilde{\lambda}$ is a zero of $\omega(\lambda)$ if and only if $\tilde{\lambda}$ is an eigenvalue of (5.1.6) and (5.1.7).

The next theorem, which will be used in the proof of the main theorem in Section 2, is also interesting in its own right.

THEOREM 5.1

Let $\phi(x,\lambda)$ be defined as above. Then

$$(\lambda-\lambda')\int_a^b \phi(x,\lambda)\phi(x,\lambda')\,dx = \alpha(\lambda')\omega(\lambda)-\alpha(\lambda)\omega(\lambda'), \qquad (5.1.19)$$

where $\alpha(\lambda)$ is an entire function in λ with no common zeros with $\omega(\lambda)$.

Proof. From Lagrange's identity (5.1.3), we have for $y=\phi(x,\lambda)$ and $z=\phi(x,\lambda')$

$$(\lambda-\lambda')\int_a^b \phi(x,\lambda)\phi(x,\lambda')\,dx$$

$$=[\phi(x,\lambda),\phi(x,\lambda')]_{x=b}-[\phi(x,\lambda),\phi(x,\lambda')]_{x=a}, \qquad (5.1.20)$$

where $[\phi(x,\lambda),\phi(x,\lambda')]$ is defined by (5.1.4). Note that $\phi(x,\lambda)$ is a solution of the system

$$U_j(\phi(x,\lambda))=0, \quad j=1,\ldots,2n-1$$

$$U_{2n}(\phi(x,\lambda))=\omega(\lambda),$$

which as in (5.1.11) can be written in the matrix form

$$A\Phi(a,\lambda)+B\Phi(b,\lambda)=\Omega(\lambda),$$

where

$$\Phi(x,\lambda)=\begin{pmatrix}\phi^{[0]}(x,\lambda)\\ \phi^{[1]}(x,\lambda)\\ \cdot\\ \cdot\\ \cdot\\ \phi^{[2n-1]}(x,\lambda)\end{pmatrix} \quad \text{and} \quad \Omega(\lambda)=\begin{pmatrix}0\\0\\ \cdot\\ \cdot\\ \cdot\\ \omega(\lambda)\end{pmatrix}.$$

Since A is nonsingular, we have, as in Lemma 5.1.1,

$$\phi^{[k-1]}(a,\lambda)=\sum_{j=1}^{2n}\gamma_{k,j}\phi^{[j-1]}(b,\lambda)+\eta_{k,2n}\omega(\lambda), \qquad (5.1.21)$$

where $\gamma_{k,j}$ are given in Lemma 5.1.1 and $\eta_{j,k} = (A^{-1})_{j,k}$. Thus, as in the proof of Lemma 5.1.1, we obtain by using (5.1.21)

$$[\phi(x,\lambda),\phi(x,\lambda')]_{x=a} = \sum_{k=1}^{n} \begin{vmatrix} \phi^{[k-1]}(a,\lambda) & \phi^{[2n-k]}(a,\lambda) \\ \phi^{[k-1]}(a,\lambda') & \phi^{[2n-k]}(a,\lambda') \end{vmatrix}$$

$$= \sum_{j=1}^{2n} \sum_{i=j}^{2n} \zeta_{j,i} \begin{vmatrix} \phi^{[j-1]}(b,\lambda) & \phi^{[i-1]}(b,\lambda) \\ \phi^{[j-1]}(b,\lambda') & \phi^{[i-1]}(b,\lambda') \end{vmatrix}$$

$$+\alpha(\lambda)\,\omega(\lambda') - \alpha(\lambda')\,\omega(\lambda),$$

where

$$\alpha(\lambda) = \sum_{k=1}^{n} \sum_{i=1}^{2n} (\eta_{2n-k+1,2n}\gamma_{k,i} - \eta_{k,2n}\gamma_{2n-k+1,i})\,\phi^{[i-1]}(b,\lambda)$$

and $\zeta_{j,i}$ are given by (5.1.15). In view of (5.1.14)

$$[\phi(x,\lambda),\phi(x,\lambda')]_{x=a} = [\phi(x,\lambda),\phi(x,\lambda')]_{x=b} + \alpha(\lambda)\omega(\lambda') - \alpha(\lambda')\omega(\lambda),$$

which, when combined with (5.1.20), yields (5.1.19).

Clearly, $\alpha(\lambda)$ is an entire function in λ since $\phi^{[i-1]}(b,\lambda)$, $i = 1,\ldots,2n$, are. To show that $\alpha(\lambda)$ has no common zeros with $\omega(\lambda)$, let us recall that $\tilde{\lambda}$ is a zero of $\omega(\lambda)$ if and only if $\tilde{\lambda}$ is an eigenvalue of problem (5.1.6), (5.1.7) and that all the zeros of $\omega(\lambda)$ are simple. Let us denote these eigenvalues by $\{\lambda_n\}_{n=0}^{\infty}$ and put $\lambda' = \lambda_n$ in (5.1.19), then

$$(\lambda - \lambda_n) \int_a^b \phi(x,\lambda)\,\phi_n(x)\,dx = \alpha(\lambda_n)\,\omega(\lambda), \tag{5.1.22}$$

where $\phi_n(x) = \phi(x,\lambda_n)$ is the corresponding eigenfunction. By dividing both sides of (5.1.22) by $(\lambda - \lambda_n)$ and taking the limit as $\lambda \to \lambda_n$, we obtain

$$\|\phi_n\|^2 = \int_a^b |\phi_n(x)|^2\,dx = \alpha(\lambda_n)\,\omega'(\lambda_n).$$

If $\alpha(\lambda_n) = 0$, then $\phi_n(x)$ is zero almost everywhere, which is a contradiction. Hence, $\alpha(\lambda)$ cannot have a common zero with $\omega(\lambda)$. ∎

5.2 The Sampling Theorem

In this section we prove our sampling theorem which shows that Kramer's sampling expansions associated with one-dimensional nth order self-adjoint boundary-value problems are nothing more than Lagrange-type interpolation expansions. This theorem generalizes Theorem 4.1 (cf. [4]) and Theorem 1 in [5] not only to the case where the differential operator l has order ≥ 2, but also to the case where the boundary conditions are of mixed types. The role of the Wronskian in Theorem 4.1 will now be played by the function $\omega(\lambda)$ introduced in Section 5.1.

THEOREM 5.2

Consider the boundary-value problem (5.1.6), (5.1.7) and let $\phi(x,\lambda)$ be as defined in Section 5.1. If

$$f(\lambda) = \int_a^b F(x)\phi(x,\lambda)\,dx \tag{5.2.1}$$

for some $F(x) \in L^2(a,b)$, then $f(\lambda)$ is an entire function that admits the sampling representation

$$f(\lambda) = \sum_{n=0}^{\infty} f(\lambda_n) \frac{\omega(\lambda)}{(\lambda-\lambda_n)\omega'(\lambda_n)}, \tag{5.2.2}$$

where $\{\lambda_n\}_{n=0}^{\infty}$ are the eigenvalues of problem (5.1.6), (5.1.7) and $\omega(\lambda)$ is the entire function defined by (5.1.18) which, without loss of generality, may be written in the form

$$\omega(\lambda) = \begin{cases} \prod_{n=0}^{\infty}\left(1-\dfrac{\lambda}{\lambda_n}\right) & \text{if none of the eigenvalues is zero,} \\ \lambda\prod_{n=1}^{\infty}\left(1-\dfrac{\lambda}{\lambda_n}\right) & \text{if one of the eigenvalues, say } \lambda_0, \text{ is zero.} \end{cases}$$

Proof. First, since $\omega(\lambda)$ is an entire function whose only zeros are the λ_n's and they are all simple, it follows from Hadamard's factorization theorem for entire functions that $\omega(\lambda) = H(\lambda)G(\lambda)$, where

$$G(\lambda) = \begin{cases} \prod_{n=0}^{\infty}\left(1-\dfrac{\lambda}{\lambda_n}\right) & \text{if none of the eigenvalues is zero,} \\ \lambda\prod_{n=1}^{\infty}\left(1-\dfrac{\lambda}{\lambda_n}\right) & \text{if one of the eigenvalues, say } \lambda_0, \text{ is zero,} \end{cases}$$

and $H(\lambda)$ is an entire function with no zeros. Hence,

$$\frac{\omega(\lambda)}{\omega'(\lambda_n)} = \frac{H(\lambda)G(\lambda)}{H(\lambda_n)G'(\lambda_n)},$$

and (5.2.2) will now take the form

$$K(\lambda) = \sum_{n=0}^{\infty} K(\lambda_n) \frac{G(\lambda)}{(\lambda-\lambda_n)G'(\lambda_n)},$$

where $K(\lambda) = f(\lambda)/H(\lambda)$. Therefore, without loss of generality, we may assume that $\omega(\lambda) = G(\lambda)$.

The integral in (5.2.1) converges absolutely and uniformly in view of the fact that

$$\left(\int_a^b |F(x)\phi(x,\lambda)|\,dx\right)^2 \le \left(\int_a^b |F(x)|^2\,dx\right)\left(\int_a^b |\phi(x,\lambda)|^2\,dx\right) < \infty,$$

and since $\phi(x,\lambda)$ is an entire function in λ, then so is $f(\lambda)$. Since both $F(x)$ and $\phi(x,\lambda)$ are in $L^2(a,b)$, then

$$F(x) = \sum_{n=0}^{\infty} \hat{F}(n) \frac{\phi_n(x)}{\|\phi_n\|^2}, \tag{5.2.3}$$

and

$$\phi(x,\lambda) = \sum_{n=0}^{\infty} \langle\phi,\phi_n\rangle \frac{\phi_n(x)}{\|\phi_n\|^2}, \tag{5.2.4}$$

where

$$\hat{F}(n) = \int_a^b F(x)\phi_n(x)\,dx, \tag{5.2.5}$$

and

$$\langle\phi,\phi_n\rangle = \int_a^b \phi(x,\lambda)\phi_n(x)\,dx. \tag{5.2.6}$$

$\phi_n(x)$ is the eigenfunction corresponding to the eigenvalue λ_n. The two series (5.2.3) and (5.2.4) converge in the sense of $L^2(a,b)$. From (5.2.1) and Parseval's equality, it follows that

$$f(\lambda) = \sum_{n=0}^{\infty} f(\lambda_n) \frac{\langle\phi,\phi_n\rangle}{\|\phi_n\|^2}, \tag{5.2.7}$$

since $f(\lambda_n) = \hat{F}(n)$.

From (5.1.19) with $\lambda' = \lambda_n$, we obtain

$$(\lambda - \lambda_n)\langle \phi, \phi_n \rangle = \alpha(\lambda_n)\omega(\lambda), \tag{5.2.8}$$

which, upon dividing by $(\lambda - \lambda_n)$ and taking the limit as $\lambda \to \lambda_n$, yields

$$\|\phi_n\|^2 = \alpha(\lambda_n)\omega'(\lambda_n). \tag{5.2.9}$$

By combining (5.2.7), (5.2.8), and (5.2.9) we obtain (5.2.2). ∎

References

1. E. Coddington and N. Levinson, *Theory of Ordinary Differential Equations*, McGraw-Hill, New York (1955).

2. M. Naimark, *Linear Differential Operators*, Vol. I, *Elementary Theory of Linear Differential Operators*, George Harrap & Co., London, UK (1967).

3. M. Naimark, *Linear Differential Operators*, Vol. II, *Linear Differential Operators in Hilbert Space*, George Harrap & Co., London, UK (1968).

4. A. Zayed, On Kramer's sampling theorem associated with general Sturm-Liouville problems and Lagrange interpolation, *SIAM J. Appl. Math.*, 51 (1991), 575-604.

5. A. Zayed, G. Hinsen, and P. Butzer, On Lagrange interpolation and Kramer-type sampling theorems associated with Sturm-Liouville problems, *SIAM J. Appl. Math.*, 50 (1990), 893-909.

6

SAMPLING BY USING GREEN'S FUNCTION

6.0 Introduction

In Chapter 5 we obtained a Kramer-type sampling theorem for the case where the kernel function $K(x, \lambda)$ in Kramer's theorem arises from a self-adjoint boundary-value problem associated with nth order differential operator and boundary conditions of mixed types. The derivation was based on the assumption that the boundary-value problem was one dimensional, which is admittedly a rather stringent assumption.

In this chapter we shall show that Kramer's sampling theorem actually holds under less restrictive conditions than originally thought; it holds even if the boundary-value problem is *not* self-adjoint. We shall require that neither the eigenfunctions be generated from one single function as Kramer did nor that the boundary-value problem be one dimensional as we did in the previous chapter; we only require that the eigenvalues be simple poles of the Green's function of the problem. Moreover, we shall show that even under these general conditions, Kramer's sampling series is nothing more than a Lagrange-type interpolation series. This again may lend some credence to Campbell's conjecture on the equivalence of the Whittaker-Shannon-Kotel'nikov (WSK) and Kramer sampling theorems when the latter is associated with self-adjoint boundary-value problems with nth ($n > 2$) order differential operators. Nevertheless, it should be stressed that whether Campbell's conjecture is true or not is not yet known.

The technique we shall use is different from those used in [6-8] and is based on utilizing the Green's function of the boundary-value problem to construct the interpolating functions. One of the main advantages of this technique is that it allows a boundary-value problem like (3.2.11), (3.2.12) to have a Kramer-type sampling theorem associated with it; a result that was impossible to obtain under the original formulation of Kramer's sampling theorem; see Section 3.2. The idea of using Green's function in sampling theory goes back to Haddad, Yao and Thomas [2], who outlined different ways of constructing sampling series expansions. We shall show that their result is indeed a special case of ours.

6.1 Preliminaries*

Consider the differential operator

$$L = p_0(x)\frac{d^n}{dx^n} + p_1(x)\frac{d^{n-1}}{dx^{n-1}} + \dots + p_n(x), \tag{6.1.1}$$

where $p_0, p_1, \dots, p_n$ are continuous, complex-valued functions defined on a finite real interval $I = [a, b]$, with $p_0(x) \neq 0$ for any $x \in I$. The adjoint differential operator L^* of L is defined by

$$L^* = (-1)^n \left(\frac{d^n}{dx^n}\right)(\overline{p}_0(x)\,\cdot) + (-1)^{n-1}\left(\frac{d^{n-1}}{dx^{n-1}}\right)(\overline{p}_1(x)\,\cdot) \cdots + \overline{p}_n(x). \tag{6.1.2}$$

This is to be interpreted as follows: if g is any function on I such that $\overline{p}_k g (k = 0, 1, \dots, n)$ has $n - k$ derivatives on I, then

$$L^* g = (-1)^n \frac{d}{dx^n}(\overline{p}_0 g) + (-1)^{n-1}\frac{d^{n-1}}{dx^{n-1}}(\overline{p}_1 g) + \dots + \overline{p}_n g\,.$$

The equation $L^* y = 0$ is called the adjoint equation of $Ly = 0$.

From now on we shall assume that $p_k \in C^{n-k}(I)$, $k = 0, 1, \dots, n$. It can be shown that ([1, p. 86] or [4, p. 9]) if $u, v \in C^n(I)$, then

$$\int_{x_1}^{x_2} \left(\overline{v}Lu - u\overline{L^* v}\right) dx = [uv]_{x=x_2} - [uv]_{x=x_1}, \tag{6.1.3}$$

for any $x_1, x_2 \in I$, where

$$[uv] = \sum_{m=1}^{n} \sum_{\substack{j+k=m-1 \\ j \geq 0, k \geq 0}} (-1)^j u^{(k)} (p_{n-m}\overline{v})^{(j)}\,.$$

Equation (6.1.3) is known as Lagrange's formula or Lagrange's identity.

---

*Some material in Sections 6.1 and 6.2 is based on an article by the author that will appear in *J. Math. Anal. Appls.* and is printed with permission from Academic Press.

The operator L is said to be self-adjoint if $L^* = L$. It is easy to see that if the coefficient functions $p_k (k = 0, 1, \ldots, n)$ are real-valued, then L is self-adjoint if and only if L has an even order $n = 2m (m \geq 1)$ and can be written in the form

$$Ly = (p_0 y^{(m)})^{(m)} + (p_1 y^{(m-1)})^{(m-1)} + \ldots + (p_{m-1} y^{(1)})^{(1)} + p_m y \, .$$

However, we shall not assume that here.

Let

$$U_j(y) = \sum_{k=1}^{n} \alpha_{j,k} y^{(k-1)}(a) + \beta_{j,k} y^{(k-1)}(b) \, , \quad j = 1, 2, \ldots, m (m \leq 2n)$$

be linear forms in $y^{(k-1)}(a)$, $y^{(k-1)}(b)$, $k = 1, 2, \ldots, n$, and assume that they are linearly independent. This is equivalent to saying that the rank of the matrix

$$[A : B] = \begin{pmatrix} \alpha_{1,1} & \cdots\cdots & \alpha_{1,n} & \beta_{1,1} & \cdots\cdots & \beta_{1,n} \\ \vdots & & \vdots & \vdots & & \vdots \\ \vdots & & \vdots & \vdots & & \vdots \\ \alpha_{m,1} & \cdots\cdots & \alpha_{m,n} & \beta_{m,1} & \cdots\cdots & \beta_{m,n} \end{pmatrix} ,$$

is m.

To any m linearly independent forms $U_1, \ldots, U_m$, it is always possible to adjoint $2n - m$ linearly independent forms $U_{m+1}, \ldots, U_{2n}$ such that the combined system $U_1, \ldots, U_{2n}$ constitutes $2n$ linearly independent forms. This is equivalent to imbedding the matrix $[A : B]$ in a $2n \times 2n$ nonsingular matrix; see [1, p. 287].

The boundary forms $U_{m+1}, \ldots, U_{2n}$ are said to be complementary boundary forms to the forms $U_1, \ldots, U_m$. It is also known ([1, p. 288; 4, p. 9]) that given any boundary forms $U_1, \ldots, U_m$ and their complementary forms $U_{m+1}, \ldots, U_{2n}$, we can find a unique system of linearly independent boundary forms $V_1, \ldots, V_{2n}$ such that

$$\int_a^b (Ly)\bar{z} \, dx = U_1 V_{2n} + \ldots + U_m V_{2n-m+1} + U_{m+1} V_{2n-m} + \ldots$$

$$\ldots + U_{2n} V_1 + \int_a^b y(\overline{L^* z}) \, dx \, . \tag{6.1.4}$$

The boundary conditions

$$V_j(y) = 0 \, , \quad j = 1, 2, \ldots, 2n - m \, ; \quad y \in C^{n-1}(y)$$

associated with the boundary forms $V_1, \dots, V_{2n-m}$ are called the adjoint boundary conditions of the boundary conditions

$$U_j(y) = 0, \quad j = 1,2,\dots,m; \quad y \in C^{n-1}(I)$$

associated with the boundary forms $U_1, \dots, U_m$.

Now consider the boundary-value problems:

$$\Pi_m: \begin{cases} Ly = \lambda y, & x \in I \\ U_j(y) = 0, & j = 1,2,\dots,m, \end{cases} \qquad \begin{matrix} (6.1.5) \\ (6.1.6) \end{matrix}$$

and

$$\Pi^*_{2n-m}: \begin{cases} L^*y = \lambda y, & x \in I \\ V_j(y) = 0, & j = 1,2,\dots,2n-m, \end{cases} \qquad \begin{matrix} (6.1.7) \\ (6.1.8) \end{matrix}$$

where L and L^* are given by (6.1.1) and (6.1.2) respectively. The boundary-value problem (6.1.7), (6.1.8) is called the adjoint boundary-value problem of the boundary-value problem (6.1.5), (6.1.6).

It follows immediately from (6.1.4) that if y and z are two functions satisfying the boundary conditions (6.1.6) and (6.1.8) respectively, then

$$\int_a^b (Ly)\bar{z}\,dx = \int_a^b y(\overline{L^*z})\,dx,$$

or $\langle Ly, z\rangle = \langle y, L^*z\rangle$, where $\langle y, z\rangle = \int_a^b y(x)\overline{z(x)}\,dx.$

The boundary-value problem Π_m is said to be self-adjoint if L is self-adjoint and the boundary forms U_j, $j = 1,2,\dots,m$, are equivalent to their adjoints V_j, $j = 1,\dots,2n-m$. Evidently, for Π_m to be self-adjoint, it is necessary that $m = n$.

From now on we shall assume that $m = n$, and denote the boundary-value problems Π_n, Π_n^* by Π and Π^*, respectively, but we shall not assume that problem Π is self-adjoint. Since both problems Π and Π^* always have the trivial solution $y \equiv 0$, we shall be concerned only with the nontrivial solutions.

Let $y_1(x,\lambda), \dots, y_n(x,\lambda)$ denote the fundamental system of solutions of equation (6.1.5) that satisfy the initial conditions

$$y_j^{(\nu-1)}(a,\lambda) = \begin{cases} 0 & \text{for } j \neq \nu \\ 1 & \text{for } j = \nu \end{cases}, \quad j, \nu = 1,2,\dots,n. \qquad (6.1.9)$$

For $y(x,\lambda) = \sum_{i=1}^{n} c_i y_i(x,\lambda)$ to be a solution of Π, we must have

$$\sum_{i=1}^{n} c_i U_j(y_i) = 0, \quad j = 1,\ldots,n .$$

Hence, to have a nontrivial solution of Π, the rank of the matrix

$$U(\lambda) = \begin{bmatrix} U_1(y_1)\ldots\ldots & U_1(y_n) \\ \ldots & \ldots \\ U_n(y_1)\ldots\ldots & U_n(y_n) \end{bmatrix}$$

must be less than n, or equivalently $\Delta(\lambda) = \det(U(\lambda)) = 0$.

$\Delta(\lambda)$ is an entire function in λ since so are $y_1(x,\lambda),\ldots, y_n(x,\lambda)$; see [4, p. 13]. Thus, unless identically zero, $\Delta(\lambda)$ can have at most countably many zeros with no finite limit point. The zeros of $\Delta(\lambda)$ are the eigenvalues of problem Π. The set of all eigenfunctions belonging to the same eigenvalue is a finite-dimensional vector space of dimension less than or equal to n. The dimension of this space is called the multiplicity of this eigenvalue. It is known [4, p. 15] that if λ_0 is a zero of $\Delta(\lambda)$ with multiplicity m, then the multiplicity of λ_0 as an eigenvalue is less than or equal to m. In particular, if λ_0 is a simple zero of $\Delta(\lambda)$, i.e., $m = 1$, then the number of linearly independent eigenfunctions corresponding to λ_0 is one. In this case, we say that λ_0 is a simple eigenvalue. Generally speaking, the eigenvalues of Π are not necessarily simple.

It is a fact that if λ_0 is an eigenvalue of problem Π with multiplicity m, then $\bar{\lambda}_0$ is also an eigenvalue of problem Π^* with the same multiplicity. Furthermore, the eigenfunctions of problem Π corresponding to the eigenvalue λ_0 are orthogonal to the eigenfunctions of problem Π^* corresponding to the eigenvalue μ_0, provided that $\lambda_0 \neq \bar{\mu}_0$ [4, p. 21].

Finally, it is known that the Green's function $G(x,y,\lambda)$ of problem Π (or Π^*), which is a meromorphic function in λ, does not necessarily have simple poles; however, throughout the rest of this chapter we shall assume that it does.

6.2 The Sampling Theorem

Consider the boundary-value problem Π, and let $\{y_i(x,\lambda)\}_{i=1}^{n}$ be the fundamental system of solutions of (6.1.5) that satisfy (6.1.9). Let W be their Wronskian

$$W(x) = \begin{vmatrix} y_1^{(n-1)}(x) & y_2^{(n-1)}(x) & \dots & y_n^{(n-1)}(x) \\ y_1^{(n-2)}(x) & y_2^{(n-2)}(x) & \dots & y_n^{(n-2)}(x) \\ \dots & \dots & \dots & \\ y_1(x) & y_2(x) & \dots & y_n(x) \end{vmatrix}. \tag{6.2.1}$$

Set

$$g(x,\xi) = \pm\frac{1}{W(\xi)}\begin{vmatrix} y_1(x) & y_2(x) & \dots & y_n(x) \\ y_1^{(n-2)}(\xi) & y_2^{(n-2)}(\xi) & \dots & y_n^{(n-2)}(\xi) \\ \dots & \dots & \dots & \\ y_1(\xi) & y_2(\xi) & \dots & y_n(\xi) \end{vmatrix}, \tag{6.2.2}$$

where the positive sign is taken if $x > \xi$, and the negative sign if $x < \xi;\ x,\xi \in I$.

It is shown in [4, p. 37] that the Green's function $G(x,\xi,\lambda)$ of problem Π can be given by

$$G(x,\xi,\lambda) = \frac{(-1)^n}{\Delta(\lambda)}H(x,\xi,\lambda), \tag{6.2.3}$$

where

$$\Delta(\lambda) = \begin{vmatrix} U_1(y_1) & U_1(y_2) & \dots & U_1(y_n) \\ U_2(y_1) & U_2(y_2) & \dots & U_2(y_n) \\ \dots & \dots & \dots & \\ U_n(y_1) & U_n(y_2) & \dots & U_n(y_n) \end{vmatrix}, \tag{6.2.4}$$

and

$$H(x,\xi,\lambda) = \begin{vmatrix} y_1(x) & y_2(x) & \dots & y_n(x) & g(x,\xi) \\ U_1(y_1) & U_1(y_2) & \dots & U_1(y_n) & U_1(g) \\ U_2(y_1) & U_2(y_2) & \dots & U_2(y_n) & U_2(g) \\ \dots & \dots & \dots & \dots & \dots \\ U_n(y_1) & U_n(y_2) & \dots & U_n(y_n) & U_n(g) \end{vmatrix}. \tag{6.2.5}$$

For fixed x and ξ, $G(x,\xi,\lambda)$ is a meromorphic function in λ since both $\Delta(\lambda)$ and $H(x,\xi,\lambda)$ are entire functions in λ. The poles of G are not necessarily simple; however, as indicated in the preceding section, we shall always assume that they are. Clearly, if $\lambda = \lambda_0$ is a pole of G, then it must be a zero of $\Delta(\lambda)$. However, a simple pole of G may not necessarily be a simple zero of Δ.

Let us denote the poles of G, or equivalently the eigenvalues of Π, by $\{\lambda_k\}_{k=0}^{\infty}$. Although all are assumed to be simple poles of G, some or all of the λ_k's may not be simple zeros of $\Delta(\lambda)$; hence, they may not be simple eigenvalues of Π, i.e., their multiplicities as eigenvalues may be greater than one.

We shall now assume that the asymptotic behavior of the eigenvalues $\{\lambda_k\}_{k=0}^{\infty}$ of problem Π is given by $\lambda_k^{1/n} = O(k\pi/(b-a))$ as $|k| \to \infty$. As a consequence of this, for $n > 1$, the canonical product

$$P(\lambda) = \begin{cases} \prod_{k=1}^{\infty}\left(1-\dfrac{\lambda}{\lambda_k}\right), & \text{if none of the eigenvalues is zero}, \\[2ex] \lambda\prod_{k=2}^{\infty}\left(1-\dfrac{\lambda}{\lambda_k}\right), & \text{if one of the eigenvalues, say } \lambda_1, \text{ is zero}, \end{cases} \tag{6.2.6}$$

is well defined since $\sum_{k=2}^{\infty} 1/\lambda_k < \infty$. For $n = 1$, if the product in (6.2.6) does not converge, we set

$$P(\lambda) = \begin{cases} \prod_{k=1}^{\infty}\left(1-\dfrac{\lambda}{\lambda_k}\right)\exp(\lambda/\lambda_k), & \text{if none of the eigenvalues is zero}, \\[2ex] \lambda\prod_{k=2}^{\infty}\left(1-\dfrac{\lambda}{\lambda_k}\right)\exp(\lambda/\lambda_k), & \text{if one of the eigenvalues, say } \lambda_1, \text{ is zero}. \end{cases} \tag{6.2.7}$$

Such an asymptotic behavior is guaranteed, for example, if the boundary conditions of problem Π are regular (cf. [4, p. 64]). The definition of regular boundary conditions is rather technical and will not be stated here; but an interested reader may consult [4, p. 56] for details. The class of regular boundary conditions is rather large and contains most of the standard boundary conditions, such as the Sturm-Liouville type for even-order differential operators. In this type, half the boundary conditions contain only the function values of y and its derivatives at $x = a$, and the other half only the values at $x = b$. Boundary conditions of periodic type, which are conditions in the form

$$U_j(y) = y^{(j)}(a) - y^{(j)}(b) = 0, \quad j = 0, 1, \ldots, n-1,$$

are also regular.

To simplify the proofs, we shall assume, without loss of generality, that none of the eigenvalues is zero; otherwise we choose a fixed number c and consider the boundary-value problem:

$$Ly = (\lambda - c)y$$

$$U_j(y) = 0\,; \quad j = 1, 2, \ldots, n\,,$$

which for c large enough, does not have zero as an eigenvalue.

Since the boundary-value problem Π^* has at most countably many eigenfunctions $\psi_k(x)$, and each has at most countably many zeros, it is possible to choose $\xi_0 \in [a, b]$ such that $\psi_k(\xi_0) \neq 0$ for all k.

Now fix $\xi_0 \in [a, b]$ and define

$$\Phi(x, \lambda) = P(\lambda) G(x, \xi_0, \lambda)\,, \tag{6.2.8}$$

where G is given by (6.2.3) and P is given by (6.2.6) or (6.2.7) depending on whether n is larger than or equal to 1. Since G has only a simple pole at $\lambda = \lambda_k$, $k = 1, 2, \ldots$, $\Phi(x, \lambda)$ is an entire function in λ.

THEOREM 6.1

Let

$$f(\lambda) = \int_a^b \overline{F}(x) \Phi(x, \lambda)\, dx\,, \tag{6.2.9}$$

for some $F \in L^2(a, b)$. Then f is an entire function of order not exceeding $1/n$ (n is the order of the differential operator L), which admits the following sampling representation

$$f(\lambda) = \sum_{k=1}^{\infty} f(\lambda_k) \frac{P(\lambda)}{(\lambda - \lambda_k) P'(\lambda_k)}\,. \tag{6.2.10}$$

Moreover, if problem Π is self-adjoint, or if F satisfies the adjoint boundary conditions (6.1.8), then the series converges uniformly on compact subsets of the complex λ-plane.

Proof. $\Phi(x, \lambda)$ is in $L^2(a, b)$ as a function of x because so is $G(x, y, \lambda)$. Then it follows from the Cauchy-Schwarz inequality that

$$|f(\lambda)|^2 \le \left(\int_a^b |F(x)|^2\, dx \right) \left(\int_a^b |\Phi(x, \lambda)|^2\, dx \right) < \infty\,;$$

hence $f(\lambda)$ is well defined. Moreover, $H(x, \xi_0, \lambda)$ and $\Delta(\lambda)$ are entire functions in λ of order $1/n$ because of (6.2.4), (6.2.5) and Theorem 1, p. 48 in [4]. Since $P(\lambda)$ is an entire function of order not exceeding $1/n$, as can be seen from Theorem 4, p. 12 in [3], it follows that $\Phi(x,\lambda)$, and consequently $f(\lambda)$ are entire functions of order not exceeding $1/n$.

Let λ_p be an eigenvalue of problem Π with multiplicity v_p and $\phi_{p,1}(x), \phi_{p,2}(x), \ldots, \phi_{p,v_p}(x)$ be the corresponding eigenfunctions. Similarly, let $\psi_{p,1}(x), \psi_{p,2}(x), \ldots, \psi_{p,v_p}(x)$ be the eigenfunctions corresponding to the eigenvalue $\bar{\lambda}_p$ of problem Π^*. It is known [1, p. 310-311] that each $\phi_{p,i}(x)$; $i = 1, 2, \ldots, v_p$, is orthogonal to all except one of the eigenfunctions of Π^*. The exception is one of the eigenfunctions corresponding to the eigenvalue $\bar{\lambda}_p$, and will be denoted by $\psi_{p,i}(x)$. Moreover, $\phi_{p,i}$ and $\psi_{p,i}$ can be normalized so that $\langle \phi_{p,i}, \psi_{p,i} \rangle = 1$.

It is now possible to rearrange the eigenfunctions of both problems Π and Π^* in two sequences $\{\phi_i\}_{i=1}^{\infty}$ and $\{\psi_i\}_{i=1}^{\infty}$ respectively such that $\langle \phi_i, \psi_j \rangle = \delta_{i,j}$, and if ϕ_i has eigenvalue λ_p, then ψ_i has eigenvalue $\bar{\lambda}_p$.

Since F and Φ are in $L^2(a,b)$, we have

$$F(x) = \sum_{p=1}^{\infty} \langle F, \phi_p \rangle \psi_p(x), \tag{6.2.11}$$

where

$$\langle F, \phi_p \rangle = \int_a^b F(x) \bar{\phi}_p(x)\, dx, \tag{6.2.12}$$

and

$$\Phi(x,\lambda) = \sum_{p=1}^{\infty} \langle \Phi, \psi_p \rangle \phi_p(x).$$

By Parseval's equality we have,

$$f(\lambda) = \sum_{p=1}^{\infty} \langle \Phi, \psi_p \rangle \overline{\langle F, \phi_p \rangle}.$$

Since some of these eigenfunctions ϕ_p belong to the same eigenvalue, we may write this series in the form

$$f(\lambda) = \sum_{p=1}^{\infty} \sum_{j=1}^{v_p} \langle \Phi, \psi_{p,j} \rangle \overline{\langle F, \phi_{p,j} \rangle}, \tag{6.2.13}$$

where $\phi_{p,j}$, $j = 1,2,\ldots,\nu_p(1 \le \nu_p \le n)$ are the eigenfunctions corresponding to the eigenvalue λ_p, which has multiplicity ν_p. But from (6.2.8), (6.2.9), (6.2.12) and Theorem 5.1 in [1], (cf. [4, p. 39]), we obtain

$$f(\lambda_p) = \lim_{\lambda \to \lambda_p} f(\lambda) = \lim_{\lambda \to \lambda_p} P(\lambda) \int_a^b \overline{F}(x) G(x,\xi_0,\lambda)\, dx$$

$$= -P'(\lambda_p) \sum_{j=1}^{\nu_p} \overline{\psi}_{p,j}(\xi_0) \int_a^b \overline{F}(x) \phi_{p,j}(x)\, dx$$

$$= -P'(\lambda_p) \sum_{j=1}^{\nu_p} \overline{\psi}_{p,j}(\xi_0) \overline{\langle F, \phi_{p,j} \rangle}. \tag{6.2.14}$$

In addition, we have for $1 \le j \le \nu_p$

$$\langle \Phi, \psi_{p,j} \rangle = P(\lambda) \int_a^b G(x,\xi_0,\lambda) \overline{\psi_{p,j}(x)}\, dx = -P(\lambda) \frac{\overline{\psi_{p,j}(\xi_0)}}{(\lambda - \lambda_p)}. \tag{6.2.15}$$

Therefore, by combining (6.2.15), (6.2.14) and (6.2.13), we obtain (6.2.10).

Finally, if problem Π is self-adjoint, then $\psi_k = \phi_k$ for all k and $F \in L^2(a,b)$ implies that $\sum_{k=1}^{\infty} |\langle F, \phi_k \rangle|^2 < \infty$; hence we can apply the same technique used in Theorem 4.1 (cf. [6 and 8]) to show that the series (6.2.10) converges uniformly on compact sets.

For any compact subset K of the complex λ-plane, we have

$$\sup_{\lambda \in K} \left| \frac{P(\lambda)}{(\lambda - \lambda_k)} \right| \le C(K)$$

for all k. To show this, let $\Lambda = \{\lambda_k\}_{k=1}^{\infty}$ and $\tilde{\Lambda} = \{\lambda_{i_1}, \ldots, \lambda_{i_p}\}$ be the set of those eigenvalues that lie inside K. There are only a finite number of them since Λ has no finite limit point. Let $\rho = dist(K, \Lambda - \tilde{\Lambda})$, hence

$$\sup_{\lambda \in K} \left| \frac{P(\lambda)}{(\lambda - \lambda_k)} \right| \le \frac{1}{\rho} \sup_{\lambda \in K} |P(\lambda)| = \frac{1}{\rho} \|P\|_K,$$

for all $\lambda_k \in \Lambda - \tilde{\Lambda}$, where $\|P\|_K = \sup_{\lambda \in K} |P(\lambda)|$. Since $P(\lambda)$ has zeros at $\lambda = \lambda_{i_1}, \ldots, \lambda_{i_p}$, it follows that $h_j(\lambda) = P(\lambda)/(\lambda - \lambda_j)$ is an analytic function in K for each $j = i_1, \ldots, i_p$. If we set $\mathbf{A} = \max(\|h_{i_1}\|_K, \ldots, \|h_{i_p}\|_K)$, then we can choose $C(K)$ as the maximum of $\mathbf{A}$ and $(1/\rho)\|P\|_K$.

Now if Π is not self-adjoint but F satisfies the adjoint boundary-conditions (6.1.8) of $U_j(y) = 0$, $j = 1,2,\ldots,n$, it follows from a minor modification of Theorem 4, Ch. 5 in [4], that the series (6.2.11) converges uniformly to $F(x)$. Thus, for any compact subset K, we have in view of (6.2.14)

$$\left| f(\lambda) - \sum_{p=1}^{N-1} f(\lambda_p) \frac{P(\lambda)}{(\lambda - \lambda_p)P'(\lambda_p)} \right| \leq \sum_{p=N}^{\infty} \left| \frac{f(\lambda_p)}{P'(\lambda_p)} \right| \left| \frac{P(\lambda)}{(\lambda - \lambda_p)} \right|$$

$$\leq C(K) \sum_{p=N}^{\infty} \sum_{j=1}^{\nu_p} \left| \left\langle F, \phi_{p,j} \right\rangle \right| \left| \psi_{p,j}(\xi_0) \right| .$$

The proof is now complete due to the fact that the last series is independent of λ and goes to zero as $N \to \infty$ since the series (6.2.11) converges uniformly. ∎

If all the eigenvalues are simple, e.g., they are simple zeros of $\Delta(\lambda)$, then we may, without loss of generality, take $P(\lambda) = \Delta(\lambda)$. For, in this case both P and Δ are entire functions with the same simple zeros, then in view of the Hadamard factorization theorem for entire functions, $\Delta(\lambda) = P(\lambda)H(\lambda)$, where $H(\lambda)$ is an entire function with no zeros. Therefore, $\Delta'(\lambda_k) = P'(\lambda_k)H(\lambda_k)$, and consequently (6.2.10) becomes

$$f(\lambda) = \sum_{k=1}^{\infty} f(\lambda_k) \frac{\Delta(\lambda)H(\lambda_k)}{(\lambda - \lambda_k)H(\lambda)\Delta'(\lambda_k)},$$

which yields the sampling series

$$K(\lambda) = \sum_{k=1}^{\infty} K(\lambda_k) \frac{\Delta(\lambda)}{(\lambda - \lambda_k)\Delta'(\lambda_k)},$$

where $K(\lambda) = f(\lambda)H(\lambda)$.

To illustrate some of these ideas, let us consider the following boundary-value problems:

1) $$\begin{cases} -iy' = \lambda y, & x \in [0,1] \\ y(0) = y(1); & \end{cases}$$

2) $$\begin{cases} iy' = \lambda y, & x \in [-1,1] \\ y(-1) = y(1); & \end{cases}$$

3) $$\begin{cases} -y'' = \lambda y, & x \in [0,1] \\ y(0) = y(1), & y'(0) = y'(1). \end{cases}$$

For problem (1) we have $\Delta(\lambda)=1-e^{i\lambda}$; hence, the eigenvalues are $\lambda_k=2k\pi, k=0,\pm1,\pm2,\ldots$, and $P(\lambda)=2\sin(\lambda/2), H(\lambda)=-i\exp(i\lambda/2)$, while for problem (2)

$$(1/2i)\Delta(\lambda)=\sin\lambda=P(\lambda).$$

On the other hand, for problem (3) we have $\Delta(\lambda)=2(1-\cos\sqrt{\lambda})$. Hence, the eigenvalues are $\lambda_k=(2k\pi)^2$ and $P(\lambda)=2\sqrt{\lambda}\sin(\sqrt{\lambda}/2)$. The eigenvalues are simple zeros of $P(\lambda)$, but double zeros of $\Delta(\lambda)$, except for λ_0. Nevertheless, the Green's function has simple poles at $\lambda=\lambda_k$ (cf. [5, pp. 427-429] and [1, p. 202]), and consequently Theorem 6.1 does apply; see Section 3.2, in particular (3.2.11) and (3.2.12).

It is evident from Theorem 6.1 that different boundary-value problems with identical eigenvalues give rise to identical sampling series expansions. Examples of such expansions are given in the following important corollaries.

Corollary 6.1.1. If the eigenvalues $\{\lambda_k\}$ of the boundary-value problem Π are given by $\lambda_k=[k\pi/(b-a)]^2$, $k=0,1,2,\ldots$, then the sampling series (6.2.10) takes on the form

$$f(\lambda)=\frac{f(0)\sin(\sqrt{\lambda}(b-a))}{(b-a)\sqrt{\lambda}}+\sum_{k=1}^{\infty}f\left(\left(\frac{k\pi}{b-a}\right)^2\right)\frac{2\sqrt{\lambda}\sin((b-a)(\sqrt{\lambda}-\sqrt{\lambda_k}))}{(b-a)(\lambda-\lambda_k)}.$$

In particular, if $a=0$, $b=\pi$, we have

$$f(\lambda)=f(0)\frac{\sin\pi\sqrt{\lambda}}{\pi\sqrt{\lambda}}+\sum_{k=1}^{\infty}f(k^2)\frac{2\sqrt{\lambda}\sin(\pi(\sqrt{\lambda}-k))}{\pi(\lambda-k^2)},$$

or

$$\tilde{f}(t)=\tilde{f}(0)\frac{\sin\pi t}{\pi t}+\sum_{k=1}^{\infty}\tilde{f}(k)\frac{2t\sin(\pi(t-k))}{\pi(t^2-k^2)},$$

where $\tilde{f}(t)=f(\lambda)$, $\sqrt{\lambda}=t$.

Proof. In this case

$$P(\lambda)=\lambda\prod_{k=1}^{\infty}\left(1-\frac{\lambda(b-a)^2}{k^2\pi^2}\right)=\frac{\sqrt{\lambda}}{(b-a)}\sin(\sqrt{\lambda}(b-a)),$$

and

$$P'(\lambda_k) = \begin{cases} \frac{1}{2}(-1)^k, & \text{for} \quad k = 1,2,\dots \\ 1 \quad , & \text{for} \quad k = 0 . \end{cases}$$

∎

Corollary 6.1.2. If the eigenvalues $\{\lambda_k\}$ of problem Π are given by $\lambda_k = k\pi/(b-a)$, $k = 0, \pm 1, \pm 2, \dots$, then the sampling expansion (6.2.10) takes the form

$$f(\lambda) = \sum_{k=-\infty}^{\infty} f\left(\frac{k\pi}{b-a}\right) \frac{\sin\left[(b-a)\left(\lambda - \frac{k\pi}{(b-a)}\right)\right]}{\left(\lambda - \frac{k\pi}{(b-a)}\right)(b-a)} .$$

In particular, if $a = 0$ and $b = \pi$, we have

$$f(\lambda) = \sum_{k=-\infty}^{\infty} f(k) \frac{\sin \pi(\lambda - k)}{\pi(\lambda - k)} .$$

Proof. In this case

$$P(\lambda) = \lambda \prod_{\substack{k=-\infty \\ k \neq 0}}^{\infty} \left(1 - \frac{\lambda(b-a)}{k\pi}\right) = \lambda \prod_{k=1}^{\infty} \left(1 - \frac{\lambda^2(b-a)^2}{k^2\pi^2}\right) = \frac{\sin \lambda(b-a)}{(b-a)} ,$$

and hence

$$P'(\lambda_k) = (-1)^k .$$

∎

The eigenvalues in Corollaries 6.1.1 and 6.1.2 arise from boundary-value problems whose associated differential operators are of 2nd and 1st orders respectively.

Theorem 6.1 and Corollary 6.1.2 give a new insight into why the WSK sampling theorem works the way it does. The kernel of the Fourier transform $\Phi(x,\lambda) = e^{i\lambda x}$ is a multiple of the Green's function $G(x,\xi_0,\lambda)$ of the boundary-value problem

$$(-i)y' = \lambda y , \quad x \in [-\sigma, \sigma]$$

$$y(-\sigma) = y(\sigma) ,$$

whose eigenvalues are $\lambda_k = k\pi/\sigma$, $k = 0, \pm 1, \pm 2, \dots$; hence

$$\left(\frac{1}{2i\sigma}\right)\Delta(\lambda) = P(\lambda) = \frac{\sin\sigma\lambda}{\sigma}, \quad P'\left(\frac{k\pi}{\sigma}\right) = (-1)^k,$$

and

$$f(\lambda) = \sum_{k=-\infty}^{\infty} f\left(\frac{k\pi}{\sigma}\right)\frac{\sin\sigma\left(\lambda - \frac{k\pi}{\sigma}\right)}{\sigma\left(\lambda - \frac{k\pi}{\sigma}\right)}.$$

Corollary 6.1.3. If the eigenvalues of problem Π are given by $\lambda_k = k^4$, $k = 0, 1, 2, \dots$, then the sampling series (6.2.10) takes the form

$$f(\lambda) = f(0)\,\frac{\sin(\sqrt[4]{\lambda}\,\pi)\sinh(\sqrt[4]{\lambda}\,\pi)}{\pi^2\sqrt{\lambda}}$$

$$+\sum_{k=1}^{\infty} f(k^4)\,\frac{4k\sqrt{\lambda}\sinh(\sqrt[4]{\lambda}\,\pi)\sin\pi(\sqrt[4]{\lambda} - k)}{\pi(\lambda - k^4)\sinh k\pi}, \tag{6.2.16}$$

if zero is an eigenvalue, and

$$f(\lambda) = \sum_{k=1}^{\infty} f(k^4)\,\frac{4k\sqrt{\lambda}\sinh(\sqrt[4]{\lambda}\,\pi)\sin\pi(\sqrt[4]{\lambda} - k)}{\pi(\lambda - k^4)\sinh\pi k}, \tag{6.2.17}$$

if zero is not an eigenvalue.

Proof. We only prove (6.2.16) since the proof for (6.2.17) is similar.

$$P(\lambda) = \lambda\prod_{k=1}^{\infty}\left(1 - \frac{\lambda}{k^4}\right) = \lambda\prod_{k=1}^{\infty}\left(1 - \frac{\sqrt{\lambda}}{k^2}\right)\left(1 + \frac{\sqrt{\lambda}}{k^2}\right)$$

$$= \frac{\sqrt{\lambda}}{\pi^2}\sin(\sqrt[4]{\lambda}\,\pi)\sinh(\sqrt[4]{\lambda}\,\pi),$$

and

$$P'(k) = \begin{cases} \dfrac{(-1)^k}{4\pi k}\sinh k\pi & \text{if} \quad k = 1, 2, \dots \\ 1 & \text{if} \quad k = 0 \end{cases}.$$

∎

Although $\sqrt[4]{\lambda_k} = \pm k, \pm ik$, the reader can verify that regardless of the choice of $\sqrt[4]{\lambda_k}$, $P'(k)$ is invariant.

If we set $G(\lambda) = f(\lambda)/(\sqrt{\lambda}\sinh \sqrt[4]{\lambda}\,\pi)$ in (6.2.17), we obtain

$$G(\lambda) = \sum_{k=1}^{\infty} G(k^4)\, \frac{4k^3 \sin\pi(\sqrt[4]{\lambda} - k)}{\pi(\lambda - k^4)},$$

which yields

$$\tilde{G}(t) = \sum_{k=1}^{\infty} \tilde{G}(k)\, \frac{(4k^3)\sin\pi(t-k)}{\pi(t^4 - k^4)}, \tag{6.2.18}$$

where

$$\tilde{G}(t) = G(\lambda), \quad \lambda = t^4 .$$

Corollary 6.1.4. If the eigenvalues are given by $\lambda_k = k^n$, then for $\tilde{f}(t) = f(\lambda)$, $\lambda = t^n$, we have

$$\tilde{f}(t) = \sum_{k=1}^{\infty} \tilde{f}(k)\, \frac{n(k)^{n-1}\tilde{P}(t)}{(t^n - k^n)\tilde{P}'(k)} ,$$

where $\tilde{P}(t) = P(\lambda)$.

Proof.
$$\left.\frac{dP}{d\lambda}\right|_{\lambda=k^n} = \frac{1}{nk^{n-1}} \left.\frac{d\tilde{P}}{dt}\right|_{t=k} .$$

▮

We conclude this section with some examples.

***Example* 1.** Consider the boundary-value problem

$$y^{(4)} = \lambda y , \quad x \in [0, \pi]$$

and

$$y^{(1)}(0) = 0 = y^{(3)}(0), \quad y^{(1)}(\pi) = 0 = y^{(3)}(\pi) .$$

With some easy computations, one can show that

$$\Delta(\lambda) = \sqrt{\lambda}\sin(\pi\sqrt[4]{\lambda})\sinh(\pi\sqrt[4]{\lambda}) ;$$

hence the eigenvalues are $\lambda_k = k^4$, $k = 0, 1, 2, \ldots$ and

$$P(\lambda) = \frac{\sqrt{\lambda}}{\pi^2}\sin(\pi\sqrt[4]{\lambda})\sinh(\pi\sqrt[4]{\lambda}) .$$

The sampling series expansion in this case is the same as in Corollary 6.1.3.

Example 2. Consider the boundary-value problem

$$y^{(4)} = \lambda y, \quad x \in [0, \pi],$$

and

$$y(0) = 0 = y''(0), \quad y(\pi) = 0 = y''(\pi).$$

It is readily seen that

$$\Delta(\lambda) = -\frac{1}{\sqrt{\lambda}} \sin(\pi \sqrt[4]{\lambda}) \sinh(\pi \sqrt[4]{\lambda}),$$

hence, the eigenvalues are $\lambda_k = k^4$, $k = 1, 2, \ldots$, and $P(\lambda) = (-1/\pi^2)\,\Delta(\lambda)$. The sampling series is the same as in Example 1 except for the absence of the term that corresponds to the eigenvalue zero.

6.3 The Haddad-Yao-Thomas Sampling Theorem

We close this chapter by showing that the sampling theorem obtained by Haddad, Yao and Thomas [2] is a special case of Theorem 6.1. In [2], the authors considered a self-adjoint boundary-value problem of the form (6.1.5), (6.1.6) and its associated nonhomogeneous problem

$$(L - \nu) y = F(x), \tag{6.3.1}$$

together with (6.1.6), where $F \in L^2(a, b)$. If we denote the Green's function of this problem by $G(x, \xi, \nu)$, then the solution of (6.3.1) and (6.1.6) is given by

$$y(x, \nu) = \int_a^b G(x, \xi, \nu) F(\xi)\, d\xi. \tag{6.3.2}$$

Moreover, we have

$$F(x) = \frac{1}{2\pi i} \lim_{m \to \infty} \oint_{C_m} [y(x, \nu)]\, d\nu = \frac{1}{2\pi i} \oint_C [y(x, \nu)]\, d\nu$$

$$= \frac{1}{2\pi i} \oint_C \left(\int_a^b [G(x, \xi, \nu)] F(\xi)\, d\xi \right) d\nu, \tag{6.3.3}$$

where $\{C_m\}$ is a sequence of positively oriented contours such that C_m encloses the eigenvalues $\lambda_1, \lambda_2, \ldots, \lambda_m$ and $\lim_{m \to \infty} \oint_{C_m} = \oint_C$.

In addition, the authors also assumed that there exists a function $\psi(x, \lambda)$ that satisfies the homogeneous equation (6.1.5) and which generates the eigenfunctions $\{\psi_k\}$ of the problem, when the parameter λ is replaced by the eigenvalues λ_k. Their main result essentially states that if

$$f(\lambda) = \int_a^b F(x)\,\psi(x,\lambda)\,dx\,, \tag{6.3.4}$$

then

$$f(\lambda) = \sum_{k=1}^{\infty} f(\lambda_k)\,\Psi_k(\lambda)\,, \tag{6.3.5}$$

where

$$\Psi_k(\lambda) = \frac{1}{\|\psi_k\|^2}\int_a^b \psi(x,\lambda)\,\psi_k(x)\,dx = \frac{\langle\psi,\psi_k\rangle}{\|\psi_k\|^2}\,. \tag{6.3.6}$$

To derive (6.3.5) from (6.2.10), first let us observe that in (6.2.8) through (6.2.10) we have suppressed the fixed parmeter ξ_0, but if we let ξ_0 run over the interval $[a,b]$, then (6.2.9), (6.2.10) can be written in the form

$$f(x,\nu) = \int_a^b \Phi(x,\xi,\nu)\,F(\xi)\,d\xi\,, \tag{6.3.7}$$

and

$$f(x,\nu) = \sum_{k=1}^{\infty} f(x,\lambda_k)\,S_k(\nu)\,, \tag{6.3.8}$$

where
$$S_k(\nu) = \frac{P(\nu)}{(\nu-\lambda_k)\,P'(\lambda_k)}\,.$$

From (6.3.2) and (6.3.7), it is easy to see that

$$y(x,\nu) = \frac{f(x,\nu)}{P(\nu)}\,,$$

and hence by combining (6.3.3), (6.3.4) and (6.3.8) we obtain

$$f(\lambda) = \frac{1}{2\pi i}\int_a^b \psi(x,\lambda)\left(\oint_C \left[\frac{f(x,\nu)}{P(\nu)}\right]d\nu\right)dx$$

$$= \sum_{k=1}^{\infty}\frac{1}{P'(\lambda_k)}\int_a^b f(x,\lambda_k)\,\psi(x,\lambda)\,dx\,.$$

From this, we further obtain

$$f(\lambda_n) = \sum_{k=1}^{\infty} \frac{1}{P'(\lambda_k)} \int_a^b f(x, \lambda_k)\, \psi_n(x)\, dx\,,$$

and by expanding ψ in terms of the eigenfunctions $\{\psi_n(x)\}$, we also obtain

$$f(\lambda) = \sum_{k=1}^{\infty} \sum_{n=1}^{\infty} \frac{\langle \psi, \psi_n \rangle}{P'(\lambda_k)\, \|\psi_n\|^2} \int_a^b f(x, \lambda_k)\, \psi_n(x)\, dx\,.$$

Putting these last two equations together, we finally obtain

$$f(\lambda) = \sum_{n=1}^{\infty} f(\lambda_n)\, \Psi_n(\lambda)\,,$$

where

$$\Psi_n(\lambda) = \frac{\langle \psi, \psi_n \rangle}{\|\psi_n\|^2}\,,$$

which is (6.3.5).

References

1. E. Coddington and N. Levinson, *Theory of Ordinary Differential Equations*, McGraw-Hill, New York (1955).

2. A. Haddad, K. Yao and J. Thomas, General methods for the derivation of sampling theorems, *IEEE Trans. Inform. Theory*, IT-13 (1967), 227-230.

3. B. Ja. Levin, *Distribution of Zeros of Entire Functions*, Transl, Math. Monographs Ser., Vol. 5, Amer. Math. Soc., Providence, RI (1964).

4. M. Naimark, *Linear Differential Operators*, Vol. 1, *Elementary Theory of Linear Differential Operators*, George Harrap & Co., London, U.K. (1967).

5. I. Stakgold, *Green's Functions and Boundary Value Problems*, John Wiley & Sons, New York (1979).

6. A. Zayed, On Kramer's sampling theorem associated with general Sturm-Liouville problems and Lagrange interpolation, *SIAM J. Appl. Math.*, 51 (1991), 575-604.

7. A. Zayed, M. A. El-Sayed and M. H. Annaby, On Lagrange interpolations and Kramer's sampling theorem associated with self-adjoint boundary-value problems, *J. Math. Anal. Appls.*, 158, 1 (1991), 269-284.

8. A. Zayed, G. Hinsen and P. Butzer, On Lagrange interpolation and Kramer-type sampling theorems associated with Sturm-Liouville problems, *SIAM J. Appl. Math.*, 50, 3 (1990), 893-909.

7

SAMPLING THEOREMS AND SPECIAL FUNCTIONS

7.0 Introduction

One of the most recent mathematical applications of Sampling Theory is in the field of Special Functions, where sampling theorems have been proved to be a very useful tool in summing infinite series. The main aim of this chapter is to report on some advances in this direction. We will show how sampling theorems can be used to derive numerous summation formulae, some of which we believe to be new. Among all available methods to sum up infinite series, the sampling theory technique is the easiest and the most straightforward.

We begin by giving a brief historical background of the relationship between Sampling Theory and Special Functions, in particular, of the relationship between sampling theorems and infinite series involving special functions. Following this, we will present some of the new results that we have recently obtained.

7.1 Historical Background

The connection between Sampling Theory and Special Functions is, in fact, an old one and can be traced back to the work of J. M. Whittaker [23, p. 71] who suggested a sampling series expansion for the Hankel transform of a function with compact support. But insomuch as using sampling theorems as a tool to sum infinite series involving special functions, the first connection of which we are aware appears to be in the work of K. S. Krishnan [18], who in 1948 obtained summation formulae for a number of infinite series, such as

$$\sum_{n=-\infty}^{\infty} \frac{\sin[a\sqrt{n^2\alpha^2+\lambda^2}]}{\sqrt{n^2\alpha^2+\lambda^2}} = \frac{\pi}{\alpha} J_0(\lambda a), \quad 0 < \alpha \le 2\pi/a, \qquad (7.1.1)$$

using the equivalence of these series with some integral representations of suitably chosen functions. The summation formulae he obtained, including (7.1.1), were of the form

$$\sum_{n=-\infty}^{\infty} f(n\alpha+\theta) = \sum_{n=-\infty}^{\infty} f(n\alpha) = \frac{1}{\alpha}\int_{-\infty}^{\infty} f(t)\,dt\,, \tag{7.1.2}$$

where the Fourier transform of f has support in $[-a,a]$ and $0<\alpha\le 2\pi/a$.

Krishnan's work was inspired by an idea that was suggested to him and his collaborator, A. B. Bhatia, by Professor Norbert Wiener in regard to one of their earlier papers on light scattering in homogeneous medi and thermal elastic waves. In that paper, Bhatia and Krishnan [1] needed the sum

$$\sum_{n=-\infty}^{\infty} \frac{\sin^2(na+b)}{(na+b)^2} = \frac{\pi}{a}, \quad 0<a\le\pi, \quad b \text{ is a constant},$$

which they derived by integrating other known summation formulae and manipulating the results. In a footnote, the authors added another more elegant proof that is based on the Poisson summation formula (cf. formula (2.2.3)) and attributed it to Professor Wiener. Capitalizing on Wiener's idea, Krishnan used the Poisson summation formula and some integrals of Ramanujan to derive his summation formulae.

Neither in the work of Krishnan nor in that of Bhatia and Krishnan was a link between their summation formulae and the Whittaker-Shannon-Kotel'nikov (WSK) sampling theorem ever mentioned. This is not surprising because the WSK sampling theorem was not very well known at the time. But such a linkage can now be established. For example, we can easily see that Krishnan's formulae can be derived from the WSK sampling series because of the fact that (7.1.2) is an immediate consequence of the Poisson summation formula, which is known to be equivalent to the WSK sampling theorem [3]. In fact, by integrating the relation

$$f(t) = \sum_{k=-\infty}^{\infty} f\left(\frac{k\pi}{\sigma}\right) \frac{\sin(\sigma t - k\pi)}{(\sigma t - k\pi)}$$

over the whole real line, we obtain (7.1.2) with $\alpha = \pi/\sigma$, provided that f is integrable.

Apart from these indirect connections between sampling theorems and infinite sums, it seems that the first direct connection between these two topics appeared in the work of A. Jerri [16], who was probably the first to explicitly use the language of sampling theory to derive summation formulae for infinite series. In his Ph.D. thesis [16], Jerri used the integral representation [22, p. 186]

$$\frac{\pi\Gamma(a-1)}{2^{a-2}\Gamma((a+x)/2)\Gamma((a-x)/2)} = \int_{-\pi/2}^{\pi/2} (\cos u)^{a-2} e^{ixu} du\,, \quad a > 1\,, \qquad (7.1.3)$$

to obtain the sampling series expansion

$$\frac{1}{\Gamma((a+x)/2)\Gamma((a-x)/2)} =$$

$$\sum_{n=-\infty}^{\infty} \frac{1}{\Gamma((a+2n)/2)\Gamma((a-2n)/2)} \frac{\sin[(x-2n)\pi/2]}{(x-2n)\pi/2}, \quad a > 1 \qquad (7.1.4)$$

which yields the familiar relation

$$\frac{1}{\Gamma(x/2)\Gamma(1-(x/2))} = \frac{\sin(\pi x/2)}{\pi},$$

when $a = 2$.

One can easily discern that (7.1.3) is the usual representation of a band-limited function (cf. (2.1.1)) and that (7.1.4) is its WSK sampling series.

Some years later, another connection between the WSK sampling theorem and the sums of infinite series involving special functions was made by Higgins in [12], where he derived a sampling theorem similar to the WSK sampling theorem, but for the Bessel-Hankel transform of functions with support in [0,1]. This means functions f of the form

$$f(x) = \int_0^1 \sqrt{xt}\, J_\nu(xt) F(t)\, dt\,,$$

where $F \in L^2(0,1)$. He then utilized his sampling theorem to obtain the summation formula

$$x^{\nu-\mu} J_\mu(x) = 2J_\nu(x) \sum_{n=1}^{\infty} \frac{x_n^{\nu-\mu+1} J_\mu(x_n)}{J_\nu'(x_n)(x^2 - x_n^2)}, \quad \mu > \nu + \frac{1}{2}, \qquad (7.1.5)$$

where $\{x_n\}$ are the positive zeros of $J_\nu(x)$.

Recently J. L. Brown [2] also used sampling theorems, in particular, the WSK sampling theorem to derive some well-known summation formulae, mainly, Mittag-Leffler expansions of some elementary trigonometric functions. To the best of my knowledge, the method of using sampling theorems to sum infinite series involving special functions was not fully investigated until 1991 (cf. [24]). In [24], the author obtained, by using sampling theorem techniques, some unusual and more complicated summation formulae that include (7.1.1), (7.1.4) and (7.1.5) as special cases. Though unusual and complicated, these formulae are easy to prove when sampling theorems are used.

Before we get involved in technicalities, let us have another look at the WSK sampling theorem. From the Mittag-Leffler expansion of meromorphic functions [17, p. 37], we recall that if $F(z)$ is a meromorphic function whose poles are all simple and located at the points $\{z_n\}_{n=0}^{\infty}$ with corresponding residues $\{a_n\}_{n=0}^{\infty}$, then

$$F(z) = \sum_{n=0}^{\infty} \frac{a_n}{(z - z_n)} + h(z), \tag{7.1.6}$$

provided that the series converges absolutely and uniformly on any compact subset of the complex z-plane not containing the poles of F, where h is an entire function. The series in (7.1.6) is called the principle part of F. In particular, if f and g are two entire functions with no common zeros and if g has simple zeros at the points $\{z_n\}_{n=0}^{\infty}$, then

$$\frac{f(z)}{g(z)} = \sum_{n=0}^{\infty} \frac{f(z_n)}{(z - z_n)g'(z_n)} + h(z),$$

provided that the series converges absolutely and uniformly on any compact subset of the complex z-plane not containing the zeros of g. Or equivalently,

$$f(z) = \sum_{n=0}^{\infty} \frac{f(z_n)g(z)}{(z - z_n)g'(z_n)} + H(z), \tag{7.1.7}$$

where $H = gh$ is an entire function. The last series is a Lagrange-type interpolation series which, unfortunately, does not always converge to f since the function h is not, in general, equal to zero. For example, it is known that the Gamma function

$$\Gamma(z) = \int_0^{\infty} t^{z-1} e^{-t} dt, \quad Re(z) > 0$$

is a meromorphic function in z whose poles, which are located at the points $z_n = -n$, $n = 0, 1, 2, \ldots$, are all simple and having corresponding residues $(-1)^n/n!$. Nevertheless, the Mittag-Leffler expansion of $\Gamma(z)$ is given by

$$\Gamma(z) = \sum_{n=0}^{\infty} \frac{(-1)^n}{n!(z+n)} + h(z),$$

where $h(z)$ is the incomplete Gamma function $h(z) = \int_1^\infty t^{z-1} e^{-t}\,dt$, which is an entire function in z. It is readily seen that the principle part of the Γ-function $\sum_{n=0}^{\infty} (-1)^n/(n!\,(z+n))$ is equal to $\int_0^1 t^{z-1} e^{-t}\,dt$.

Formula (7.1.7), which may also be derived by using the residue theorem and contour integration (cf. [11]), yields the WSK sampling theorem when $H = 0$, $g(z) = \sin \pi z$ and f is band-limited. More generally, when $H = 0$, (7.1.7) yields the Lagrange-type interpolation series discussed in previous chapters. But showing that $H = 0$ in a specific problem is usually not an easy task; however, when sampling theorems are used, this difficulty seems to disappear. Finally, we note that (7.1.1), (7.1.4) and (7.1.5) are special cases of (7.1.7) with $H = 0$.

We close this section by pointing out that some other work on sampling theorems and special functions, not necessarily in the context of summing up infinite series, can be found in [14-16, 19, 20, 25, 26] and Chapters 3, 4.

7.2 New Summation Formulae Involving Trigonometric* and Bessel Functions

Using symbolic manipulation programs and a fair amount of ingenuity, William Gosper recently obtained new and unusual summation formulae for series involving trigonometric functions. Some of these formulae have just been mathematically verified by M. Ismail and R. Zhang (see [9, 13]), who have also generalized some of these summation formulae to series involving the Bessel function of the first kind. Their proofs involve several special function relations together with some techniques borrowed from the theory of Fourier series and integrals and Mittag-Leffler expansions of some meromorphic functions.

In this section we shall show that some of these new summation formulae, due to Gosper, Ismail and Zhang, can actually be obtained from already known results in sampling theory in an easy and straightforward fashion. In fact, we shall show that other summation formulae, which we believe to be new, can be obtained just as easily.

---

*Some material presented in this chapter is based on the author's article that appeared in the *Proc. Amer. Math. Soc.*, 117 (1993) and is printed with permission from the Amer. Math. Soc.

We begin with

$$\sum_{n=0}^{\infty} \frac{(-1)^n}{\left(n+\frac{1}{2}\right)} \frac{\sin\sqrt{b^2+\pi^2\left(n+\frac{1}{2}\right)^2}}{\sqrt{b^2+\pi^2\left(n+\frac{1}{2}\right)^2}} = \frac{\pi}{2}\frac{\sin b}{b}, \tag{7.2.1}$$

which is due to Gosper and its generalization

$$\sum_{n=0}^{\infty} \frac{(-1)^n}{\left(n+\frac{1}{2}\right)} \frac{J_\nu\left(\sqrt{b^2+\pi^2\left(n+\frac{1}{2}\right)^2}\right)}{\left[\sqrt{b^2+\pi^2\left(n+\frac{1}{2}\right)^2}\right]^\nu} = \frac{\pi}{2} b^{-\nu} J_\nu(b), \tag{7.2.2}$$

which is due to Ismail and Zhang (see [9]); J_ν is the Bessel function of the first kind. Formula (7.2.2) is a generalization of (7.1.1) since the former reduces to the latter when $\nu = 1/2$.

The following result can be obtained from Theorem 4.1 or from Theorem 1 and Corollary 4 in [26] by a simple change of variables: if for some $g \in L^2(0, a)$

$$f(t) = \int_0^a g(x)\cos(xt)\,dx\,, \tag{7.2.3}$$

then

$$f(t) = \sum_{k=-\infty}^{\infty} f\left(\left(k+\frac{1}{2}\right)\frac{\pi}{a}\right) \frac{\sin\left(at-\left(k+\frac{1}{2}\right)\pi\right)}{\left(at-\left(k+\frac{1}{2}\right)\pi\right)}, \tag{7.2.4}$$

uniformly on compact subsets of $\Re$. In particular,

$$f(0) = \sum_{k=-\infty}^{\infty} f\left(\left(k+\frac{1}{2}\right)\frac{\pi}{a}\right) \frac{(-1)^k}{\left(k+\frac{1}{2}\right)\pi}. \tag{7.2.5}$$

By putting $g(x) = (a^2-x^2)^{\nu/2} J_\nu(b\sqrt{a^2-x^2})$ in (7.2.3) and using formula (50) in Section 1.13 of [4], which states that

$$\int_0^a (a^2-x^2)^{\nu/2} J_\nu(b\sqrt{a^2-x^2})\cos(tx)\,dx = \sqrt{\frac{\pi}{2}}\, a^{\nu+1/2} b^\nu J_{\nu+\frac{1}{2}}(a\sqrt{b^2+t^2})(b^2+t^2)^{-\frac{1}{2}\left(\nu+\frac{1}{2}\right)},$$

for $Re\,\nu > -1,\ b > 0$, we obtain that

$$f(t) = \sqrt{\frac{\pi}{2}}\, a^{\nu+1/2} b^{\nu} J_{\nu+\frac{1}{2}}(a\sqrt{b^2+t^2})(b^2+t^2)^{-\frac{1}{2}(\nu+\frac{1}{2})}.$$

Hence, (7.2.4) becomes

$$\frac{J_{\nu+\frac{1}{2}}(a\sqrt{b^2+t^2})}{[b^2+t^2]^{\frac{1}{2}(\nu+\frac{1}{2})}} = \sum_{k=-\infty}^{\infty} \frac{J_{\nu+\frac{1}{2}}\left(\sqrt{a^2b^2+\left(k+\frac{1}{2}\right)^2\pi^2}\right)}{\left[b^2+\left(k+\frac{1}{2}\right)^2\frac{\pi^2}{a^2}\right]^{\frac{1}{2}(\nu+\frac{1}{2})}} \frac{\sin\left(at-\left(k+\frac{1}{2}\right)\pi\right)}{\left(at-\left(k+\frac{1}{2}\right)\pi\right)},$$

which reduces to (7.2.1) when $t = 0,\ a = 1$ and $\nu = 0$. It also reduces to (7.2.2) when $t = 0,\ a = 1$ and ν is replaced by $\nu - \frac{1}{2}$. In both cases observe that $\sum_{k=-\infty}^{\infty} = 2\sum_{k=0}^{\infty}$.

One interesting consequence of (7.2.2) is the following formula which is obtained by taking the limit of both sides as $b \to 0$

$$\left(\frac{\pi}{4}\right)^{\nu+1} \frac{1}{\Gamma(\nu+1)} = \sum_{k=0}^{\infty} \frac{(-1)^k J_{\nu}\left[\pi\left(k+\frac{1}{2}\right)\right]}{(2k+1)^{\nu+1}},$$

which, in turn, gives for $\nu = 1/2$ the well known formula

$$\sum_{k=0}^{\infty} 1/(2k+1)^2 = \pi^2/8.$$

Now we can employ the above argument to produce new summation formulae which we believe to be new.

In view of formula 6.739, p. 762 in [10], we have

$$\int_0^{\pi} J_{2\nu}(b\sqrt{\pi^2-x^2})(\pi^2-x^2)^{-\frac{1}{2}}\cos(tx)\,dx = \frac{\pi}{2} J_{\nu}\left[\frac{\pi}{2}(\sqrt{b^2+t^2}+t)\right] J_{\nu}\left[\frac{\pi}{2}(\sqrt{b^2+t^2}-t)\right].$$

Thus, by putting $g(x,b) = J_{2\nu}(b\sqrt{\pi^2-x^2})(\pi^2-x^2)^{-\frac{1}{2}}$, which is in $L^2(0,\pi)$ for $\nu > 1/4$, in (7.2.3), formulae (7.2.4) and (7.2.5) after some simple computations become

$$J_{\nu}\left[\frac{\pi}{2}(\sqrt{b^2+t^2}+t)\right] J_{\nu}\left[\frac{\pi}{2}(\sqrt{b^2+t^2}-t)\right] =$$

$$\sum_{k=-\infty}^{\infty} J_{\nu}\left[\frac{\pi}{2}\left(\sqrt{b^2+\left(k+\frac{1}{2}\right)^2}+\left(k+\frac{1}{2}\right)\right)\right] J_{\nu}\left[\frac{\pi}{2}\left(\sqrt{b^2+\left(k+\frac{1}{2}\right)^2}-\left(k+\frac{1}{2}\right)\right)\right] \frac{\sin\pi\left(t-\left(k+\frac{1}{2}\right)\right)}{\pi\left(t-\left(k+\frac{1}{2}\right)\right)},$$

and

$$\left[J_\nu\left(\frac{\pi b}{2}\right)\right]^2 = 2\sum_{k=0}^{\infty} J_\nu\left[\frac{\pi}{2}\left(\sqrt{b^2+\left(k+\frac{1}{2}\right)^2}+\left(k+\frac{1}{2}\right)\right)\right]J_\nu\left[\frac{\pi}{2}\left(\sqrt{b^2+\left(k+\frac{1}{2}\right)^2}-\left(k+\frac{1}{2}\right)\right)\right]\frac{(-1)^k}{\pi\left(k+\frac{1}{2}\right)}.$$

Upon replacing b by $2b$ in the last equation, we obtain for $\nu = \frac{1}{2}$

$$[\sin\pi b]^2 = 2\sum_{k=0}^{\infty}\sin\left[\pi\left(\sqrt{b^2+\left(\frac{k}{2}+\frac{1}{4}\right)^2}+\left(\frac{k}{2}+\frac{1}{4}\right)\right)\right]\sin\left[\pi\left(\sqrt{b^2+\left(\frac{k}{2}+\frac{1}{4}\right)^2}-\left(\frac{k}{2}+\frac{1}{4}\right)\right)\right]\frac{(-1)^k}{\pi\left(k+\frac{1}{2}\right)}$$

which, in turn, gives as a special case for $b = 1/2$

$$\frac{\pi}{4} = \sum_{k=0}^{\infty}\sin\left[\pi\left(\sqrt{\frac{1}{4}+\left(\frac{k}{2}+\frac{1}{4}\right)^2}+\left(\frac{k}{2}+\frac{1}{4}\right)\right)\right]\sin\left[\pi\left(\sqrt{\frac{1}{4}+\left(\frac{k}{2}+\frac{1}{4}\right)^2}-\left(\frac{k}{2}+\frac{1}{4}\right)\right)\right]\frac{(-1)^k}{(2k+1)}.$$

We also obtain with the use of the identity

$$2\sin A\,\sin B = \cos(A-B)-\cos(A+B)$$

that

$$1-\cos 2\pi b = \sum_{k=0}^{\infty}\cos\left[\pi\sqrt{4b^2+\left(k+\frac{1}{2}\right)^2}\right]\frac{(-1)^{k+1}}{\pi\left(\frac{k}{2}+\frac{1}{4}\right)}, \tag{7.2.6}$$

which gives for $b = 1$ and $1/4$

$$\sum_{k=0}^{\infty}\cos\left[\pi\sqrt{4+\left(k+\frac{1}{2}\right)^2}\right]\frac{(-1)^{k+1}}{(2k+1)} = 0,$$

and

$$\frac{\pi}{4} = \sum_{k=0}^{\infty}\cos\left[\pi\sqrt{\frac{1}{4}+\left(k+\frac{1}{2}\right)^2}\right]\frac{(-1)^{k+1}}{(2k+1)}.$$

By differentiating (7.2.6) with respect to b, we obtain (7.2.1) once more.

Another summation formula involving the Bessel function, which we believe to be new, can be obtained as follows: From formula 6.681-1, p. 738 in [10], we obtain

$$\int_0^{\pi} J_{2\nu}\left(2b\cos\frac{x}{2}\right)\cos(tx)\,dx = \pi J_{\nu+t}(b)J_{\nu-t}(b), \tag{7.2.7}$$

Hence, by putting $g(x) = J_{2\nu}\left(2b\cos\frac{x}{2}\right)$, which is in $L^2(0,\pi)$ for $\nu > -\frac{1}{4}$, in (7.2.3), we then have $f(b,t) = \pi J_{\nu+t}(b)J_{\nu-t}(b)$, and consequently (7.2.4), (7.2.5) yield

$$J_{\nu+t}(b)J_{\nu-t}(b) = \sum_{k=-\infty}^{\infty} J_{\nu+\left(k+\frac{1}{2}\right)}(b)J_{\nu-\left(k+\frac{1}{2}\right)}(b)\,\frac{\sin\pi\left(t-\left(k+\frac{1}{2}\right)\right)}{\pi\left(t-\left(k+\frac{1}{2}\right)\right)}, \qquad (7.2.8)$$

and

$$[J_\nu(b)]^2 = \frac{4}{\pi}\sum_{k=0}^{\infty} J_{\nu+\left(k+\frac{1}{2}\right)}(b)J_{\nu-\left(k+\frac{1}{2}\right)}(b)\,\frac{(-1)^k}{(2k+1)}. \qquad (7.2.9)$$

Similar summation formulae involving the modified Bessel functions of the first and second kinds $\mathrm{I}_\nu(b)$ and $\mathrm{K}_\nu(b)$ respectively, can also be obtained since they have integral representations similar to the one given in (7.2.7).

As a special case of (7.2.9), we obtain for $\nu = 1/2$ that

$$[\sin b]^2 = \sum_{k=0}^{\infty} bJ_{k+1}(b)J_k(b)\,\frac{2}{2k+1}.$$

Summation formulae involving other types of special functions can also be obtained as the next example shows.

From formula 6.685, p. 741 in [10], we have

$$\int_0^\pi\left(\sec\frac{x}{2}\right)\mathrm{K}_{2\nu}\left(b\sec\frac{x}{2}\right)\cos(tx)\,dx = \frac{\pi}{b}\mathrm{W}_{t,\nu}(b)\,\mathrm{W}_{-t,\nu}(b),$$

where $\mathrm{W}_{\lambda,\mu}(z)$ is the Whittaker function, which is related to the confluent hypergeometric function $\Psi(a,c;z)$ by

$$\mathrm{W}_{\lambda,\mu}(z) = e^{-z/2}z^{c/2}\Psi(a,c;z) \quad \text{where} \quad \lambda = -a+\frac{c}{2} \quad \text{and} \quad \mu = \frac{c-1}{2}.$$

By putting $g(b,x) = \left(\sec\frac{x}{2}\right)\mathrm{K}_{2\nu}\left(b\sec\frac{x}{2}\right)$ in (7.2.3), we obtain that $f(b,t) = \frac{\pi}{b}\mathrm{W}_{t,\nu}(b)\mathrm{W}_{-t,\nu}(b)$; hence, formulae (7.2.4) and (7.2.5) take the form

$$\mathrm{W}_{t,\nu}(b)\mathrm{W}_{-t,\nu}(b) = \sum_{k=-\infty}^{\infty}\mathrm{W}_{\left(k+\frac{1}{2}\right),\nu}(b)\mathrm{W}_{-\left(k+\frac{1}{2}\right),\nu}(b)\,\frac{\sin\pi\left(t-\left(k+\frac{1}{2}\right)\right)}{\pi\left(t-\left(k+\frac{1}{2}\right)\right)},$$

and

$$[\mathrm{W}_{o,\nu}(b)]^2 = \frac{4}{\pi}\sum_{k=0}^{\infty} \mathrm{W}_{\left(k+\frac{1}{2}\right),\nu}(b)\mathrm{W}_{-\left(k+\frac{1}{2}\right),\nu}(b)\,\frac{(-1)^k}{(2k+1)}.$$

But since $\mathrm{W}_{o,\nu}(b) = \sqrt{\frac{b}{\pi}}\mathrm{K}_\nu\left(\frac{b}{2}\right)$ (cf. 9.235, p. 1062 in [10]), we then have

$$b\left[\mathrm{K}_\nu\left(\frac{b}{2}\right)\right]^2 = \frac{4}{\pi}\sum_{k=0}^{\infty} \mathrm{W}_{\left(k+\frac{1}{2}\right),\nu}(b)\mathrm{W}_{-\left(k+\frac{1}{2}\right),\nu}(b)\,\frac{(-1)^k}{(2k+1)}.$$

More formulae of similar types can be obtained in the same fashion.

7.3 Summation Formulae Involving the Zeros of the Bessel Function

Now we turn our attention to a different type of summation formulae. These formulae, which involve the zeros of the Bessel function, will also be obtained from already known results in sampling theory.

The following result, which has been known for a while (cf. [26, corollary 5, p. 904]), does not follow directly from the original version of Kramer's sampling theorem since the kernel of the integral transform $K(x,t)$ arises from a *singular* Sturm-Liouville problem. Nevertheless, it follows from Theorem 4.2 and Example B1 in Section 4.4 (cf. [25]).

If

$$f(t) = \int_0^1 g(x)\sqrt{x}\,J_\nu(tx)\,dx\,, \quad \text{for some} \quad g\in L^2(0,1) \tag{7.3.1}$$

then,

$$f(t) = \sum_{k=1}^{\infty} f(t_{k,\nu})\,\frac{2t_{k,\nu}J_\nu(t)}{J_{\nu+1}(t_{k,\nu})[(t_{k,\nu})^2 - t^2]}, \tag{7.3.2}$$

where $t_{k,\nu}$ is the kth positive zero of $J_\nu(t)$.

In view of formula 6.59-16, p. 703, in [10], we have

$$\int_0^1 x^{\frac{1}{2}-\nu}(1-x^2)^{\mu-1}\sqrt{x}\,J_\nu(tx)\,dx = \frac{2^{1-\nu}t^{-\mu}}{\Gamma(\nu)}\mathrm{s}_{\mu+\nu-1,\mu-\nu}(t)\,, \quad Re\,\mu>0$$

where $\mathrm{s}_{\mu,\nu}(z)$ is the Lommel function, which is defined in terms of the hypergeometric function by

$$\mathrm{s}_{\mu,\nu}(z) = \frac{z^{\mu+1}}{(\mu-\nu+1)(\mu+\nu+1)}\,{}_1F_2\left(1;\frac{\mu-\nu+3}{2},\frac{\mu+\nu+3}{2};-\frac{z^2}{4}\right), \tag{7.3.3}$$

where $\mu\pm\nu$ is not a negative integer (cf. 8574-3, p. 985 in [10]).

If we set $g(x) = x^{\frac{1}{2}-\nu}(1-x^2)^{\mu-1}$, which is in $L^2(0,1)$ for $\nu < 1$ and $\mu > 1/2$, in (7.3.1), we then have $f(t) = ((2^{1-\nu}t^{-\mu})/\Gamma(\nu))\, s_{\mu+\nu-1,\mu-\nu}(t)$. Thus, (7.3.2) now yields

$$t^{-\mu} s_{\mu+\nu-1,\mu-\nu}(t) = \sum_{k=1}^{\infty} \frac{2(t_{k,\nu})^{1-\mu} s_{\mu+\nu-1,\mu-\nu}(t_{k,\nu}) J_\nu(t)}{J_{\nu+1}(t_{k,\nu})[(t_{k,\nu})^2 - t^2]}. \tag{7.3.4}$$

But from (7.3.3), we have

$$s_{\mu+\nu-1,\mu-\nu}(t) = \frac{t^{\nu+\mu}}{4\mu\nu}\, {}_1F_2\left(1;\nu+1,\mu+1;-\frac{t^2}{4}\right).$$

Hence, by taking the limit in (7.3.4) as $t \to 0$, we obtain after some simplifications

$$2^{\nu-3}\,\frac{\Gamma(\nu)}{\mu} = \sum_{k=1}^{\infty} \frac{s_{\mu+\nu-1,\mu-\nu}(t_{k,\nu})}{J_{\nu+1}(t_{k,\nu})\,(t_{k,\nu})^{\mu+1}}, \tag{7.3.5}$$

which we believe is new.

In particular, if $\nu = 1/2$, we have $t_{k,\nu} = k\pi$; hence (7.3.5) becomes

$$\frac{\sqrt{\pi}}{\sqrt{2^5}\,\mu} = \sum_{k=1}^{\infty} \frac{s_{\mu-\frac{1}{2},\mu-\frac{1}{2}}(k\pi)}{J_{\frac{3}{2}}(k\pi)\,(k\pi)^{\mu+1}}. \tag{7.3.6}$$

But since (cf. [10, 8.573-5, p. 986])

$$s_{\nu,\nu}(z) = \Gamma\left(\nu+\frac{1}{2}\right)\sqrt{\pi}\, 2^{\nu-1} \mathbf{H}_\nu(z),$$

where $\mathbf{H}_\nu(z)$ is the Struve function defined by (cf. [10, 8550-1, p. 982])

$$\mathbf{H}_\nu(z) = \sum_{m=0}^{\infty} (-1)^m \frac{(z/2)^{2m+\nu+1}}{\Gamma\left(m+\frac{3}{2}\right)\Gamma\left(\nu+m+\frac{3}{2}\right)},$$

equation (7.3.6) can now be written in the form

$$\frac{1}{2^{\mu+1}\Gamma(\mu+1)} = \sum_{k=1}^{\infty} \frac{\mathbf{H}_{\mu-\frac{1}{2}}(k\pi)}{J_{\frac{3}{2}}(k\pi)\,(k\pi)^{\mu+1}}. \tag{7.3.7}$$

It is easy to see that $J_{\frac{3}{2}}(k\pi) = \sqrt{2/k}\ (-1)^{k+1}/\pi$; thus, (7.3.7) can be reduced to

$$\frac{(\pi)^{\mu}}{2^{\mu+\frac{1}{2}}\Gamma(\mu+1)} = \sum_{k=1}^{\infty} \frac{(-1)^{k+1}\mathbf{H}_{\mu-\frac{1}{2}}(k\pi)}{(k)^{\mu+\frac{1}{2}}}; \tag{7.3.8}$$

see formula (45), p. 70 in [5]. The series in (7.3.8) is called a Schlömilch-type series.

In view of the fact that $\mathbf{H}_{\frac{1}{2}}(z) = \sqrt{2/\pi z}\ (1-\cos z)$, (cf. [10, 8.552-6, p. 983]), we obtain, from (7.3.8) for $\mu = 1$, the well known formula

$$\sum_{k=1}^{\infty} 1/(2k-1)^2 = \pi^2/8.$$

Now let us start with Sonine's second integral ((4) in sec. 7.7 of [5])

$$\int_0^1 x^{\mu+1}(1-x^2)^{\nu/2}J_{\nu}(b\sqrt{1-x^2})J_{\mu}(tx)\,dx = \frac{t^{\mu}b^{\nu}}{(t^2+b^2)^{(\mu+\nu+1)/2}}\,J_{\nu+\mu+1}(\sqrt{t^2+b^2}),$$

$t, b > 0$; $Re\,\nu$, $Re\,\mu > -1$.

By setting $g(b,x) = x^{\mu+\frac{1}{2}}(1-x^2)^{\nu/2}J_{\nu}(b\sqrt{1-x^2})$ in (7.3.1), we then have

$$f(b,t) = \frac{t^{\mu}b^{\nu}}{(t^2+b^2)^{(\nu+\mu+1)/2}}\,J_{\nu+\mu+1}(\sqrt{t^2+b^2}),$$

and consequently (7.3.2) gives

$$\frac{t^{\mu}}{2J_{\mu}(t)\,(t^2+b^2)^{(\nu+\mu+1)/2}}\,J_{\nu+\mu+1}(\sqrt{t^2+b^2}) =$$

$$\sum_{k=1}^{\infty} \frac{(t_{k,\mu})^{1+\mu}}{J_{\mu+1}(t_{k,\mu})[(t_{k,\mu})^2-t^2]}\,\frac{J_{\nu+\mu+1}\left(\sqrt{(t_{k,\mu})^2+b^2}\right)}{[(t_{k,\mu})^2+b^2]^{(\nu+\mu+1)/2}}, \tag{7.3.9}$$

which is formula (4.1) in [9].

By taking the limit in the above equation as $t \to 0$, we obtain

$$\frac{2^{\nu-1}\Gamma(\nu+1)}{b^{\nu+\mu+1}}\,J_{\nu+\mu+1}(b) = \sum_{k=1}^{\infty} \frac{(t_{k,\mu})^{\mu-1}}{J_{\mu+1}(t_{k,\mu})}\,\frac{J_{\nu+\mu+1}\left(\sqrt{(t_{k,\mu})^2+b^2}\right)}{[(t_{k,\mu})^2+b^2]^{(\mu+\nu+1)/2}},$$

from which, upon taking the limit once more as $b \to 0$, we obtain

$$\frac{\Gamma(\nu+1)}{2^{\mu+2}\Gamma(\nu+\mu+2)} = \sum_{k=1}^{\infty} \frac{J_{\nu+\mu+1}(t_{k,\mu})}{J_{\mu+1}(t_{k,\mu})\,(t_{k,\mu})^{\nu+2}}.$$

By taking the limit in (7.3.9) as $b \to 0$, we obtain

$$\frac{J_{\nu+\mu+1}(t)}{2J_{\mu}(t)t^{\nu+1}} = \sum_{k=1}^{\infty} \frac{J_{\nu+\mu+1}(t_{k,\mu})}{J_{\mu+1}(t_{k,\mu})\,(t_{k,\mu})^{\nu}\,[(t_{k,\mu})^2 - t^2]},$$

which can also be obtained from (7.3.1), (7.3.2) and Sonine's first integral (cf. (5) - 7.7.2 in [5]).

7.4 Summation Formulae Involving the Γ-Function

In Section 4.4, Example C2, the following sampling formula has been proved (cf. (4.4.8), (4.4.10)): if

$$f(t) = \int_0^{\pi} g(\theta)\left(\sin\frac{\theta}{2}\right)^{\alpha+\frac{1}{2}} \left(\cos\frac{\theta}{2}\right)^{\beta+\frac{1}{2}} \mathrm{R}_{t-\gamma}^{(\alpha,\beta)}(\cos\theta)\,d\theta,$$

$$\alpha,\beta > -1 \quad \text{and} \quad 2\gamma = \alpha+\beta+1, \tag{7.4.1}$$

then

$$f(t) = \begin{cases} \displaystyle\sum_{k=0}^{\infty} f(k+\gamma)\, \frac{(-1)^{k+1}2(k+\gamma)\,\Gamma(k+2\gamma)}{\Gamma(\gamma+1)\,\Gamma(\gamma-t)[t^2-(k+\gamma)^2]\Gamma(k+1)} & \text{if } \gamma \neq 0 \\ \displaystyle f(0)\,\frac{\sin\pi t}{\pi t} + \sum_{k=1}^{\infty} f(k)\, \frac{2t\sin[\pi(t-k)]}{\pi[t^2-k^2]} & \text{if } \gamma = 0, \end{cases} \tag{7.4.2}$$

where $\mathrm{R}_t^{(\alpha,\beta)}(z) = {}_2\mathrm{F}_1\left(-t, t+2\gamma; \alpha+1; \frac{1-z}{2}\right)$ is the Jacobi function, which reduces to the Jacobi polynomial of degree n when $t = n$.

By substituting $x = \sin^2(\theta/2)$ in (7.4.1), we obtain

$$f(t) = \int_0^1 \hat{g}(x)\, x^{\left(\frac{\alpha}{2}-\frac{1}{4}\right)} (1-x)^{\left(\frac{\beta}{2}-\frac{1}{4}\right)}\, {}_2\mathrm{F}_1(-t+\gamma, t+\gamma; \alpha+1; x)\,dx, \tag{7.4.3}$$

where $\hat{g}(x) = g(2\sin^{-1}\sqrt{x})$.

Also, by using the notation $(a)_k = \Gamma(a+k)/\Gamma(a)$ and the relations

$$(x+k) = \frac{x(x+1)_k}{(x)_k}, \quad (x-k) = -\frac{\Gamma(1-x)(1-x)_k}{\Gamma(-x)(-x)_k} \quad \text{and} \quad \Gamma(z)\Gamma(1-z) = \pi\csc(\pi z),$$

we can transform (7.4.2) into the form

$$f(t) = \sum_{k=0}^{\infty} f(k+\gamma)A_k(t,\gamma), \tag{7.4.4}$$

where

$$A_k(t,\gamma) = \frac{(-1)^k\Gamma(2\gamma+1)(2\gamma)_k(\gamma+1)_k(\gamma-t)_k(\gamma+t)_k}{\Gamma(\gamma+1+t)\Gamma(\gamma+1-t)(\gamma)_k(\gamma+1+t)_k(\gamma+1-t)_k\Gamma(k+1)},$$

for $\gamma \neq 0$. Similar expression can be obtained for $\gamma = 0$.

In view of formula 7.51-4, p. 849 in [10], we have

$$\int_0^1 x^{\alpha}(1-x)^{\delta-1}{}_2F_1(-t+\gamma,t+\gamma;\alpha+1;x)\,dx = \frac{\Gamma(\alpha+1)\Gamma(\delta)\Gamma(\delta-\beta)}{\Gamma(h+t)\Gamma(h-t)},$$

where $h = (\alpha-\beta+1)/2+\delta$ and $2\gamma = \alpha+\beta+1$.

This last integral is in the form (7.4.3) for $\hat{g}(x) = (x)^{(2\alpha+1)/4}(1-x)^{(4\delta-2\beta-3)/4}$, which is in $L^2(0,1)$ for $\alpha > -1$, and $\delta > (2\beta+1)/4$. Therefore, by substituting $f(t) = (\Gamma(\alpha+1)\Gamma(\delta)\Gamma(\delta-\beta))/(\Gamma(h+t)\Gamma(h-t))$ in (7.4.4), we obtain after some computations that for $\gamma \neq 0$,

$$\frac{\Gamma(h+\gamma)\Gamma(h-\gamma)\Gamma(1-t+\gamma)\Gamma(1+t+\gamma)\Gamma(\gamma)}{\Gamma(h+t)\Gamma(h-t)\Gamma(2\gamma)\Gamma(\gamma+1)}$$

$$= \sum_{k=0}^{\infty} \frac{2(\gamma-t)_k(\gamma+t)_k(\gamma+1)_k(2\gamma)_k(1-h+\gamma)_k}{(1-t+\gamma)_k(1+t+\gamma)_k(\gamma)_k(h+\gamma)_k\Gamma(k+1)}$$

$$= 2\,{}_5F_4\left(\begin{matrix}2\gamma,1+\gamma,\gamma-t,\gamma+t,1-h+\gamma; \\ \gamma,1+\gamma+t,1+\gamma-t,h+\gamma;\end{matrix}\;1\right),$$

which is formula (III.12), p. 244 in [21] for $a = 2\gamma$, $b = \gamma - t$, $c = \gamma + t$, $d = 1-h+\gamma$.

Finally, we conclude this section with a summation formula involving the hypergeometric function ${}_3F_2$. This summation formula gives an expansion of a hypergeometric function in terms of hypergeometric polynomials.

In view of formula 7.51-5, p. 849 in [10], we have

$$\int_0^1 x^{\delta-1}(1-x)^{\sigma-1}{}_2F_1(-t+\gamma,t+\gamma;\alpha+1;x)\,dx =$$

$$\frac{\Gamma(\delta)\Gamma(\sigma)}{\Gamma(\delta+\sigma)}\,{}_3F_2(-t+\gamma,t+\gamma,\delta;\alpha+1,\delta+\sigma;1)$$

Re $\delta > 0$, *Re* $\sigma > \max\{0,\beta\}$.

This last integral is in the form (7.4.3) for

$$\hat{g}(x) = x^{(4\delta-2\alpha-3)/4}(1-x)^{(4\sigma-2\beta-3)/4},$$

which is in $L^2(0,1)$ for $\delta > (2\alpha+1)/4$, and $\sigma > (2\beta+1)/4$.

Now with $f(t) = \{\Gamma(\delta)\Gamma(\sigma)/\Gamma(\delta+\sigma)\}\ {}_3F_2(-t+\gamma, t+\gamma, \delta; \alpha+1, \delta+\sigma; 1)$, we obtain from (7.4.4) the summation formula

$$\begin{aligned}&{}_3F_2(-t+\gamma, t+\gamma, \delta; \alpha+1, \delta+\sigma; 1) = \\ &\sum_{k=0}^{\infty} {}_3F_2(-k, k+2\gamma, \delta; \alpha+1, \delta+\sigma; 1) A_k(t,\gamma),\end{aligned} \tag{7.4.5}$$

which, upon using the relation

$${}_2F_1(a,b;c;1) = \frac{\Gamma(c)\Gamma(c-b-a)}{\Gamma(c-a)\Gamma(c-b)}$$

and carrying out some computations, can be transformed once more, for $b_1 = \alpha+1$ and $b_2 = \delta+\sigma$, into

$$\begin{aligned}{}_3F_2(-t+\gamma, t+\gamma, \delta; \alpha+1, \delta+\sigma; 1) &= \sum_{k=0}^{\infty} \frac{(-1)^k(\gamma+t)_k(\gamma-t)_k}{(2\gamma+k)_k\Gamma(k+1)} \\ &\quad {}_2F_1(k-t+\gamma, k+t+\gamma; 1+2k+2\gamma; 1)\,{}_3F_2(-k, k+2\gamma, \delta; \alpha+1, \delta+\sigma; 1) \\ &= \sum_{k=0}^{\infty} \frac{(-1)^k(\gamma+t)_k(\gamma-t)_k(b_1)_k(b_2)_k}{(2\gamma+k)_k\Gamma(k+1)(b_1)_k(b_2)_k} \\ &\quad {}_4F_3\left(\begin{matrix}k+b_1, k+b_2, k+\gamma-t, k+\gamma-t \\ k+b_1, k+b_2, 1+2k+2\gamma\end{matrix}\ \middle|\ 1\right) {}_3F_2\left(\begin{matrix}-k, k+2\gamma, \delta \\ b_1, b_2\end{matrix}\ \middle|\ 1\right)\end{aligned} \tag{7.4.6}$$

Formula (7.4.6) is a special case of formula (1.2) in [7] for $r = 1$, $s = 0$, $p = 2 = q$, $c_1 = \delta$, $b_1 = \alpha+1$, $b_2 = \delta+\sigma$, $a_1 = \gamma-t$, $a_2 = \gamma+t$, $\beta = \delta+\sigma$ and α replaced by $\alpha+1$, γ replaced by 2γ; see also [8].

7.5 Summation Formula Involving the Kampé de Fériet Function

In this section we derive a summation formula analogous to (7.4.5), but involving the Kampé de Fériet function. We will use the following contracted notation for the generalized hypergeometric function

$${}_pF_q\left(\begin{matrix}a_1, \dots, a_p; \\ b_1, \dots, b_q;\end{matrix}\ z\right) = {}_pF_q\left(\begin{matrix}a_P; \\ b_Q;\end{matrix}\ z\right) = {}_pF_q(a_P; b_Q; z) = \sum_{k=0}^{\infty} \frac{(a_P)_k}{(b_Q)_k}\frac{z^k}{k!}$$

where $(a_P)_k = \prod_{j=1}^{p}(a_j)_k$, $(b_Q)_k = \prod_{j=1}^{q}(b_j)_k$ and $(c)_k = \Gamma(c+k)/\Gamma(c)$.

Let $\hat{g}(x) = x^{\delta-\alpha/2-3/4}(1-x)^{\sigma-\beta/2-3/4}\left(\sum_{n=0}^{\infty} h_n x^n\right)$ be in $L^2(0,1)$ and assume that the series converges uniformly on [0, 1]. Then by substituting this in (7.4.3), interchanging the summation and integration signs and using formulae (2.3.2), (2.3.3) in [6], we obtain

$$f(t) = \frac{\Gamma(\delta)\Gamma(\sigma)}{\Gamma(\delta+\sigma)} \sum_{n=0}^{\infty} h_n \frac{(\delta)_n}{(\delta+\sigma)_n} \, {}_3F_2\left(\begin{matrix} -t+\gamma, t+\gamma, \delta+n \\ \alpha+1, \delta+\sigma+n \end{matrix} \; 1\right).$$

Hence, from (7.4.4) we have

$$\sum_{n=0}^{\infty} h_n \frac{(\delta)_n}{(\delta+\sigma)_n} \, {}_3F_2\left(\begin{matrix} -t+\gamma, t+\gamma, \delta+n \\ \alpha+1, \delta+\sigma+n \end{matrix} \; 1\right) =$$

$$\sum_{k=0}^{\infty} \sum_{n=0}^{\infty} h_n \frac{(\delta)_n}{(\delta+\sigma)_n} \, {}_3F_2\left(\begin{matrix} -k, k+2\gamma, \delta+n \\ \alpha+1, \delta+\sigma+n \end{matrix} \; 1\right) A_k(t,\gamma) .$$

As a special case of this, if we take $h_n = \{(a_P)_n/(b_Q)_n\}\ \{y^n/n!\}$ with $p \le q+1$, we obtain after some calculations that

$$F^{1:2;p}_{1:1;q}\left(\begin{matrix} \delta: -t+\gamma, t+\gamma; (a_P); \\ \delta+\sigma: \alpha+1; (b_Q); \end{matrix} \; 1, y\right) =$$

$$\sum_{k=0}^{\infty} F^{1:2;p}_{1:1;q}\left(\begin{matrix} \delta: -k, k+2\gamma; (a_P); \\ \delta+\sigma: \alpha+1; (b_Q); \end{matrix} \; 1, y\right) A_k(t,\gamma), \tag{7.5.1}$$

where

$$F^{p:q;k}_{l:m;n}\left(\begin{matrix} (a_P):(b_Q);(c_K); \\ (\alpha_L):(\beta_M);(\gamma_N); \end{matrix} \; x, y\right) = \sum_{r,s=0}^{\infty} \frac{\prod_{i=1}^{p}(a_i)_{r+s} \prod_{i=1}^{q}(b_i)_r \prod_{i=1}^{k}(c_i)_s}{\prod_{i=1}^{l}(\alpha_i)_{r+s} \prod_{i=1}^{m}(\beta_i)_r \prod_{i=1}^{n}(\gamma_i)_s} \frac{x^r}{r!} \frac{y^s}{s!}$$

is the Kampé de Fériet function. To the best of the author's knowledge formula (7.5.1) is new.

References

1. A. B. Bhatia and K. S. Krishnan, Light-scattering in homogeneous media regarded as reflexion from appropriate thermal elastic waves, *Proc. Royal. Soc. London*, 192 (1948), 181-195.

2. J. L. Brown, Summation of certain series using the Shannon sampling theorem, *IEEE Trans. on Education*, 33 (1990), 337-340.

3. P. L. Butzer, N. Hauss and R. L. Stens, The sampling theorem and its unique role in various branches of mathematics, Manuscript (1990).

4. A. Erdelyi, W. Magnus, F. Oberhettinger and F. Tricomi, *Tables of Integral Transforms*, Vol. 1, McGraw-Hill, New York (1954).

5. A. Erdelyi, W. Magnus, F. Oberhettinger and F. Tricomi, *Higher Transcendental Functions*, Vol.2, McGraw-Hill, New York (1953).

6. H. Exton, *Handbook of Hypergeometric Integral*, John Wiley & Sons Publ., New York (1978).

7. J. Fields and M. Ismail, Polynomial expansions, *Math. Comp.*, 29, 131 (1975), 894-902.

8. J. Fields and J. Wimp, Expansions of hypergeometric functions in hypergeometric polynomials, *Math. Comp.*, 15 (1961), 390-395.

9. R. Gosper, M. Ismail and R. Zhang, On some strange summation formulas of Gosper, to appear in Ill. J. of Math.

10. I. Gradshteyn and I. Ryzhik, *Tables of Integrals, Series and Products*, Academic Press, New York (1965).

11. J. R. Higgins, Sampling theorems and the contour integral method, *J. Applicable Analysis*, 41 (1991), 155-169.

12. ______, An interpolation series associated with the Bessel-Hankel transform, *J. London Math. Soc.*, 5 (1972), 707-714.

13. M. Ismail, Y. Takeuchi and R. Zhang, Pages from the computer files of R. William Gosper, Manuscript (1992).

14. A. Jerri, A note on sampling expansion for a transform with parabolic cylindrical kemel, *Inform. Sci.*, 26 (1982), 155-158.

15. ______, On the application of some interpolating functions in physics, *J. Res. Nat. Bur. Standards* Sec. B (80), (B) 3 (1969).

16. ______, On Extension of the Generalized Sampling Theorem, Ph.D. Thesis, Oregon State University (1967), 37-38.

17. K. Knopp, *Theory of Functions*, Part II, Dover Publication, New York (1947).

18. K. S. Krishnan, On the equivalence of certain infinite series and the corresponding integrals, *J. Indian Math. Soc.* (N.S.), 12 (1948), 79-88.

19. F. C. Mehta, Sampling expansion for band-limited signals through some special functions, *J. Cybernetics*, 5 (1975), 61-68.

20. S. D. Selvaratnam, Shannon-Whittaker Sampling Theorems, Ph.D. thesis, University of Wisconsin-Milwaukee (1987).

21. L. Slater, *Generalized Hypergeometric Functions*, Cambridge Univ. Press, Cambridge, England (1966).

22. E. C. Titchmarsh, *Introduction to the Theory of Fourier Integrals*, Oxford University Press, Oxford, England (1937).

23. J. M. Whittaker, *Interpolatory Function Theory*, Cambridge University Press, Cambridge, England (1935).

24. A. Zayed, A proof of new summation formulae by using sampling theorems, *Proc. Amer. Math. Soc.*, 117 (1993), 699-710.

25. ______, On Kramer's sampling theorem associated with general Sturm-Liouville problems and Lagrange interpolation, *SIAM J. Appl. Math.*, 51, 2 (1991), 575-604.

26. A. Zayed, G. Hinsen and P. Butzer, On Lagrange interpolation and Kramer-type sampling theorems associated with Sturm-Liouville problems, *SIAM J. Appl. Math.*, 50, 3 (1990), 893-909.

8

KRAMER'S SAMPLING THEOREM AND LAGRANGE-TYPE INTERPOLATION IN N-DIMENSIONS

8.0 Introduction

In Chapters 8 and 9 we shall extend some of the results obtained in Chapters 4 through 6 to higher dimensions. We begin this chapter by generalizing Kramer's sampling theorem to $N (N \geq 1)$ dimensions and then demonstrate, as in the one-dimensional case, how the kernel function $K(x, \lambda)$ and the sampling points $\{\lambda_n\}_{n=0}^{\infty}$ arise naturally when we solve certain Dirichlet or Neumann-type boundary-value problems. We then investigate the relationship between this generalization of Kramer's theorem on the one hand and N-dimensional versions of both the Whittaker-Shannon-Kotel'nikov (WSK) sampling theorem (Theorem 3.6) and the Paley-Wiener interpolation theorem for band-limited functions (Theorems 3.7 and 3.8) on the other. It will be shown that the sampling series associated with this generalization of Kramer's theorem is again nothing more than a Lagrange-type interpolation series.

Some of the results in the one-dimensional analysis do not easily carry over to N-dimensions. For example, the equivalence of the WSK and Kramer sampling theorems in N-dimensions is still an open question.

We close this chapter by giving some applications to show how the techniques that will be developed in Sections 2 and 3 can be used to find the sums of multivariate infinite series as done in Chapter 7 for the one-dimensional case.

8.1 Preliminaries

In this short section we introduce some of the notation that will be used throughout this chapter.

Let N be a positive integer and $\underline{x} \in \Re^N$ stand for $\underline{x} = (x_1, \ldots, x_N)$ where $x_i \in \Re$, $i = 1, \ldots, N$; $\lambda \underline{x} = (\lambda x_1, \ldots, \lambda x_N)$ and $\underline{x} \cdot \underline{y} = x_1 y_1 + \ldots + x_N y_N$ where

$\underline{y} = (y_1, \ldots, y_N)$. Moreover, let $d\underline{x}$ denote $dx_1 dx_2 \ldots dx_N$, hence

$$\int_{\mathfrak{R}^N} \ldots d\underline{x} = \int_{-\infty}^{\infty} \ldots d\underline{x}$$

will denote

$$\int_{x_1=-\infty}^{\infty} \int_{x_2=-\infty}^{\infty} \cdots \int_{x_N=-\infty}^{\infty} \ldots dx_1\, dx_2 \ldots dx_N .$$

We set

$$\Delta = \frac{\partial^2}{\partial x_1^2} + \ldots + \frac{\partial^2}{\partial x_N^2},$$

and for any function q of N variables, $q(\underline{x})$ will mean $q(x_1, \ldots, x_N)$.

If $\mathcal{G}$ denotes a finite region in $\mathfrak{R}^N$, then its boundary will be denoted by $\partial \mathcal{G}$. An N-dimensional rectangle $\mathcal{B} = [\underline{a}, \underline{b}]$ is defined by $\mathcal{B} = \{\underline{x} \mid a_i \le x_i \le b_i ; i = 1, \ldots, N\}$, where $\underline{a} = (a_1, \ldots, a_N)$, $\underline{b} = (b_1, \ldots, b_N)$ and $L^k(\mathfrak{R}^N)$ is defined as $\left\{ f(\underline{x}): \int_{-\infty}^{\infty} | f(\underline{x}) |^k d\underline{x} < \infty \right\}$, $1 \le k < \infty$.

8.2 An N-Dimensional Kramer's Sampling Theorem*

First, let us begin by generalizing Kramer's sampling theorem to N-dimensions ($N \ge 1$). The obvious generalization of Kramer's sampling theorem goes as follows: Let $\mathcal{G}$ be a bounded region in $\mathfrak{R}^N$ with a smooth boundary. Assume that there exists a function $K(\underline{x} ; \underline{\lambda})$ of $2N$ variables with $\underline{x} \in \mathcal{G}$, $\underline{\lambda} = (\lambda_1, \ldots, \lambda_N) \in \mathfrak{R}^N$ that is continuous in $\underline{\lambda}$ and a sequence of vectors $\underline{\lambda}_{\underline{k}} = (\lambda_{1,k_1}, \ldots, \lambda_{N,k_N})$, $\underline{k} = (k_1, \ldots, k_N)$ where k_i is a nonnegative integer, $i = 1, \ldots, N$, such that $\{\psi_{\underline{k}}(\underline{x}) = K(\underline{x} ; \underline{\lambda}_{\underline{k}})\}$ is a complete orthogonal family of functions in $L^2(\mathcal{G})$.

If

$$f(\underline{\lambda}) = \int_{\mathcal{G}} F(\underline{x}) K(\underline{x} ; \underline{\lambda})\, d\underline{x}\,, \quad \text{for some} \quad F \in L^2(\mathcal{G}), \tag{8.2.1}$$

*Some material presented in Sections 8.2 and 8.3 is based on the author's article that appeared in the *J. Multidimensinal Systems and Signal Processing*, 3 (1992) and is printed with permission from Kluwer Academic Publishers.

then

$$f(\underline{\lambda}) = \sum_{\underline{k}} f(\underline{\lambda}_{\underline{k}}) S^{*}_{\underline{k}}(\underline{\lambda})$$

$$= \sum_{k_1,\ldots,k_N=0}^{\infty} f\left(\lambda_{1,k_1},\ldots,\lambda_{N,k_N}\right) S^{*}_{(k_1,\ldots,k_N)}(\underline{\lambda}), \tag{8.2.2}$$

where

$$S^{*}_{(k_1,\ldots,k_N)}(\underline{\lambda}) = \frac{\left\{\int_{\mathcal{G}} K(\underline{x};\underline{\lambda})\,\psi_{\underline{k}}(\underline{x})\,d\,\underline{x}\right\}}{\|\psi_{\underline{k}}(\underline{x})\|^{2}}. \tag{8.2.3}$$

The proof is exactly the same as in the one-dimensional case.

Now we demonstrate a prototype situation where the function $K(\underline{x};\underline{\lambda})$ and the vectors $\underline{\lambda}_{\underline{k}}$ are generated in a natural way.

Consider the partial differential equation

$$\Delta\psi + \{\lambda - q(\underline{x})\}\psi = 0, \tag{8.2.4}$$

where $q(\underline{x})$ is assumed to be continuous on some N-dimensional rectangle $\mathcal{B} = [\underline{a},\underline{b}]$ with $\underline{a} = (a_1,\ldots,a_N)$ and $\underline{b} = (b_1,\ldots,b_N)$.

Let $\psi(\underline{x};\lambda)$ be the solution of (8.2.4) which satisfies the following $2N$ boundary conditions:

$$\cos\alpha_i\psi(x_1,\ldots,x_{i-1},a_i,x_{i+1},\ldots,x_N) + \sin\alpha_i\frac{\partial}{\partial x_i}\psi(x_1,\ldots,x_{i-1},a_i,x_{i+1},\ldots,x_N) = 0, \tag{8.2.5}$$

$$\cos\beta_i\psi(x_1,\ldots,x_{i-1},b_i,x_{i+1},\ldots,x_N) + \sin\beta_i\frac{\partial}{\partial x_i}\psi(x_1,\ldots,x_{i-1},b_i,x_{i+1},\ldots,x_N) = 0, \tag{8.2.6}$$

$i = 1,2,\ldots,N$. Let Π denote the boundary-value problem (8.2.4) through (8.2.6).

Under the assumption that $q(\underline{x}) = \sum_{i=1}^{N} q_i(x_i)$, equation (8.2.4) becomes separable and $\psi(\underline{x},\lambda)$ can be assumed to be in the form $\psi(\underline{x};\lambda) = \prod_{i=1}^{N}\psi_i(x_i;\lambda)$. For simplicity, we may sometimes suppress the parameter λ and write $\psi(\underline{x})$ and $\psi_i(x_i)$ instead. It then follows from (8.2.4) through (8.2.6) that

$$\psi_i''(x_i) - q_i(x_i)\psi_i(x_i) = -\lambda_i\psi_i(x_i), \quad a_i \le x_i \le b_i \tag{8.2.7}$$

and

$$\cos\alpha_i\psi_i(a_i) + \sin\alpha_i\psi_i'(a_i) = 0\,, \tag{8.2.8}$$

$$\cos\beta_i\psi_i(b_i) + \sin\beta_i\psi_i'(b_i) = 0\,, \tag{8.2.9}$$

where $i = 1,\ldots,N$ and $\lambda = \sum_{i=1}^{N}\lambda_i$. For each fixed i, let Π_i denote the boundary-value problem (8.2.7) through (8.2.9).

It is known that each one of these N Sturm-Liouville boundary-value problems Π_i has a discrete spectrum [8, 10]. Let us denote the eigenvalues of the ith problem by $\{\lambda_{i,k_i}\}_{k_i=0}^{\infty}$ and let $\psi_i(x_i)$ be the solution of (8.2.7) that satisfies $\psi_i(a_i) = \sin\alpha_i$ and $\psi_i'(a_i) = -\cos\alpha_i$. Hence, $\psi_i(x_i)$ is a solution of (8.2.7) and (8.2.8), and the eigenvalues $\{\lambda_{i,k_i}\}_{k_i=0}^{\infty}$ are the zeros of the function $\cos\beta_i\psi_i(b_i\,;\lambda_i) + \sin\beta_i\psi_i'(b_i\,;\lambda_i)$.

Evidently, $\psi_{i,k_i}(x_i) = \psi_i(x_i\,;\lambda_{i,k_i})$ is an eigenfunction of Π_i corresponding to the eigenvalue λ_{i,k_i}. Since $\psi_i(x_i) \in L^2(I_i)$, where $I_i = [a_i, b_i]$ for $i = 1,\ldots,N$, $\psi(\underline{x}\,;\underline{\lambda}) = \prod_{i=1}^{N}\psi_i(x_i\,;\lambda_i)$ is in $L^2(\mathcal{B})$, where $\underline{\lambda} = (\lambda_1,\ldots,\lambda_N)$. Hence,

$$\psi_{\underline{k}}(\underline{x}) = \psi_{1,k_1}(x_1)\psi_{2,k_2}(x_2)\ldots\psi_{N,k_N}(x_N)\,, \tag{8.2.10}$$

is an eigenfunction of Π corresponding to the eigenvalue $\lambda_{\underline{k}} = \sum_{i=1}^{N}\lambda_{i,k_i}$, where $\underline{k} = (k_1,\ldots,k_N)$ and k_i is a nonnegative integer.

The converse is also true: in fact, it can be shown [11, p. 114] that all the eigenvalues and eigenfunctions of problem Π arise this way. Clearly, the set of eigenvalues $E = \{\lambda_{\underline{k}}\}$ of Π is countable. However, unlike the one-dimensional case, the eigenvalues are not necessarily simple, i.e., an eigenvalue may be repeated finitely many times. Consequently, to any eigenvalue there may correspond finitely many linearly independent eigenfunctions, which by a standard orthogonalization process, can be chosen to be orthogonal on $\mathcal{B}$.

The set E can now be written in the form $E = \{\lambda_K\}_{K=0}^{\infty}$ where each eigenvalue is repeated j times if and only if it has j eigenfunctions. Thus, each eigenvalue λ_K is associated with exactly one eigenfunction $\psi_K(\underline{x})$. The set of functions $\{\psi_K(\underline{x})\}_{K=0}^{\infty}$ is orthogonal since eigenfunctions belonging to different eigenvalues are orthogonal and, by the above construction, eigenfunctions belonging to the same eigenvalue are also orthogonal. It also follows from the above construction that there is a one-to-one correspondence between the different representations of the eigenvalues of the boundary-value problem Π, i.e.,

$$\lambda_K \leftrightarrow \lambda_{\underline{k}} \leftrightarrow \left(\lambda_{1,k_1},\ldots,\lambda_{N,k_N}\right)$$

where $\underline{k} = (k_1,\ldots,k_N)$, $\lambda_{\underline{k}} = \sum_{i=1}^{N} \lambda_{i,k_i}$ and K is that unique nonnegative integer that corresponds to the vector $\underline{k}$. This is guaranteed by the fact that the set $\mathbf{N}$ of nonnegative integers is isomorphic to $\mathbf{N}^N$. We shall identify λ_K with $\lambda_{\underline{k}}$.

Now we can state our sampling theorem which shows that if the kernel $K(\underline{x};\lambda)$ in (8.2.1) is associated with the boundary-value problem Π, then Kramer's sampling series (8.2.2) is nothing more than a Lagrange-type interpolation series given by a multidimensional analogue of the Paley-Wiener interpolation theorem for band-limited functions.

THEOREM 8.1

Let $\underline{\lambda} = (\lambda_1,\ldots,\lambda_N)$, $\lambda_i \in \Re$ and $\psi(\underline{x};\underline{\lambda}) = \prod_{i=1}^{N} \psi_i(x_i;\lambda_i)$. If

$$f(\underline{\lambda}) = \int_{\mathcal{B}} F(\underline{x})\psi(\underline{x},\underline{\lambda})\,d\underline{x}, \quad \text{for some} \quad F \in L^2(\mathcal{B}), \qquad (8.2.11)$$

then $f(\underline{\lambda})$ is an entire function of order 1/2 and type σ_i in each of the variables λ_i with $0 \le \sigma_i \le b_i - a_i$, $i = 1,\ldots,N$, and has the following sampling series representation:

$$f(\lambda_1,\ldots,\lambda_N) = \sum_{k_1,\ldots,k_N=0}^{\infty} f\left(\lambda_{1,k_1},\ldots,\lambda_{N,k_N}\right) \prod_{i=1}^{N} \frac{G_i(\lambda_i)}{(\lambda_i - \lambda_{i,k_i})G_i{}'(\lambda_{i,k_i})}, \qquad (8.2.12)$$

where

$$G_i(\lambda_i) = \begin{cases} \prod_{k_i=0}^{\infty}\left(1 - \dfrac{\lambda_i}{\lambda_{i,k_i}}\right) & \text{if none of the eigenvalues } \lambda_{i,k_i} \text{ is zero} \\ \lambda_i \prod_{k_i=1}^{\infty}\left(1 - \dfrac{\lambda_i}{\lambda_{i,k_i}}\right) & \text{if one of the eigenvalues, say } \lambda_{i,0}, \text{ is zero.} \end{cases} \qquad (8.2.13)$$

The series converges uniformly on any compact subset of the complex λ_i-plane.

Proof. That $f(\underline{\lambda})$ is an entire function of order 1/2 and type σ_i with $0 \le \sigma_i \le b_i - a_i$ in the variable λ_i follows from the relation

$$|f(\underline{\lambda})| \le \|F\|_2 \left(\prod_{i=1}^{N} \max_{a_i \le x_i \le b_i} |\psi_i(x_i;\lambda_i)| \right) \sqrt{V},$$

and the fact that $\psi_i(x_i;\lambda_i)$ has these properties (cf. [10, p. 10]), where V denotes the volume of $\mathcal{B}$.

Since both F and ψ are in $L^2(\mathcal{B})$, we have

$$F(\underline{x}) = \sum_{K=0}^{\infty} \hat{F}(K) \frac{\psi_K(\underline{x})}{\|\psi_K\|^2}, \tag{8.2.14}$$

and

$$\psi(\underline{x};\underline{\lambda}) = \sum_{K=0}^{\infty} \frac{\langle \psi, \psi_K \rangle}{\|\psi_K\|^2} \psi_K(\underline{x}), \tag{8.2.15}$$

where

$$\hat{F}(K) = \int_{\mathcal{B}} F(\underline{x}) \psi_K(\underline{x}) d\underline{x}, \tag{8.2.16}$$

$$\langle \psi, \psi_K \rangle = \int_{\mathcal{B}} \psi(\underline{x};\underline{\lambda}) \psi_K(\underline{x}) d\underline{x}. \tag{8.2.17}$$

The convergence in (8.2.14) and (8.2.15) is in the sense of $L^2(\mathcal{B})$.

From (8.2.11), (8.2.14), (8.2.15) and Parseval's equality, we obtain

$$f(\underline{\lambda}) = \sum_{K=0}^{\infty} \hat{F}(K) \frac{\langle \psi, \psi_K \rangle}{\|\psi_K\|^2}. \tag{8.2.18}$$

But from (8.2.10), (8.2.11) and (8.2.16), we also have

$$\begin{aligned} f(\lambda_{1,k_1}, \ldots, \lambda_{N,k_N}) &= \int_{\mathcal{B}} F(\underline{x}) \psi(\underline{x};\lambda_{1,k_1}, \ldots, \lambda_{N,k_N}) d\underline{x} \\ &= \int_{\mathcal{B}} F(\underline{x}) \prod_{i=1}^{N} \psi_i(x_i;\lambda_{i,k_i}) d\underline{x} \\ &= \int_{\mathcal{B}} F(\underline{x}) \psi_{\underline{k}}(\underline{x}) d\underline{x} = \int_{\mathcal{B}} F(\underline{x}) \psi_K(\underline{x}) d\underline{x} \\ &= \hat{F}(K), \end{aligned}$$

where K is that unique integer determined by the vector $\underline{k} = (k_1, \ldots, k_N)$. Thus,

$$f(\underline{\lambda}) = \sum_{k_1, \ldots, k_N = 0}^{\infty} f\left(\lambda_{1,k_1}, \ldots, \lambda_{N,k_N}\right) \frac{\langle \psi, \psi_{\underline{k}} \rangle}{\| \psi_{\underline{k}} \|^2}. \tag{8.2.19}$$

It is easy to see from (8.2.10) and (8.2.17) that for $\underline{k} = (k_1, \ldots, k_N)$

$$\langle \psi, \psi_{\underline{k}} \rangle = \prod_{i=1}^{N} \left(\int_{x_i = a_i}^{b_i} \psi_i(x_i; \lambda_i) \psi_i(x_i; \lambda_{i,k_i}) \, dx_i \right)$$

$$= \prod_{i=1}^{N} \left(\int_{a_i}^{b_i} \psi_i(x_i; \lambda_i) \psi_{i,k_i}(x_i) \, dx_i \right),$$

and in addition

$$\| \psi_{\underline{k}} \|^2 = \prod_{i=1}^{N} \| \psi_{i,k_i} \|^2.$$

Therefore,

$$\frac{\langle \psi, \psi_{\underline{k}} \rangle}{\| \psi_{\underline{k}} \|^2} = \prod_{i=1}^{N} \frac{\int_{a_i}^{b_i} \psi_i(x_i; \lambda_i) \psi_{i,k_i}(x_i) \, dx_i}{\| \psi_{i,k_i} \|^2}. \tag{8.2.20}$$

As is shown in the proof of Theorem 4.1 (see also [17 and 19]), for each fixed i

$$\frac{\int_{a_i}^{b_i} \psi_i(x_i; \lambda_i) \psi_{i,k_i}(x_i) \, dx_i}{\| \psi_{i,k_i} \|^2} = \frac{G_i(\lambda_i)}{(\lambda_i - \lambda_{i,k_i}) G_i'(\lambda_{i,k_i})}, \tag{8.2.21}$$

where $G_i(\lambda_i)$ is the Wronskian $\psi_i(x_i; \lambda_i) \chi_i'(x_i; \lambda_i) - \psi_i'(x_i; \lambda_i) \chi_i(x_i; \lambda_i)$ of the two functions $\psi_i(x_i; \lambda_i)$, $\chi_i(x_i; \lambda_i)$, and $\chi_i(x_i; \lambda_i)$ is the solution of (8.2.7) that satisfies $\chi_i(b_i; \lambda_i) = \sin \beta_i$ and $\chi_i'(b_i; \lambda_i) = -\cos \beta_i$.

This Wronskian, which is known to be independent of x_i, is an entire function of λ_i whose zeros are exactly the eigenvalues of problem Π_i. It was shown in Theorem 4.1 that, as far as the sampling theorem is concerned, $G_i(\lambda_i)$ can be taken as the canonical product of its zeros, i.e., as in (8.2.13). Therefore, by combining (8.2.21), (8.2.20) and (8.2.19), we obtain (8.2.12). Finally, the convergence of (8.2.12) can be proved as in Theorem 4.1. ▌

The relationship between our generalization of Kramer's theorem on the one hand and the N-dimensional versions of both the WSK (Theorem 3.6) and the Paley-Wiener interpolation theorems (Theorem 3.7), on the other, can now be stated. Clearly, the N-dimensional version of the WSK theorem is a special case of Theorem 8.1 (cf. Examples 1 and 2 in Section 3), whereas the series (8.2.12) is a special case of the series given by the N-dimensional version of the Paley-Wiener-Parzen interpolation theorem (Theorem 3.7) only if

$$f(\lambda_1,\ldots,\lambda_N) = f_1(\lambda_1)\ldots f_N(\lambda_N)\,.$$

On the other hand, if $N = 2$, then the series (8.2.12) is a special case of the series given by Theorem 3.8, when the sampling points $\{y_{nm}\}$ are independent of n.

It is worth noting ([10], p. 19) that the sampling points in (8.2.12) satisfy the following estimates:

$$\left(k_i - \frac{1}{2}\right)\frac{\pi}{b_i - a_i} < t_{i,k_i} < \left(k_i + \frac{1}{2}\right)\frac{\pi}{(b_i - a_i)},$$

where $t_{i,k_i} = \sqrt{\lambda_{i,k_i}}$. Hence, there exists $C > 0$, independent of i, such that

$$\sup_{|k_i| \geq C} \left| t_{i,k_i} - \frac{k_i\pi}{(b_i - a_i)} \right| < \frac{\pi}{4(b_i - a_i)}$$

for all $i = 1,\ldots,N$.

One question naturally arises: Given sequences of points $\{t_{i,k_i}\}$ satisfying the above estimates, can we construct the kernel $\psi(\underline{x};\underline{\lambda})$ in Theorem 8.1 to obtain a sampling series of type (8.2.12)?

The answer is "yes." This question is known in the theory of boundary-value problems as the inverse problem, i.e., constructing a boundary-value problem for a given set of eigenvalues; for more details we refer the reader to [19, Lemma 2] and [8].

8.3 Examples

***Example* 1.** Consider the partial differential equation

$$\Delta\psi(\underline{x}) = -\lambda\psi(\underline{x}), \quad \underline{x} \in \mathcal{B} = [\underline{0},\underline{b}]$$

together with the boundary conditions (8.2.5) and (8.2.6) for $\alpha_i = 0 = \beta_i$, i.e.,

$$\psi(x_1,\ldots,x_{i-1},0,x_{i+1},\ldots,x_N) = 0$$

$$\psi(x_1,\ldots,x_{i-1},b_i,x_{i+1},\ldots,x_N) = 0$$

$i = 1,2,\ldots,N.$

It is easy to see that $\psi_i(x_i;\lambda_i) = \sin(t_i x_i)/t_i$ and $\lambda_{i,k_i} = (k_i\pi/b_i)^2$, where $t_i = \sqrt{\lambda_i}$, $k_i = 1,2,\ldots;\ i = 1,2,\ldots,N$. Hence, $\psi(\underline{x},\underline{\lambda}) = (\sin t_1x_1)\ldots(\sin t_Nx_N)$,

$$G_i(\lambda_i) = \prod_{k_i=1}^{\infty}\left(1-\frac{\lambda_i}{(k_i\pi/b_i)^2}\right) = \frac{\sin b_it_i}{b_it_i},$$

and

$$G_i'\left(\lambda_{i,k_i}\right) = \frac{(-1)^{k_i}b_i^2}{2(k_i\pi)^2}.$$

Therefore, Theorem 8.1 now takes on the form:
If

$$f(\lambda_1,\ldots,\lambda_N) = \int_{x_1=0}^{b_1}\ldots\int_{x_N=0}^{b_N} F(x_1,\ldots,x_N)\,\frac{(\sin t_1x_1)}{t_1}\cdots\frac{(\sin t_Nx_N)}{t_N}\,dx_1\ldots dx_N,$$

for some $F(x_1,\ldots,x_N) \in L^2(\mathcal{B})$, then $f(\underline{\lambda})$ is an entire function of order 1/2 that admits the sampling expansion

$$f(\lambda_1,\ldots,\lambda_N) = 2^N \sum_{k_1,\ldots,k_N=0}^{\infty} f\left(\lambda_{1,k_1},\ldots,\lambda_{N,k_N}\right)\prod_{i=1}^{N}\frac{\lambda_{i,k_i}\sin b_i\left(\sqrt{\lambda_i}-\sqrt{\lambda_{i,k_i}}\right)}{\sqrt{\lambda_i}\,b_i(\lambda_i-\lambda_{i,k_i})}.$$

Upon setting $t_i = \sqrt{\lambda_i}$, $t_{i,k_i} = \sqrt{\lambda_{i,k_i}}$ and $\tilde{f}(t_1,\ldots,t_N) = f(\lambda_1,\ldots,\lambda_N)$, we obtain

$$\tilde{f}(t_1,\ldots,t_N) = 2^N \sum_{k_1,\ldots,k_N=0}^{\infty} \tilde{f}\left(t_{1,k_1},\ldots,t_{N,k_N}\right)\prod_{i=1}^{N}\frac{t_{i,k_i}^2\sin b_i(t_i-t_{i,k_i})}{t_ib_i\left(t_i^2-t_{i,k_i}^2\right)},$$

from which we further obtain

$$g(t_1,\ldots,t_N) = 2^N \sum_{k_1,\ldots,k_N=0}^{\infty} g\left(t_{1,k_1},\ldots,t_{N,k_N}\right)\prod_{i=1}^{N}\frac{t_{i,k_i}\sin b_i(t_i-t_{i,k_i})}{b_i\left(t_i^2-t_{i,k_i}^2\right)},$$

where $g(t_1, \ldots, t_N) = (t_1 \ldots t_N)\tilde{f}(t_1, \ldots, t_N)$. g is an entire function of exponential type σ_i with $0 \leq \sigma_i \leq b_i$ in each of the variables t_i. Since $\tilde{f}(t_1, \ldots, t_N)$ is an even function in each variable t_i, g is an odd function in each one of these variables. And hence, with straightforward calculations, we can obtain the well-known relation

$$g(t_1, \ldots, t_N) = \sum_{k_1, \ldots, k_N = -\infty}^{\infty} g\left(t_{1,k_1}, \ldots, t_{N,k_N}\right) \frac{\sin b_1(t_1 - t_{1,k_1}) \ldots \sin b_N(t_N - t_{N,k_N})}{b_1(t_1 - t_{1,k_1}) \ldots b_N(t_N - t_{N,k_N})},$$

which is sometimes called the rectangular cardinal series, (cf. [6, 9, p. 453]).

***Example* 2.** Consider the differential equation

$$\Delta\psi(\underline{x}) = -\lambda\psi(\underline{x}), \quad \underline{x} \in \mathcal{B} = [\underline{0}, \underline{b}]$$

together with the boundary conditions (8.2.5) and (8.2.6) for $\alpha_i = \pi/2 = \beta_i$, i.e.,

$$\frac{\partial}{\partial x_i}\psi(x_1, \ldots, x_{i-1}, 0, x_{i+1}, \ldots, x_N) = 0,$$

and

$$\frac{\partial}{\partial x_i}\psi(x_1, \ldots, x_{i-1}, b_i, x_{i+1}, \ldots, x_N) = 0,$$

$i = 1, 2, \ldots, N$.

It is easy to see that $\psi_i(x_i; \lambda_i) = \cos\sqrt{\lambda_i}\, x_i$ and $\lambda_{i,k_i} = (k_i\pi/b_i)^2$, where $k_i = 0, 1, \ldots$. Zero is an admissible eigenvalue since the corresponding eigenfunction is not identically zero. In this case, for each $i = 1, \ldots, N$,

$$G_i(\lambda_i) = \lambda_i \prod_{k_i=1}^{\infty} \left(1 - \frac{\lambda_i b_i^2}{(k_i\pi)^2}\right) = \frac{\sqrt{\lambda_i}\sin b_i\sqrt{\lambda_i}}{b_i},$$

and

$$G_i{}'\left(\lambda_{i,k_i}\right) = \begin{cases} \frac{1}{2}(-1)^{k_i}, & k_i = 1, 2, \ldots \\ 1 \quad, & k_i = 0 \end{cases}.$$

Hence, Theorem 8.1 now takes on the form: if

$$f(\lambda_1,\ldots,\lambda_N)=\int_{x_1=0}^{b_1}\cdots\int_{x_N=0}^{b_N}F(x_1,\ldots,x_N)(\cos\sqrt{\lambda_1}\,x_1)\ldots(\cos\sqrt{\lambda_N}\,x_N)\,dx_1\ldots dx_N\,,$$

for some $F(\underline{x})\in L^2(\mathcal{B})$, then

$$f(\lambda_1,\ldots,\lambda_N)=\sum_{k_1,\ldots,k_N=0}^{\infty}f\left(\lambda_{1,k_1},\ldots,\lambda_{N,k_N}\right)H\left(\lambda_1,\ldots,\lambda_N;\lambda_{1,k_1},\ldots,\lambda_{N,k_N}\right),$$

where

$$H\left(\lambda_1,\ldots,\lambda_N;\lambda_{1,k_1},\ldots,\lambda_{N,k_N}\right)=\prod_{i=1}^{N}H_i\left(\lambda_i;\lambda_{i,k_i}\right),$$

and

$$H_i\left(\lambda_i;\lambda_{i,k_i}\right)=\begin{cases}\dfrac{2\sqrt{\lambda_i}\sin b_i\left(\sqrt{\lambda_i}-\sqrt{\lambda_{i,k_i}}\right)}{b_i(\lambda_i-\lambda_{i,k_i})} & \text{if}\quad k_i\neq 0\\[2ex] \dfrac{\sin b_i\sqrt{\lambda_i}}{b_i\sqrt{\lambda_i}} & \text{if}\quad k_i=0\end{cases}.$$

Thus,

$$\begin{aligned}f(\lambda_1,\ldots,\lambda_N)&=f(0,0,\ldots,0)\left(\frac{\sin b_1\sqrt{\lambda_1}}{b_1\sqrt{\lambda_1}}\right)\cdots\left(\frac{\sin b_N\sqrt{\lambda_N}}{b_N\sqrt{\lambda_N}}\right)\\&+2^N\sum_{k_1,\ldots,k_N=1}^{\infty}f\left(\lambda_{1,k_1},\ldots,\lambda_{N,k_1}\right)\prod_{i=1}^{N}\frac{\sqrt{\lambda_i}\sin b_i\left(\sqrt{\lambda_i}-\sqrt{\lambda_{i,k_i}}\right)}{b_i(\lambda_i-\lambda_{i,k_i})}\\&+\sum f\left(\lambda_{1,k_1},\ldots,\lambda_{N,k_N}\right)H\left(\lambda_1,\ldots,\lambda_N;\lambda_{1,k_1},\ldots,\lambda_{N,k_N}\right),\end{aligned}$$

where in the last summation at least one, but not all of the k_i's, is zero.

As a special case of this, we obtain for $N=2$, $b_i=\pi$, $t_i=\sqrt{\lambda_i}$, $i=1,2$ and $\tilde{f}(t_1,t_2)=f(\lambda_1,\lambda_2)$ that: if

$$\tilde{f}(t_1,t_2)=\int_0^{\pi}\int_0^{\pi}F(x_1,x_2)\cos t_1x_1\cos t_2x_2\,dx_1dx_2\,,$$

for some $F(x_1,x_2)\in L^2(\mathcal{B})$ where $\mathcal{B}=[0,\pi]\times[0,\pi]$, then

$$\tilde{f}(t_1,t_2)=\tilde{f}(0,0)\frac{\sin\pi t_1}{\pi t_1}\frac{\sin\pi t_2}{\pi t_2}+\sum_{k_2=1}^{\infty}\tilde{f}(0,k_2)\frac{\sin\pi t_1}{\pi t_1}\frac{2t_2\sin\pi(t_2-k_2)}{\pi(t_2^2-k_2^2)}$$

$$+\sum_{k_1=1}^{\infty}\tilde{f}(k_1,0)\frac{\sin\pi t_2}{\pi t_2}\frac{2t_1\sin\pi(t_1-k_1)}{\pi(t_1^2-k_1^2)}$$

$$+\sum_{k_1,k_2=1}^{\infty}\tilde{f}(k_1,k_2)\frac{2t_1\sin\pi(t_1-k_1)}{\pi(t_1^2-k_1^2)}\frac{2t_2\sin\pi(t_2-k_2)}{\pi(t_2^2-k_2^2)},$$

or equivalently

$$\tilde{f}(t_1,t_2)=\tilde{f}(0,0)\frac{\sin\pi t_1}{\pi t_1}\frac{\sin\pi t_2}{\pi t_2}+\frac{\sin\pi t_1}{\pi t_1}\sum_{\substack{k_2=-\infty\\k_2\neq 0}}^{\infty}\tilde{f}(0,k_2)\frac{\sin\pi(t_2-k_2)}{\pi(t_2-k_2)}$$

$$+\frac{\sin\pi t_2}{\pi t_2}\sum_{\substack{k_1=-\infty\\k_1\neq 0}}^{\infty}\tilde{f}(k_1,0)\frac{\sin\pi(t_1-k_1)}{\pi(t_1-k_1)}$$

$$+\sum_{\substack{k_1,k_2=-\infty\\k_1\neq 0\neq k_2}}^{\infty}\tilde{f}(k_1,k_2)\frac{\sin\pi(t_1-k_1)}{\pi(t_1-k_1)}\frac{\sin\pi(t_2-k_2)}{\pi(t_2-k_2)}.$$

Although we have proved our results under the assumption that $q(\underline{x})$ is continuous on $\mathcal{B}$, i.e., under the assumption that the boundary-value problem Π is regular, they may hold for some singular problems, provided that the spectrum of problem Π remains discrete. This is demonstrated in the following example:

***Example* 3.** Consider the differential equation

$$\Delta\psi(\underline{x})+\left\{\lambda-\sum_{i=1}^{N}\frac{\left(\alpha_i^2-\frac{1}{4}\right)}{x_i^2}\right\}\psi(\underline{x})=0,$$

where $\underline{x}\in\mathcal{B}=[\underline{0},\underline{b}]$, $\underline{b}=(b_1,\ldots,b_N)$, $b_i>0$, $\alpha_i>-\frac{1}{2}$, together with the boundary conditions

$$\psi(x_1,\ldots,x_{i-1},0,x_{i+1},\ldots,x_N)=0,$$

$$\psi(x_1,\ldots,x_{i-1},b_i,x_{i+1},\ldots,x_N)=0,$$

for all $i=1,\ldots,N$.

We can verify that $\psi_i(x_i;\lambda_i)$ may be taken as

$$\psi_i(x_i;\lambda_i) = \sqrt{x_i}\,\lambda_i^{-\alpha_i/2} J_{\alpha_i}(\sqrt{\lambda_i}\,x_i)\,,$$

where $J_\alpha(z)$ is the Bessel function of the first kind and order α; see (4.4.2) for the one-dimensional version. Moreover, $\lambda_{i,k_i} = (\alpha_{i,k_i}/b_i)^2$, where α_{i,k_i} is the k_ith positive zero of $J_{\alpha_i}(z)$, $k_i = 1,2,\ldots$. Thus,

$$G_i(\lambda_i) = \prod_{k_i=1}^{\infty}\left(1-\frac{\lambda_i b_i^2}{\alpha_{i,k_i}^2}\right),$$

$i = 1,\ldots,N$ and in view of the relation [12, p. 498]

$$J_\alpha(b\sqrt{\lambda}) = \frac{(b\sqrt{\lambda})^\alpha}{2^\alpha\Gamma(\alpha+1)}\,G(\lambda)\,,$$

we obtain that if

$$f(\underline{\lambda}) = \int_{x_1=0}^{b_1}\cdots\int_{x_N=0}^{b_N} F(x_1,\ldots,x_N)\sqrt{x_1\ldots x_N}\,J_{\alpha_1}(\sqrt{\lambda_1}\,x_1)\cdots$$

$$J_{\alpha_N}(\sqrt{\lambda_N}\,x_N)\,dx_1\ldots dx_N\,, \qquad (8.3.1)$$

for some $F(\underline{x}) \in L^2(\mathcal{B})$, then

$$f(\lambda_1,\ldots,\lambda_N) = 2^N \sum_{k_1,\ldots,k_N=1}^{\infty} f\left(\lambda_{1,k_1},\ldots,\lambda_{N,k_N}\right)\prod_{i=1}^{N}\frac{\sqrt{\lambda_{i,k_i}}\,J_{\alpha_i}(b_i\sqrt{\lambda_i})}{b_i(\lambda_i-\lambda_{i,k_i})J'_{\alpha_i}\left(b_i\sqrt{\lambda_{i,k_i}}\right)}\,. \qquad (8.3.2)$$

The integral in (8.3.1) is the N-dimensional finite Hankel transform and (8.3.2) is its sampling series expansion.

The extension of Theorem 8.1 to the case where the region $\mathcal{G}$ is not necessarily a rectangular region or the differential equation (8.2.4) is not separable, is not that obvious and needs further investigation.

8.4 Applications to Special Functions

In this section we show how Theorem 8.1 and a variation thereof can be used in the area of Special Functions, in particular, in summing up infinite series in several variables, as done in Chapter 7 for infinite series in one variable; see [13, 14 and 18].

Although the summation formulae we are about to derive may look somewhat complicated, their proofs, by using sampling theorems, are rather easy and straightforward; this is one of the most interesting features of our technique.

For simplicity we shall restrict ourselves throughout the rest of this section to the case where $N = 2$ and E is the rectangle $E = [0,a] \times [0,b]$.

***Example* 1.** Now consider the Dirichlet problem:

$$\Delta\psi(\underline{x}) = -\lambda\psi(\underline{x}), \qquad \underline{x} = (x,y) \in E$$

with

$$\psi(\underline{x}) = 0 \qquad \text{on} \qquad \partial E .$$

It is easy to see that the eigenvalues of this problem are

$$\lambda_{m,n} = (m\,\pi/a)^2 + (n\,\pi/b)^2$$

and the corresponding eigenfunctions are

$$\psi_{m,n}(x,y) = \sin(m\,\pi\,x/a)\sin(n\,\pi\,y/b) .$$

From Example 1 in Section 8.3, one can verify the following sampling theorem (cf. [16, 19]): If

$$f(u,v) = \int_0^b \int_0^a F(x,y)\sin ux \sin vy\,dx\,dy , \tag{8.4.1}$$

for some $F \in L^2(E)$, then

$$f(u,v) = 4 \sum_{m,n=1}^{\infty} f\left(\frac{m\pi}{a}, \frac{n\pi}{b}\right) \frac{(m\pi/a)\sin(au - m\pi)}{a[u^2 - (m\pi/a)^2]} \frac{(n\pi/b)\sin(bv - n\pi)}{b[v^2 - (n\pi/b)^2]} .$$

In particular, if $a = b$, we have

$$f(u,v) = 4 \sum_{m,n=1}^{\infty} f\left(\frac{m\,\pi}{b}, \frac{n\,\pi}{b}\right) \frac{(m\,\pi/b)\sin(bu - m\,\pi)}{b[u^2 - (m\,\pi/b)^2]} \frac{(n\,\pi/b)\sin(bv - n\,\pi)}{b[v^2 - (n\,\pi/b)^2]} . \tag{8.4.2}$$

Now we show how this sampling theorem can be used to find summation formulae for some infinite series in two variables.

Let $F(x,y) = xy(b^2 - x^2 - y^2)^{\mu - 3/2}\chi_{\mathbf{D}}$, $\mu > \frac{1}{2}$, where $\chi_{\mathbf{D}}$ is the characteristic function of the region $\mathbf{D} = \{(x,y) : x^2 + y^2 \le b^2,\ x,y \ge 0\}$. Then, with the aid of formulae (2.3-9) and (2.13-51) in [2], we obtain from (8.4.1) for $a = b$

$$f(u,v) = \int_0^b \int_0^b F(x,y)\sin ux \sin vy\,dx\,dy$$

$$= \int_0^b y\left(\int_0^A x(A^2-x^2)^{\mu-3/2}\sin ux\,dx\right)\sin vy\,dy$$

$$= d_\mu u^{-\mu+1} \int_0^b y(b^2-y^2)^{\mu/2} J_\mu(u\sqrt{b^2-y^2})\sin vy\,dy$$

$$= c_\mu \frac{uv}{(u^2+v^2)^{\mu/2+3/4}} J_{\mu+3/2}(b\sqrt{u^2+v^2}), \tag{8.4.3}$$

where

$$A = \sqrt{b^2-y^2}, \qquad d_\mu = 2^{\mu-2}\sqrt{\pi}\,\Gamma(\mu-1/2),$$
$$c_\mu = \pi 2^{\mu-5/2}\Gamma(\mu-1/2)b^{\mu+3/2}.$$

By substituting (8.4.3) into (8.4.2), we obtain the summation formula

$$\frac{uv}{(u^2+v^2)^{\mu/2+3/4}} J_{\mu+3/2}(b\sqrt{u^2+v^2}) = \sum_{m,n=1}^{\infty}\left\{\frac{(m\,n\,\pi^2)/b^2}{[(m\pi/b)^2+(n\pi/b)^2]^{\mu/2+3/4}}\right.$$
$$\left.\times J_{\mu+3/2}\left(b\sqrt{(m\pi/b)^2+(n\pi/b)^2}\right)\right\}$$
$$\times \frac{(m\pi/b)\sin(bu-m\pi)}{b[u^2-(m\pi/b)^2]}\,\frac{(n\pi/b)\sin(bv-n\pi)}{b[v^2-(n\pi/b)^2]}.$$

In particular, if $b = \pi$, we have

$$\frac{1}{[u^2+v^2]^{\mu/2+3/4}} J_{\mu+3/2}(\pi\sqrt{u^2+v^2}) =$$

$$\frac{\sin\pi u}{\pi u}\,\frac{\sin\pi v}{\pi v}\sum_{m,n=1}^{\infty}\left\{\frac{(mn)^2(-1)^{m+n}}{[m^2+n^2]^{\mu/2+3/4}} J_{\mu+3/2}(\pi\sqrt{m^2+n^2})\right\}\frac{1}{(u^2-m^2)}\,\frac{1}{(v^2-n^2)}. \tag{8.4.4}$$

This is reminiscent of some of the summation formulae recently obtained by Gosper, Ismail and Zhang [3], and Ismail, Takeuchi and Zhang [5] for infinite series in one variable; see also Section 7.2.

As a special case of (8.4.4), we obtain by taking the limit as $u, v \to 0$

$$\frac{(\pi/2)^{\mu+3/2}}{\Gamma(\mu+5/2)} = \sum_{m,n=1}^{\infty} \frac{(-1)^{m+n}}{[m^2+n^2]^{\mu/2+3/4}} J_{\mu+3/2}(\pi\sqrt{m^2+n^2}) .$$

Example 2. Consider the Neumann boundary-value problem:

$$\Delta\psi(\underline{x}) = -\lambda\psi(\underline{x}), \quad \underline{x} \in E$$

$$\frac{\partial\psi}{\partial n} = 0 \quad \text{on} \quad \partial E ,$$

where E is the same as in Example 1.

It is easy to see that the eigenvalues and eigenfunctions of this problem are $\lambda_{m,n} = (m\pi/a)^2 + (n\pi/b)^2$ and $\psi_{m,n}(x,y) = \cos(m\pi x/a)\cos(n\pi y/b)$.

From Example 2 in Section 8.3 (cf. [16, 17]): If

$$f(u,v) = \int_0^b \int_0^a F(x,y) \cos ux \cos vy \, dx \, dy , \tag{8.4.5}$$

for some $F \in L^2(E)$, then

$$\begin{aligned} f(u,v) &= f(0,0)\frac{\sin au}{au}\frac{\sin bv}{bv} + \frac{\sin au}{au}\sum_{n=1}^{\infty} f\left(0,\frac{n\pi}{b}\right)\frac{2v\sin(bv-n\pi)}{b[v^2-(n\pi/b)^2]} \\ &+ \frac{\sin bv}{bv}\sum_{m=1}^{\infty} f\left(\frac{m\pi}{a},0\right)\frac{2u\sin(au-m\pi)}{a[u^2-(m\pi/a)^2]} \\ &+ \sum_{m,n=1}^{\infty} f\left(\frac{m\pi}{a},\frac{n\pi}{b}\right)\frac{2u\sin(au-m\pi)}{a[u^2-(m\pi/a)^2]}\,\frac{2v\sin(bv-n\pi)}{b[v^2-(n\pi/b)^2]} . \end{aligned} \tag{8.4.6}$$

For simplicity we shall take $a = b = \pi$.

Now let Δ be the region bounded by the lines $x = 0$, $y = x$ and $y = \pi$ and let χ_Δ be its characteristic function.

Let $F(x,y) = J_0(c\sqrt{y^2-x^2})\chi_\Delta$, where $c > 0$. Then, with the aid of formula 1.13-47 in [2], we obtain from (8.4.5)

$$f(u,v) = \int_0^\pi \left(\int_0^y J_0(c\sqrt{y^2-x^2}) \cos ux\,dx \right) \cos vy\,dy$$

$$= \int_0^\pi \frac{\sin(y\sqrt{c^2+u^2})}{\sqrt{c^2+u^2}} \cos vy\,dy$$

$$= \frac{1}{2w}\left\{ \frac{1-\cos\pi(w+v)}{w+v} + \frac{1-\cos\pi(w-v)}{w-v} \right\}$$

$$= \frac{1}{w}\left\{ \frac{\sin^2\left(\frac{\pi}{2}(w+v)\right)}{w+v} + \frac{\sin^2\left(\frac{\pi}{2}(w-v)\right)}{w-v} \right\},$$

where $w = \sqrt{c^2+u^2}$. More explicitly,

$$f(u,v) = \frac{1}{\sqrt{c^2+u^2}}\left\{ \frac{\sin^2\left(\frac{\pi}{2}(\sqrt{c^2+u^2}+v)\right)}{\sqrt{c^2+u^2}+v} + \frac{\sin^2\left(\frac{\pi}{2}(\sqrt{c^2+u^2}-v)\right)}{\sqrt{c^2+u^2}-v} \right\}.$$

Therefore, by substituting this in (8.4.6), we obtain the summation formula:

$$\frac{1}{\sqrt{c^2+u^2}}\left\{ \frac{\sin^2\left(\frac{\pi}{2}(\sqrt{c^2+u^2}+v)\right)}{(\sqrt{c^2+u^2}+v)} + \frac{\sin^2\left(\frac{\pi}{2}(\sqrt{c^2+u^2}-v)\right)}{(\sqrt{c^2+u^2}-v)} \right\} = \frac{2\sin^2(\pi c/2)}{c^2}\,\frac{\sin\pi u}{\pi u}\,\frac{\sin\pi v}{\pi v}$$

$$+\frac{\sin\pi u}{\pi u}\sum_{n=1}^{\infty}\frac{1}{c}\left\{ \frac{\sin^2\left(\frac{\pi}{2}(c+n)\right)}{(c+n)} + \frac{\sin^2\left(\frac{\pi}{2}(c-n)\right)}{(c-n)} \right\}\frac{2v\sin\pi(v-n)}{\pi[v^2-n^2]}$$

$$+\frac{\sin\pi v}{\pi v}\sum_{m=1}^{\infty}\frac{2\sin^2\left(\frac{\pi}{2}\sqrt{c^2+m^2}\right)}{(c^2+m^2)}\,\frac{2u\sin\pi(u-m)}{\pi[u^2-m^2]}$$

$$+\sum_{m,n=1}^{\infty}\frac{1}{\sqrt{c^2+m^2}}\left\{ \frac{\sin^2\left[\frac{\pi}{2}(\sqrt{c^2+m^2}+n)\right]}{(\sqrt{c^2+m^2}+n)} + \frac{\sin^2\left[\frac{\pi}{2}(\sqrt{c^2+m^2}-n)\right]}{(\sqrt{c^2+m^2}-n)} \right\}$$

$$\times\frac{2u\sin\pi(u-m)}{\pi[u^2-m^2]}\,\frac{2v\sin\pi(v-n)}{\pi[v^2-n^2]}.$$

By taking the limit as $c \to 0$, we obtain

$$\frac{1}{u}\left\{\frac{\sin^2\left[\frac{\pi}{2}(u+v)\right]}{(u+v)}+\frac{\sin^2\left[\frac{\pi}{2}(u-v)\right]}{(u-v)}\right\}=\frac{\pi^2}{2}\frac{\sin\pi u}{\pi u}\frac{\sin\pi v}{\pi v}$$

$$+\frac{\sin\pi u}{\pi u}\sum_{n=1}^{\infty}\left\{\frac{(-1)^n-1}{n^2}\right\}\frac{2v\sin\pi(v-n)}{\pi[v^2-n^2]}$$

$$+\frac{\sin\pi v}{\pi v}\sum_{m=1}^{\infty}\left\{\frac{2\sin^2(m\pi/2)}{m^2}\right\}\frac{2u\sin\pi(u-m)}{\pi[u^2-m^2]}$$

$$+\sum_{m,n=1}^{\infty}\frac{1}{m}\left\{\frac{\sin^2\left[\frac{\pi}{2}(m+n)\right]}{(m+n)}+\frac{\sin^2\left[\frac{\pi}{2}(m-n)\right]}{(m-n)}\right\}\frac{2u\sin\pi(u-m)}{\pi[u^2-m^2]}\frac{2v\sin\pi(v-n)}{\pi[v^2-n^2]},$$

which, after some simplifications, yields

$$\frac{1}{u}\left\{\frac{\sin^2\left[\frac{\pi}{2}(u+v)\right]}{(u+v)}+\frac{\sin^2\left[\frac{\pi}{2}(u-v)\right]}{(u-v)}\right\}=4\frac{\sin\pi u}{\pi u}\frac{\sin\pi v}{\pi v}\left\{\frac{\pi^2}{8}+\frac{1}{2}\sum_{n=1}^{\infty}\frac{(1-(-1)^n)(u^2-v^2)}{(u^2-n^2)(v^2-n^2)}\right.$$

$$\left.+\sum_{m,n=1}^{\infty}\frac{((-1)^{m+n}-1)}{(m^2-n^2)}\frac{u^2v^2}{(u^2-m^2)(v^2-n^2)}\right\}.$$

When $v=0$, we have

$$2\left(\frac{\sin(\pi u/2)}{u}\right)^2=4\frac{\sin\pi u}{\pi u}\left\{\frac{\pi^2}{8}+\sum_{n=1}^{\infty}\left(\frac{1}{(2n-1)^2-u^2}-\frac{1}{(2n-1)^2}\right)\right\},$$

which yields the Mittag-Leffler expansion of $\tan(\pi u/2)$, i.e.,

$$\pi\tan\left(\frac{\pi u}{2}\right)=\sum_{n=1}^{\infty}\frac{4u}{(2n-1)^2-u^2}.$$

***Example* 3.** From Theorem 8.1 we can obtain the following result, which is a variation of Example 2 in Section 8.3; see also Corollary 4 in [19]. If

$$\tilde{f}(\lambda_1,\ldots,\lambda_N)=\int_{x_1=0}^{b_1}\cdots\int_{x_N=0}^{b_N}F(x_1,\ldots,x_N)(\cos\sqrt{\lambda_1}x_1)\ldots$$

$$(\cos\sqrt{\lambda_N}x_N)\,dx_1\ldots dx_N, \qquad (8.4.7)$$

for some $F(x_1,\ldots,x_N)\in L^2(\mathcal{B})$, where $\mathcal{B}=[0,b_1]\times\ldots\times[0,b_N]$, then

$$\tilde{f}(\lambda_1,\ldots,\lambda_N)=\sum_{k_1,\ldots,k_N=1}^{\infty}\tilde{f}(\lambda_{1,k_1},\ldots,\lambda_{N,k_N})H(\lambda_1,\ldots,\lambda_N;\lambda_{1,k_1},\ldots,\lambda_{N,k_N}),$$

where

$$H(\lambda_1,\ldots,\lambda_N;\lambda_{1,k_1},\ldots,\lambda_{N,k_N}) = \prod_{i=1}^{N} H_i(\lambda_i;\lambda_{i,k_i}),$$

$$H_i(\lambda_i,\lambda_{i,k_i}) = \frac{2\sqrt{\lambda_{i,k_i}}(-1)^{k_i}\cos(\sqrt{\lambda_i}\,b_i)}{(\lambda_i-\lambda_{i,k_i})b_i},$$

$$\lambda_{i,k_i} = \left[\left(k_i-\frac{1}{2}\right)\frac{\pi}{b_i}\right]^2; \quad k_i = 1,2,3,\ldots, \quad i = 1,\ldots,N.$$

Or equivalently (cf. Corollary 4 [19])

$$\tilde{f}(\lambda_1,\ldots,\lambda_N) = \sum_{k_1,\ldots,k_N=-\infty}^{\infty} \tilde{f}(\lambda_{1,k_1},\ldots,\lambda_{N,k_N}) \prod_{i=1}^{N} \frac{\sin b_i\left[\sqrt{\lambda_i}-\left(k_i-\frac{1}{2}\right)\pi/b_i\right]}{b_i\left[\sqrt{\lambda_i}-\left(k_i-\frac{1}{2}\right)\pi/b_i\right]}.$$

If we set $N=2$, $u=\sqrt{\lambda_1}$, $\mathrm{v}=\sqrt{\lambda_2}$, $b_1=\pi=b_2$ and $f(u,v)=\tilde{f}(\lambda_1,\lambda_2)$, we obtain

$$f(u,\mathrm{v}) = \sum_{m,n=-\infty}^{\infty} f\left(m+\frac{1}{2},n+\frac{1}{2}\right) \times \frac{\sin\pi\left(u-\left(m+\frac{1}{2}\right)\right)}{\pi\left[u-\left(m+\frac{1}{2}\right)\right]}\,\frac{\sin\pi\left(\mathrm{v}-\left(n+\frac{1}{2}\right)\right)}{\pi\left[\mathrm{v}-\left(n+\frac{1}{2}\right)\right]}. \tag{8.4.8}$$

By following the same steps as in Example 2, we obtain

$$\frac{1}{\sqrt{b^2+u^2}}\left\{\frac{\sin^2\left(\frac{\pi}{2}(\sqrt{b^2+u^2}+\mathrm{v})\right)}{(\sqrt{b^2+u^2}+\mathrm{v})}+\frac{\sin^2\left(\frac{\pi}{2}(\sqrt{b^2+u^2}-\mathrm{v})\right)}{(\sqrt{b^2+u^2}-\mathrm{v})}\right\}$$

$$= \sum_{m,n=-\infty}^{\infty} \frac{1}{\sqrt{b^2+\left(m+\frac{1}{2}\right)^2}}\left\{\frac{\sin^2\left(\frac{\pi}{2}\left(\sqrt{b^2+\left(m+\frac{1}{2}\right)^2}+\left(n+\frac{1}{2}\right)\right)\right)}{\left(\sqrt{b^2+\left(m+\frac{1}{2}\right)^2}+\left(n+\frac{1}{2}\right)\right)}\right.$$

$$\left.+\frac{\sin^2\left(\frac{\pi}{2}\left(\sqrt{b^2+\left(m+\frac{1}{2}\right)^2}-\left(n+\frac{1}{2}\right)\right)\right)}{\left(\sqrt{b^2+\left(m+\frac{1}{2}\right)^2}-\left(n+\frac{1}{2}\right)\right)}\right\}\frac{\sin\pi\left(u-\left(m+\frac{1}{2}\right)\right)}{\pi\left(u-\left(m+\frac{1}{2}\right)\right)}\,\frac{\sin\pi\left(\mathrm{v}-\left(n+\frac{1}{2}\right)\right)}{\pi\left(\mathrm{v}-\left(n+\frac{1}{2}\right)\right)}.$$

In particular, if we take the limit as $b \to 0$, we have

$$\frac{1}{u}\left\{\frac{\sin^2\left(\frac{\pi}{2}(u+v)\right)}{u+v}+\frac{\sin^2\left(\frac{\pi}{2}(u-v)\right)}{u-v}\right\}$$

$$=\sum_{m,n=-\infty}^{\infty}\frac{1}{(m+1/2)}\left\{\frac{\cos^2\left(\frac{\pi}{2}(m+n)\right)}{(m+n+1)}+\frac{\sin^2\left(\frac{\pi}{2}(m-n)\right)}{(m-n)}\right\}\times$$

$$\frac{\sin\pi(u-(m+1/2))}{\pi(u-(m+1/2))}\,\frac{\sin\pi(v-(n+1/2))}{\pi(v-(n+1/2))}, \tag{8.4.9}$$

which, upon setting $v = 0$ and taking the limit as $u \to 0$, yields the following representation of π^4

$$\frac{\pi^4}{2}=\sum_{m,n=-\infty}^{\infty}\frac{(-1)^{m+n}}{(m+1/2)^2(n+1/2)}\left\{\frac{\cos^2\left(\frac{\pi}{2}(m+n)\right)}{(m+n+1)}+\frac{\sin^2\left(\frac{\pi}{2}(m-n)\right)}{(m-n)}\right\}.$$

By setting $v = 0$ and $u = 1/2$ in (8.4.9), we obtain the well-known formula

$$\frac{\pi}{8}-\frac{1}{3}=\sum_{k=1}^{\infty}\frac{1}{(4k+1)(4k+3)}.$$

Example 4. Let D and χ_D be as in Example 1. Consider

$$F(x,y)=(b^2-y^2)^k(b^2-y^2-x^2)^{v-1/2}C_{2k}^{v}\left(\frac{x}{\sqrt{b^2-y^2}}\right)\chi_D,$$

where $C_k^v(x)$ is the Gegenbauer polynomial of degree k and $Re\,v > -1/2$.

Set

$$g(u,y)=\int_0^{\pi}F(x,y)\cos ux\,dx$$

$$=(b^2-y^2)^k\int_0^{\sqrt{b^2-y^2}}(b^2-y^2-x^2)^{v-1/2}C_{2k}^{v}\left(\frac{x}{\sqrt{b^2-y^2}}\right)\cos ux\,dx.$$

Hence, in view of formula 3, p. 38 in [2],

$$g(u,y)=C_{k,v}\frac{(b^2-y^2)^{v/2+k}}{u^v}J_{v+2k}(u\sqrt{b^2-y^2}),$$

where

$$C_{k,\nu} = \frac{(-1)^k \pi \Gamma(2k+2\nu)}{2^\nu (2k)! \Gamma(\nu)}.$$

Therefore, in view of formula 50, p. 57 in [2], we have

$$f(u,\text{v}) = \int_0^b \int_0^b F(x,y) \cos ux \cos \text{v}y dx dy$$

$$= C_{k,\nu} u^{-\nu} \int_0^b (b^2-y^2)^{\nu/2+k} J_{\nu+2k}(u\sqrt{b^2-y^2}) \cos \text{v}y dy$$

$$= \sqrt{\frac{\pi}{2}} C_{k,\nu} \frac{b^{\nu+2k+1/2} u^{2k}}{[u^2+\text{v}^2]^{\nu/2+k+1/4}} J_{\nu+2k+1/2}(b\sqrt{u^2+\text{v}^2}).$$

By substituting this in (8.4.8), we obtain the summation formula

$$\frac{u^{2k}}{[u^2+\text{v}^2]^{\nu/2+k+1/4}} J_{\nu+2k+1/2}(b\sqrt{u^2+\text{v}^2}) =$$

$$\left(\frac{b}{\pi}\right)^{\nu+\frac{1}{2}} \sum_{m,n=-\infty}^{\infty} \left\{ \frac{\left(m+\frac{1}{2}\right)^{2k}}{\left[\left(m+\frac{1}{2}\right)^2+\left(n+\frac{1}{2}\right)^2\right]^{\nu/2+k+1/4}} J_{\nu+2k+\frac{1}{2}}\left(\pi\sqrt{\left(m+\frac{1}{2}\right)^2+\left(n+\frac{1}{2}\right)^2}\right)\right\} \times$$

$$\frac{\sin b\left[u-\left(m+\frac{1}{2}\right)\pi/b\right]}{b\left[u-\left(m+\frac{1}{2}\right)\pi/b\right]} \frac{\sin b\left[\text{v}-\left(n+\frac{1}{2}\right)\pi/b\right]}{b\left[\text{v}-\left(n+\frac{1}{2}\right)\pi/b\right]},$$

Example 5. Let D be as before and consider

$$F(x,y) = \frac{(b^2-y^2-x^2)^{\nu-1/2}}{(b^2-y^2)^{1/2(\nu+1)}} \chi_D, \quad Re\nu > -1/2.$$

Set

$$g(u,y) = \int_0^b F(x,y) \cos ux dx$$

$$= \frac{1}{(b^2-y^2)^{1/2(\nu+1)}} \int_0^{\sqrt{b^2-y^2}} (b^2-y^2-x^2)^{\nu-1/2} \cos ux dx = \frac{C_\nu}{u^\nu \sqrt{b^2-y^2}} J_\nu(u\sqrt{b^2-y^2}),$$

where

$$C_\nu = 2^{\nu-1}\sqrt{\pi}\,\Gamma(\nu+1/2).$$

Here we have used formula 8, p. 11 in [2]. Thus,

$$f(u,v) = \int_0^b \int_0^b F(x,y) \cos ux \cos vy \, dx\, dy$$

$$= C_v u^{-v} \int_0^b \frac{J_v(u\sqrt{b^2-y^2})}{\sqrt{b^2-y^2}} \cos vy \, dy \, .$$

To calculate this integral, we first make the change of variable $z = \sqrt{b^2 - y^2}$ and then use formula 3, p. 761 in [4], which states that

$$\int_0^b \frac{\cos(v\sqrt{b^2-z^2})}{\sqrt{b^2-z^2}} J_v(uz)dz = \frac{\pi}{2} J_{v/2}\left[\frac{b}{2}(\sqrt{u^2+v^2}+v)\right] J_{v/2}\left[\frac{b}{2}(\sqrt{u^2+v^2}-v)\right]$$

for $u > 0$, $Re\, v > -1$, to finally obtain that

$$f(u,v) = \frac{\pi}{2} C_v u^{-v} J_{v/2}\left[\frac{b}{2}(\sqrt{u^2+v^2}+v)\right] J_{v/2}\left[\frac{b}{2}(\sqrt{u^2+v^2}-v)\right] .$$

By substituting this into (8.4.8), we obtain the summation formula

$$u^{-v} J_{v/2}\left[\frac{b}{2}(\sqrt{u^2+v^2}+v)\right] J_{v/2}\left[\frac{b}{2}(\sqrt{u^2+v^2}-v)\right]$$

$$= \sum_{m,n=-\infty}^{\infty} \left\{ \frac{J_{v/2}\left[\frac{\pi}{2}(\sqrt{(m+1/2)^2+(n+1/2)^2}+(n+1/2))\right] J_{v/2}\left[\frac{\pi}{2}(\sqrt{(m+1/2)^2+(n+1/2)^2}-(n+1/2))\right]}{[(m+1/2)\pi/b]^v} \right\}$$

$$\times \frac{\sin b[u-(m+1/2)\pi/b]}{b[u-(m+1/2)\pi/b]} \frac{\sin b[v-(n+1/2)\pi/b]}{b[v-(n+1/2)\pi/b]} ,$$

from which the following interesting special cases are obtained:

i) for $v = 0$

$$u^{-v}\left[J_{v/2}\left(\frac{b}{2}u\right)\right]^2 = \sum_{m,n=-\infty}^{\infty} \frac{(-1)^n}{(n+1/2)\pi} \frac{\sin b[u-(m+1/2)\pi/b]}{b[u-(m+1/2)\pi/b]}$$

$$\left\{ \frac{J_{v/2}\left[\frac{\pi}{2}(\sqrt{(m+1/2)^2+(n+1/2)^2}+(n+1/2))\right] J_{v/2}\left[\frac{\pi}{2}(\sqrt{(m+1/2)^2+(n+1/2)^2}-(n+1/2))\right]}{[(m+1/2)\pi/b]^v} \right\} ,$$

ii) for $\nu = 0$, $\nu = 1$ and b replaced by $2b$,

$$\left[\frac{\sin bu}{bu}\right]^2 = \frac{1}{\pi^3}\sum_{m,n=-\infty}^{\infty} \frac{(-1)^n}{(n+1/2)(m+1/2)^2} \frac{\sin 2b[u-(m+1/2)\pi/2b]}{2b[u-(m+1/2)\pi/2b]} \times$$

$$\sin\left[\frac{\pi}{2}(\sqrt{(m+1/2)^2+(n+1/2)^2}+(n+1/2))\right]\sin\left[\frac{\pi}{2}(\sqrt{(m+1/2)^2+(n+1/2)^2}-(n+1/2))\right],$$

which for $b = \pi$ and $u \to 0$ gives another representation for π^4

$$\pi^4 = \sum_{m,n=-\infty}^{\infty} \frac{(-1)^{n+m}}{(n+1/2)(m+1/2)^3} \sin\left[\frac{\pi}{2}(\sqrt{(m+1/2)^2+(m+1/2)^2}+(n+1/2))\right]$$

$$\sin\left[\frac{\pi}{2}(\sqrt{(m+1/2)^2+(n+1/2)^2}-(n+1/2))\right],$$

iii) for $b = \pi$ and $\nu = 1/2$, all the terms in the summation over n are zeros, except for the term that corresponds to $n = 0$, which is 1; hence,

$$u^{-\nu} J_{\nu/2}\left[\frac{\pi}{2}(\sqrt{u^2+1/4}+1/2)\right] J_{\nu/2}\left[\frac{\pi}{2}(\sqrt{u^2+1/4}-1/2)\right]$$

$$= \sum_{m=-\infty}^{\infty}\left\{\frac{J_{\nu/2}\left[\frac{\pi}{2}(\sqrt{(m+1/2)^2+1/4}+1/2)\right] J_{\nu/2}\left[\frac{\pi}{2}(\sqrt{(m+1/2)^2+1/4}-1/2)\right]}{[(m+1/2)]^{\nu}}\right\} \frac{\sin\pi(u-(m+1/2))}{\pi(u-(m+1/2))}.$$

Similarly, if $b = \pi$ and $u = 1/2$, we have

$$J_{\nu/2}\left[\frac{\pi}{2}(\sqrt{\nu^2+1/4}+\nu)\right] J_{\nu/2}\left[\frac{\pi}{2}(\sqrt{\nu^2+1/4}-\nu)\right]$$

$$= \sum_{n=-\infty}^{\infty} J_{\nu/2}\left[\frac{\pi}{2}(\sqrt{(n+1/2)^2+1/4}+(n+1/2))\right] J_{\nu/2}\left[\frac{\pi}{2}(\sqrt{(n+1/2)^2+1/4}-(n+1/2))\right]$$

$$\times \frac{\sin\pi(\nu-(n+1/2))}{\pi(\nu-(n+1/2))}.$$

The above examples may suggest that our technique works only for the finite Fourier cosine and sine transforms, but a closer inspection of the process reveals that it will work for any multivariate integral transform, provided that we have a sampling expansion for it and we are able to calculate the iterated integrals in closed form. Examples involving multidimensional Hankel transforms and multivariate summation formulae involving non-uniform sample points have recently been found by the author, but will be published somewhere else.

8.5 Open Questions

In the one-dimensional case, the equivalence of the WSK and Kramer sampling theorems has long been established by Campbell [1] and Jerri [7] when the kernel function in Kramer's theorem arises from a regular Sturm-Liouville boundary-value problem. In [16] we claimed that the equivalence of the N-dimensional versions of these two theorems could be established in the same fashion as in the one-dimensional case by carrying over some of Campbell's and Jerri's arguments to N dimensions; however, our attempts in that direction were futile. Therefore, the question of whether the multidimensional versions of these two theorems are equivalent or not remains unanswered.

Finally, we would like to conclude this chapter by pointing out that one of the interesting aspects of this work is that it sheds some light on the interrelationship between Sampling Theory, on the one hand, and the Theory of Boundary-Value Problems on the other. This may not be of any practical application right now, but at least from a mathematical point of view it is rather interesting. Several questions remain unanswered; for example, what kind of a sampling theorem do we get if the differential equation (8.2.4) is not separable or the region $\mathcal{G}$ is not an N-dimensional rectangle?, is it possible to generalize Theorem 8.1 to the case where the boundary conditions are more general than those given by (8.2.5) and (8.2.6)?, under what conditions can we obtain Theorem 8.1 if problem (8.2.4) through (8.2.6) is singular?

References

1. L. Campbell, A comparison of the sampling theorems of Kramer and Whittaker, *J. SIAM*, 12 (1964), 117-130.

2. A. Erdelyi, W. Magnus, F. Oberhettinger, and F. Tricomi, *Tables of Integral Transforms*, Vol. 1, McGraw-Hill, New York (1954).

3. R. Gosper, M. Ismail and R. Zhang, On some strange summation formulas of Gosper, to appear in Ill. Journal of Math.

4. I. Gradshteyn and I. Ryzhik, *Tables of Integrals, Series and Products*, Academic Press, New York (1965).

5. M. Ismail, Y. Takeuchi and R. Zhang, Pages from the computer files of R. William Gosper, manuscript (1992).

6. A. Jerri, The Shannon sampling theorem—its various extensions and applications: A tutorial review, *Proc. IEEE*, 65, 11 (1977), 1565-1596.

7. _______, On the equivalence of Kramer's and Shannon's sampling theorems, *IEEE Trans. Inform. Theory*, Vol. IT-15 (1969), 497-499.

8. B. Levitan and I. Sargsjan, *Introduction to Spectral Theory: Self-Adjoint Ordinary Differential Operators*, Transl. Math. Monographs, Vol. 39, Amer. Math. Soc., Providence, R.I. (1975).

9. F. Reza, *An Introduction to Information Theory*, McGraw-Hill, New York (1961).

10. E. Titchmarsh, *Eigenfunction Expansions Associated with Second Order Differential Equations*, Part 1, 2nd ed., Clarendon Press, Oxford (1962).

11. _______, *Eigenfunction Expansions Associated with Second Order Differential Equations*, Part II, Clarendon Press, Oxford (1958).

12. G. Watson, *A Treatise on the Theory of Bessel Functions*, 2nd ed., Cambridge Univ. Press, Cambridge, England (1962).

13. A. Zayed, New summation formulas for multivariate infinite series by using sampling theorems, to appear in the *J. of Appl. Anal.*

14. _______, A proof of new summation formulae by using sampling theorems, *Proc. Amer. Math. Soc.*, 117 (1993), 699-710.

15. _______, Sampling theorem for functions band-limited to a disc, to appear in the *J. Complex Variables Theory Appls.*

16. A. Zayed, Kramer's sampling theorem for multidimensional signals and its relationship with Lagrange-type interpolation, *J. Multidimensional Systems Signal Processing*, 3 (1992), 323-340.

17. ________, On Kramer's sampling theorem associated with general Sturm-Liouvile problems and Lagrange interpolation, *SIAM J. Appl. Math.*, Vol. 51 (1991), 575-604.

18. ________, Sampling expansions for the continuous Bessel transform, *J. Appl. Anal.*, Vol. 27 (1988), 47-64.

19. A. Zayed, G. Hinsen and P. Butzer, On Lagrange interpolation and Kramer-type sampling theorem associated with Sturm-Liouville problems, *SIAM J. Appl. Math.*, Vol. 50 (1990), 893-909.

9

SAMPLING THEOREMS FOR MULTIDIMENSIONAL SIGNALS—THE FEICHTINGER-GRÖCHENIG SAMPLING THEORY

9.0 Introduction

The reconstruction of multidimensional signals, in particular the band-limited ones, from their values at a discrete set of sample points is of great importance in many applications, such as image processing, acoustics and geophysics. There are several results on the uniqueness of such reconstructions. For example, as we have seen in Theorem 3.9, any band-limited function $f(\underline{x})$ defined on $\Re^N$ is uniquely determined by its values at an appropriate set of points $\{\underline{x}_k\}$. However, when it comes down to the problem of how to reconstruct $f(\underline{x})$ from its values $\{f(\underline{x}_k)\}$, especially when the sample points $\{\underline{x}_k\}$ are irregularly spaced, only a few answers seem to exist and most of them do not go beyond the obvious generalizations of the one-dimensional case.

One of the new and promising answers to this difficult problem that has emerged on the stage in the last three years is the Feichtinger-Gröchenig sampling theory, to which we have alluded in Section 3.3. In a number of theorems, H. Feichtinger and K. Gröchenig derived sampling series expansions for reconstructing any member of general classes of multidimensional signals, including the band-limited ones, from the values of this member at irregularly spaced sample points.

The first section of this chapter will be devoted to the presentation of the Feichtinger-Gröchenig sampling theory. We shall not present it in its most general form, but in a form that is general enough for all practical applications and at the same time simple enough to avoid very technical mathematical details.

Section 2, on the other hand, will be dedicated to the derivation of a Kramer-type sampling theorem for integral transforms of functions with support in a simply connected bounded region in $\Re^N$. This generalizes some of the results presented in Chapter 6.

9.1 The Feichtinger-Gröchenig Sampling Theory

The starting point of the Feichtinger and Gröchenig approach is the observation that band-limited functions satisfy a reproducing formula based on the convolution operation, and the main idea of their proof is to show that the sampling theorem is equivalent to the factorization of a certain convolution operator on some spaces of functions with compact support.

We recall from Section 3.6.B that the problem of reconstructing a function f that is band-limited to $[-\sigma,\sigma]$ from its values at the regularly spaced points $\{\lambda k\}_{k\in Z}$, with $0<\lambda<2\pi/\sigma$, can be solved for example if we can find a function ϕ for which the relation

$$f(t)=\sum_{k=-\infty}^{\infty} f(\lambda k)\,\phi(t-\lambda k)$$

holds. We have also shown that, by using the Fourier transform and the Poisson summation formula, we can transform this relation to the following equivalent one

$$\hat{f}(\omega)=\frac{\sqrt{2\pi}}{\lambda}\left(\sum_{k=-\infty}^{\infty}\hat{f}\left(\omega+\frac{2\pi k}{\lambda}\right)\right)\hat{\phi}(\omega).$$

Now the easiest way to solve the latter is to assume that all the translates of $\hat{f}$ have supports disjoint from that of $\hat{\phi}$, that is $\hat{f}\left(\omega+\frac{2\pi k}{\lambda}\right)\hat{\phi}(\omega)=0$, for all $k\neq 0$. In this case, the above equation reduces to $\hat{f}(\omega)=\hat{f}(\omega)\hat{h}(\omega)$, or equivalently $f=f*h$, where $\hat{h}=(\sqrt{2\pi}/\lambda)\hat{\phi}$. By reversing the argument, we can also see that the reproducing formula $f=f*h$ would imply a sampling theorem.

Although we have derived the reproducing formula, $f=f*h$, by using the Poisson summation formula, as well as, a sampling series for band-limited functions, with regularly spaced sample points, it turns out that by reversing the argument and starting with the foregoing reproducing formula, we can obtain a more general sampling theorem that does not depend on the Poisson summation formula; hence, we can obtain a sampling series with irregularly spaced sample points.

Starting from this point of view, H. Feichtinger and K. Gröchenig in a series of articles [2-8] have been able to derive a number of sampling theorems to reconstruct any member of a large class of functions that includes multidimensional band-limited functions, from the values of this member at irregularly spaced sample points, or from some combinations thereof.

In this section we shall work with the weighted L^p-spaces

$$L_\alpha^p(\Re^N) = \left\{ f \in L_{loc}^1 : \|f\|_{p,\alpha}^p = \int_{\Re^N} |f(\underline{x})|^p (1+|\underline{x}|)^{\alpha p} d\underline{x} < \infty \right\},$$

where $1 \le p < \infty$ and $\alpha > 0$.

For $f \in L_\alpha^p(\Re^N)$, $h \in L_\alpha^1(\Re^N)$, the convolution

$$g(\underline{x}) = (f * h)(\underline{x}) = \int_{\Re^N} f(\underline{x}-\underline{y})h(\underline{y})d\underline{y}$$

is well defined and satisfies

$$\|g\|_{p,\alpha} \le \|f\|_{p,\alpha} \|h\|_{1,\alpha}; \tag{9.1.1}$$

hence $L_\alpha^p * L_\alpha^1 \subset L_\alpha^p$.

The space of functions band-limited to a fixed compact set Ω is defined as

$$B_\alpha^p(\Omega) = \{f \in L_\alpha^p : \operatorname{supp} \hat{f} \subset \Omega\},$$

where $\operatorname{supp} \hat{f}$ is the support of the Fourier transform of f. For $p > 2$, $\hat{f}$ is understood to be in the sense of tempered distributions. It is easy to see that for $f \in B_\alpha^p(\Omega)$, we have

$$|f(\underline{x})| \le C(1+|\underline{x}|)^{-\alpha}.$$

Let Ω and Ω_0 be two compact sets such that $\Omega \subset \Omega_0$. Let h be a fixed band-limited function in $L_\alpha^1(\Re^N)$ such that $\operatorname{supp} \hat{h} \subset \Omega_0$ and $\hat{h}(\underline{\omega}) = 1$ for $\underline{\omega} \in \Omega$. Then, for any $f \in B_\alpha^p(\Omega)$,

$$f = f * h,$$

and because of (9.1.1), the convolution operator

$$C_h f = f * h$$

is bounded from $L_\alpha^p(\Re^N)$ into $B_\alpha^p(\Omega_0)$ and acts as the identity on $B_\alpha^p(\Omega)$.

Definition 9.1.1.

i) The density of a set X of sampling points $\{\underline{x}_i\}$ is defined as the infimum of all $\delta > 0$ such that $\Re^N = \bigcup_i B_\delta(\underline{x}_i)$, where $B_\delta(\underline{x}_i)$ is the open ball of radius δ and center $\underline{x}_i$.

ii) The set X is said to be separated if there exists $\delta_0 > 0$ such that

$$|\underline{x}_i - \underline{x}_j| \geq \delta_0 > 0 \quad \text{for all} \quad i \neq j \,.$$

It is easy to see that if X is a separated set of sampling points $\{\underline{x}_i\}$ with density $\delta > 0$, then for any $\underline{x} \in \mathfrak{R}^N$, the ball $B_\delta(\underline{x})$ contains at least one sampling point, but at most finitely many of them. On the real line $\mathfrak{R}^1$, if $X = \{x_i\}_{i=-\infty}^{\infty}$ is a sequence of points such that $\lim_{i\to\infty} x_i = \infty$, $\lim_{i\to-\infty} x_i = -\infty$ and $\sup_i |x_{i+1} - x_i| = 2\delta$, then X has density δ. In addition, if $|x_n - n| < (1-\delta_0)/2$ for all n, then X is separated.

To any separated set X of sampling points with density $\delta > 0$, we associate a partition of unity $\Psi = \{\psi_i(\underline{x})\}$ of size δ with the following properties:

i) ψ_i is measurable and $0 \leq \psi_i \leq 1$ for all i,

ii) $\operatorname{supp} \psi_i \subset B_\delta(\underline{x}_i)$,

iii) $\sum_i \psi_i = 1$.

For example, on $\mathfrak{R}^1$, if $X = \{x_n = \delta n, \delta > 0\}_{n=-\infty}^{\infty}$, we may take Ψ as

$$\Psi = \{\psi_n(x) = \psi(x - n\delta)\}, \quad \text{where} \quad \psi(x) = \begin{cases} 1 - \dfrac{|x|}{\delta} & \text{if} \quad |x| \leq \delta \\ 0 & \text{if} \quad |x| > \delta \end{cases} .$$

More generally, if X consists of irregularly spaced points with $\sup_n |x_{n+1} - x_n| = \delta$, we may take $\Psi = \{\psi_n(x)\}_{n=-\infty}^{\infty}$

where

$$\psi_n(x) = \begin{cases} \dfrac{x - x_{n-1}}{x_n - x_{n-1}} & \text{if} \quad x_{n-1} \leq x \leq x_n \\ \dfrac{x_{n+1} - x}{x_{n+1} - x_n} & \text{if} \quad x_n \leq x \leq x_{n+1} \\ 0 & \text{otherwise}. \end{cases} \tag{9.1.2}$$

This can be easily extended to $\mathfrak{R}^2$ as follows: since the set X induces a natural triangulation on $\mathfrak{R}^2$, we define $\psi_n(x)$ to be 1 at x_n and piecewise linear on the triangles having x_n as a vertex.

Another partition of unity on $\mathfrak{R}^1$ can be chosen as follows. Let $X = \{x_n\}_{n=-\infty}^{\infty}$ be such that $\ldots < x_{n-1} < x_n < x_{n+1} < \ldots$, and $\sup_n |x_{n+1} - x_n| = \delta$. Set $y_n = (x_{n+1} + x_n)/2$ and $\psi_n(x) = \chi_{[y_{n-1}, y_n)}$. Then, in view of the fact that

$y_n - x_n < \delta/2$, $x_n - y_{n-1} < \delta/2$, it is easy to see that supp $\psi_n \subset B_\delta(x_n)$; hence $\Psi = \{\psi_n(x)\}_{n=-\infty}^{\infty}$ is a partition of unity. Other examples, using B-splines of higher degrees, can be constructed similarly.

A more general example of partitions of unity in $\mathfrak{R}^N$ is the collection of all characteristic functions of the so-called Voronoi regions $\{V_i\}$ around $\{\underline{x}_i\}$, which are defined by $V_i = \{\underline{x} \in \mathfrak{R}^N : |\underline{x} - \underline{x}_i| \leq |\underline{x} - \underline{x}_j|$, for all $j \neq i\}$.

We associate with any partition of unity Ψ, the operators

$$Sp_\Psi f(\underline{x}) = \sum_i f(\underline{x}_i)\psi_i(\underline{x}), \tag{9.1.3}$$

and

$$A f = (Sp_\Psi f) * h = \left(\sum_i f(\underline{x}_i)\psi_i\right) * h, \tag{9.1.4}$$

where $h \in L^1_\alpha(\mathfrak{R}^N)$.

The reason for the notation "Sp_Ψ" is that when Ψ consists of piecewise polynomials of fixed order, $Sp_\Psi f$ is a spline approximation of f. For example, if Ψ is as given by (9.1.2), then (9.1.3) becomes a piecewise linear interpolation of f of the form

$$Sp_\Psi f(x) = f(x_n)\psi_n(x) + f(x_{n+1})\psi_{n+1}(x) = f(x_n) + \frac{f(x_{n+1}) - f(x_n)}{x_{n+1} - x_n}(x - x_n),$$

for $x_n \leq x \leq x_{n+1}$.

To describe the local behavior of a function, the following definition will be needed.

Definition 9.1.2. For a given function f, the local maximum function $f^\#$ and the local δ-oscillation function $\text{osc}_\delta f$ are defined as

$$f^\#(\underline{x}) = \sup_{\underline{z} \in B_\delta(\underline{x})} |f(\underline{z})|,$$

and

$$\text{osc}_\delta f(\underline{x}) = \sup_{\underline{z} \in B_\delta(\underline{x})} |f(\underline{z}) - f(\underline{x})|.$$

By pulling the sup through the integrals, we can easily verify the following inequalities

$$(f * h)^\# \leq |f| * h^\# \tag{9.1.5}$$

$$\text{osc}_\delta(f * h) \leq |f| * \text{osc}_\delta h. \tag{9.1.6}$$

It is readily seen that $f^{\#}\in L^p_\alpha(\Re^N)$ if and only if f and $osc_\delta f$ are in $L^p_\alpha(\Re^N)$. Now we introduce another important space of functions. We define the space $C^p_\alpha(\Re^N)$ as the space of all continuous functions f in $L^p_\alpha(\Re^N)$ for which $f^{\#}\in L^p_\alpha(\Re^N)$, i.e.,

$$C^p_\alpha(\Re^N) = \{f \in L^p_\alpha(\Re^N) : f \text{ continuous and } f^{\#}\in L^p_\alpha(\Re^N)\}\,.$$

Clearly $C^p_\alpha(\Re^N) \subset L^p_\alpha(\Re^N)$. We provide $C^p_\alpha(\Re^N)$ with the topology generated by the norm

$$\|f\|_{C^p_\alpha} = \|f^{\#}\|_{L^p_\alpha} = \|f^{\#}\|_{p,\alpha}\,.$$

The following lemma will be needed.

Lemma 9.1.1. Let $q \in B^p_\alpha(\Omega_0)$ for some compact set Ω_0. Then $|\nabla q| \in C^p_\alpha(\Re^N)$, where ∇q is the gradient of q.

Proof. First, observe that the Schwartz space $S(\Re^N)$ of rapidly decreasing infinitely differentiable functions is a subspace of $C^1_\alpha(\Re^N)$, and that ∇q exists since q is an entire function. Moreover, $(\partial q/\partial x_k)^{\wedge} = -i\omega_k \hat{q}(\underline{\omega})$, $k = 1,2,\ldots,N$. Now choose $u_k \in S(\Re^N)$ such that $\hat{u}_k(\underline{\omega}) = -i\omega_k$ on Ω_0; hence $\partial q/\partial x_k = q * u_k$ and it follows that

$$|\nabla q| = \left(\sum_{k=1}^{N}\left(\frac{\partial q}{\partial x_k}\right)^2\right)^{1/2} \le \sum_{k=1}^{N}\left|\frac{\partial q}{\partial x_k}\right| = \sum_{k=1}^{N}|q * u_k| \le |q| * u\,,$$

where $u = \sum_{k=1}^{N}|u_k|$. Therefore, from (9.1.5) we obtain

$$|\nabla q|^{\#} \le |q| * u^{\#}\,,$$

which, in view of (9.1.1), yields

$$\|\,|\nabla q|^{\#}\|_{p,\alpha} \le \|q\|_{p,\alpha}\|u^{\#}\|_{1,\alpha} < \infty\,.$$

The last inequality follows from the fact that $u \in C^1_\alpha(\Re^N)$. ▌

It is obvious that $f \in L^p_\alpha(\Re^N)$ if $f^{\#}\in L^p_\alpha(\Re^N)$, but the converse is not always true; however, it is true if $f \in B^p_\alpha(\Omega)$, for some compact set Ω, as can be seen from the following lemma.

Lemma 9.1.2. Let $h \in B_\alpha^p(\Omega)$. Then

$$\|h^{\#}\|_{p,\alpha} \le C(\Omega)\|h\|_{p,\alpha} \tag{9.1.7}$$

and

$$\|\operatorname{osc}_\delta h\|_{p,\alpha} \le d(\delta,\Omega)\|h\|_{p,\alpha}, \tag{9.1.8}$$

for some constants C and d with $d(\delta,\Omega) = O(\delta)$ as $\delta \to 0$.

Proof. Choose $q \in B_\alpha^1(\Omega_0)$ such that $\Omega \subset \Omega_0$ and $\hat{q}(\underline{\omega}) = 1$ on Ω. Without loss of generality, we may choose q to be in $C_\alpha^1(\Re^N)$. Then $h = h * q$ and in view of (9.1.1) and (9.1.5) we obtain

$$\|h^{\#}\|_{p,\alpha} \le \|q^{\#}\|_{1,\alpha}\|h\|_{p,\alpha},$$

which leads to (9.1.7) with $C(\Omega) = \inf_q \|q^{\#}\|_{1,\alpha}$. Similarly, from (9.1.1) and (9.1.6), we obtain

$$\|\operatorname{osc}_\delta h\|_{p,\alpha} \le \|\operatorname{osc}_\delta q\|_{1,\alpha}\|h\|_{p,\alpha}. \tag{9.1.9}$$

But by the mean-value theorem

$$\operatorname{osc}_\delta q(\underline{x}) = \sup_{\underline{z}\in B_\delta(\underline{x})} |q(\underline{x}) - q(\underline{z})| \le \sup_{\underline{y}\in B_\delta(\underline{x})} |\underline{z}-\underline{x}|\,|\nabla q(\underline{y})| \le \delta \sup_{\underline{y}\in B_\delta(\underline{x})} |\nabla q(\underline{y})|\,.$$

Thus,

$$\|\operatorname{osc}_\delta q\|_{1,\alpha} \le \delta\|\,|\nabla q|^{\#}\|_{1,\alpha},$$

which, in view of Lemma 9.1.1 and (9.1.9), yields (9.1.8) with $d(\delta,\Omega) = \delta\|\,|\nabla q|^{\#}\|_{1,\alpha}$. ∎

An immediate consequence of this lemma is that $B_\alpha^p(\Omega) \subset C_\alpha^p(\Re^N)$.

Lemma 9.1.3. Given $h \in B_\alpha^1(\Omega_0)$ for some compact set Ω_0, then the operator A defined by (9.1.4) is a continuous operator from $B_\alpha^p(\Omega_0)$ into $B_\alpha^p(\Omega_0)$ and from $C_\alpha^p(\Re^N)$ into $B_\alpha^p(\Omega_0)$.

Proof. From (9.1.3), we have for any fixed $\underline{x} \in \Re^N$

$$|Sp_\psi f(\underline{x})| \le \sum_i |f(\underline{x}_i)|\psi_i(\underline{x}),$$

in which the series contains only finitely many non-zero terms since the ball $B_\delta(\underline{x})$ contains only finitely many of the sampling points $\underline{x}_i$'s. And since $\sum_i \psi_i(\underline{x}) = 1$, it is easy to see that

$$|Sp_\Psi f(\underline{x})| \le f^\#(\underline{x}).$$

Thus

$$\|Sp_\Psi f\|_{p,\alpha} \le \|f^\#\|_{p,\alpha}, \tag{9.1.10}$$

which implies that $Sp_\Psi f \in L^p_\alpha(\Re^N)$ if $f \in C^p_\alpha(\Re^N)$, in particular, if $f \in B^p_\alpha(\Omega)$ for any compact set Ω.

From (9.1.4) and (9.1.1), it follows that $Af \in L^p_\alpha(\Re^N)$ and since supp $\hat{h}$ is contained in Ω_0, then so is suppAf; hence $Af \in B^p_\alpha(\Omega_0)$. The continuity of A follows from the relation

$$\|Af\|_{p,\alpha} = \|Sp_\Psi f * h\|_{p,\alpha} \le \|Sp_\Psi f\|_{p,\alpha} \|h\|_{1,\alpha} \le \|f^\#\|_{p,\alpha} \|h\|_{1,\alpha}.$$

∎

Before we state our next lemma, let us recall that the convolution operator C_h, for $h \in B^p_\alpha(\Omega_0)$, is a continuous operator from $L^p_\alpha(\Re^N)$ into $B^p_\alpha(\Omega_0)$, in particular, from $B^p_\alpha(\Omega)$ into $B^p_\alpha(\Omega_0)$.

Lemma 9.1.4. Given $h \in B^1_\alpha(\Omega)$, there exists a minimal sampling density $\delta_0 > 0$, such that for any separated set X of sampling points with density δ_0

$$\| f * h - A\, f\|_{p,\alpha} \le C \| f\|_{p,\alpha} \tag{9.1.11}$$

for all $f \in B^p_\alpha(\Omega)$, where $0 < C < 1$. Equivalently,

$$|||\, C_h - A\, ||| < 1$$

on $B^p_\alpha(\Omega)$, where $|||-|||$ denotes the norm of the operator.

Proof. First, observe that for any fixed $\underline{x} \in \Re^N$,

$$f(\underline{x}) - Sp_\Psi f(\underline{x}) = f(\underline{x}) \sum_i \psi_i(\underline{x}) - \sum_i f(\underline{x}_i)\psi_i(\underline{x}) = \sum_i (f(\underline{x}) - f(\underline{x}_i))\psi_i(\underline{x});$$

hence, as in the proof of Lemma 9.1.3, we have

$$|f(\underline{x}) - Sp_\Psi f(\underline{x})| \le \sum_i | f(\underline{x}) - f(\underline{x}_i)|\, \psi_i(\underline{x}) \le \operatorname{osc}_\delta f(\underline{x}),$$

in which the summation contains only a finitely many non-zero terms. Therefore, we have for any fixed $\underline{x}$

$$|(f*h)(\underline{x})-(Af)(\underline{x})| = |(f*h)(\underline{x})-((Sp_\Psi f)*h)(\underline{x})|$$
$$= |((f-Sp_\Psi f)*h)(\underline{x})| \le (\mathrm{osc}_\delta f * |h|)(\underline{x}),$$

which, in view of (9.1.1) and Lemma 9.1.2, implies

$$\| f*h - Af\|_{p,\alpha} \le \| \mathrm{osc}_\delta f * |h| \|_{p,\alpha} \le \| \mathrm{osc}_\delta f\|_{p,\alpha} \| h \|_{1,\alpha}$$
$$\le d(\delta,\Omega) \| f\|_{p,\alpha} \| h \|_{1,\alpha} .$$

Since $d(\delta,\Omega) \to 0$ as $\delta \to 0$, we immediately obtain (9.1.11). ∎

Lemma 9.1.4 asserts that for a given $h \in B_\alpha^p(\Omega)$, we can find an operator A as defined by (9.1.4) such that

$$|||\, C_h - A \,||| < 1$$

on $B_\alpha^p(\Omega)$.

Lemma 9.1.5. Let $g, h \in B_\alpha^p(\Omega)$ be given and assume that $h * g = g$. Let A be such that

$$|||\, C_h - A \,||| < 1$$

on $B_\alpha^p(\Omega)$. Then the operator

$$D = \sum_{n=0}^{\infty} R^n ,$$

where $R = C_h - A$, is a well-defined continuous operator from $B_\alpha^p(\Omega)$ into $B_\alpha^p(\Omega)$ and

$$C_g = D\,A\,C_g . \tag{9.1.12}$$

Proof. Lemma 9.1.3 shows that R is a continuous operator from $B_\alpha^p(\Omega)$ into $B_\alpha^p(\Omega)$ and by assumption $\lim_{n\to\infty} |||R^n|||^{1/n} \le |||R||| < 1$; hence D is well defined and continuous on $B_\alpha^p(\Omega)$. Since $C_h = R + A$, it follows from the assumption $C_g = C_h C_g$ that

$$C_g = (R+A)C_g = R\,C_g + A\,C_g = R\,C_h C_g + A\,C_g = R\,(R+A)C_g + A\,C_g$$
$$= R^2 C_g + R\,A\,C_g + A\,C_g ,$$

which, by induction, leads to

$$C_g = R^{n+1}C_g + \left(\sum_{k=0}^{n} R^k\right) A\, C_g ,$$

where $R^0 = I$ (the identity). By letting $n \to \infty$, we obtain (9.1.12). ∎

Lemma 9.1.6. Let $\{f_n\}_{n=0}^{\infty}$ be a sequence in $B_\alpha^p(\Omega)$ converging to $f \in B_\alpha^p(\Omega)$ in the $L_\alpha^p(\Re^N)$ norm as $n \to \infty$. Then, $\lim_{n\to\infty} f_n = f$ in $C_\alpha^p(\Re^N)$. In particular, $\lim_{n\to\infty} f_n = f$ uniformly on compact sets.

Proof. Lemma 9.1.2 asserts that $B_\alpha^p(\Omega) \subseteq C_\alpha^p(\Re^N)$, hence

$$\| f_n - f \|_{C_\alpha^p} = \| (f_n^{\#} - f^{\#}) \|_{p,\alpha} .$$

Choose $h \in B_\alpha^1(\Omega_0)$ such that $\hat{h} = 1$ on Ω,with $\Omega \subset \Omega_0$. Then, by (9.1.1) and (9.1.5) we have

$$\| f_n - f \|_{C_\alpha^p} = \| (f_n - f) * h \|_{C_\alpha^p} = \| ((f_n - f) * h)^{\#} \|_{p,\alpha}$$

$$\leq \| f_n - f \|_{p,\alpha} \| h^{\#} \|_{1,\alpha} \to 0$$

as $n \to \infty$.

We also have

$$| f_n(\underline{x}) - f(\underline{x}) | = | ((f_n - f) * h)(\underline{x}) | \leq \int | f_n(\underline{t}) - f(\underline{t}) |\, | h(\underline{x} - \underline{t}) |\, d\underline{t} ,$$

for any h with support in Ω_0 and satisfying $\hat{h} = 1$ on Ω. By choosing h suitably and applying a weighted Hölder's inequality, we obtain $| f_n(\underline{x}) - f(\underline{x}) | \leq C \| f_n - f \|_{p,\alpha}$, from which the uniform convergence follows, where C is a constant that depends on h and Ω. ∎

Now we are able to state and prove the sampling theorem.

THEOREM 9.1 (Feichtinger and Gröchenig [4])

Let Ω_0 and Ω be two compact sets in $\Re^N$ such that $\Omega \subset \Omega_0$ and choose $h \in B_\alpha^1(\Omega_0)$ with the property that $\hat{h}(\underline{\omega}) = 1$ for $\underline{\omega} \in \Omega$. Then, there exist a density $\delta > 0$, depending only on Ω and h, and functions $S_i \in B_\alpha^1(\Omega_0)$ such that any $f \in B^p_\alpha(\Omega)$, $1 \le p < \infty$, $\alpha \ge 0$, can be reconstructed from any separated sampling set $X = \{\underline{x}_i\}$ with density δ as a series of the form

$$f(\underline{x}) = \sum_i f(\underline{x}_i) S_i(\underline{x}), \tag{9.1.13}$$

in which the series converges in the sense of $C^p_\alpha(\Re^N)$; hence uniformly on compact sets. The reconstruction depends continuously on the sampling values.

Proof. Choose $g \in B^1_\alpha(\Omega_0)$ such that $\Omega \subseteq \operatorname{supp} \hat{g} \subseteq \Omega_0$ and $\hat{g}(\underline{\omega}) = 1$ on Ω. Similarly, choose $h \in B^1_\alpha(\Omega_0)$ such that $\hat{h}(\underline{\omega}) = 1$ on $\operatorname{supp} \hat{g}$ so that $g = g * h$. Since $f \in B^p_\alpha(\Omega)$, we have by (9.1.10) that $Sp_\Psi f \in L^p_\alpha(\Re^N)$. It is easy to see that the partial sums of the series defining $Sp_\Psi f$ converge to $Sp_\Psi f$ in $L^p_\alpha(\Re^N)$. But since the convolution operator C_h is a continuous operator from $L^p_\alpha(\Re^N)$ into $B^p_\alpha(\Omega_0)$, we then have

$$Af = (Sp_\Psi f) * h = \left(\sum_i f(\underline{x}_i)\psi_i\right) * h = \sum_i f(\underline{x}_i)(\psi_i * h), \tag{9.1.14}$$

where the series converges in $L^p_\alpha(\Re^N)$. By Lemmas 9.1.3 and 9.1.6, it follows that the convergence in (9.1.14) is actually in the sense of $C^p_\alpha(\Re^N)$. Therefore, since $Af \in B^p_\alpha(\Omega)$, we have

$$f = f * g = C_g f = DAC_g f = DAf = D\left(\sum_i f(\underline{x}_i)(\psi_i * h)\right) = \sum_i f(\underline{x}_i) S_i,$$

where $S_i = D(\psi_i * h)$. The series converges in $C^p_\alpha(\Re^N)$; hence uniformly on compact sets by Lemma 9.1.6.

Taking the operator D inside the summation sign is justified by Lemma 9.1.5. Moreover, for all i

$$\|S_i\|_{C^1_\alpha} = \|D(\psi_i * h)\|_{C^1_\alpha} \le |||D|||\ \|(\psi_i * h)^\#\|_{1,\alpha} \le |||D|||\ \|\psi_i\|_{1,\alpha} \|h^\#\|_{1,\alpha} < \infty .$$

Finally, the relation

$$\|f\|_{p,\alpha} = \|DAf\|_{p,\alpha} \le |||D|||\ \|Sp_\Psi f * h\|_{p,\alpha} \le |||D|||\ \|h\|_{1,\alpha} \left\| \sum_i f(\underline{x}_i)\psi_i \right\|_{p,\alpha}$$

shows that the reconstruction depends continuously on the sampling points.

∎

Because S_i is in $B^1_\alpha(\Omega_0)$, it is also in $C^1_\alpha(\Re^N)$. Thus, it is easy to see that $S_i(\underline{x}) = O((1+|\underline{x}|)^{-\alpha})$ as $|\underline{x}| \to \infty$. This shows that the reconstruction (9.1.13) has better localization and convergence properties than the WSK series, in which the sampling functions are only in $L^2(\Re^N)$.

The sampling series (9.1.13) can be written in the form

$$f = \sum_{n=0}^{\infty} \phi_n ,$$

where ϕ_n can be computed by using the following iterative procedure:

$$f = DAf = \left(\sum_{n=0}^{\infty} R^n\right) Af = \sum_{n=0}^{\infty} \phi_n ,$$

where $\phi_n = R^n Af$, or equivalently

$$\phi_0 = Af = \sum_i f(\underline{x}_i)(\psi_i * h), \quad \phi_n = R\,\phi_{n-1} = \phi_{n-1} * h - \sum_i \phi_{n-1}(\underline{x}_i)(\psi_i * h) .$$

Starting with different generalizations of the operator A, Feichtinger and Gröchenig have obtained other versions and generalizations of Theorem 9.1. For instance, following similar procedure to the one above, but starting with the operator

$$A_1 f = [D_\Phi(Sp_\Psi f * h_1)] * h_2 = \sum_{i,j} f(\underline{x}_i)\left(\int \phi_j(\underline{t})(\psi_i * h_1)(\underline{t})\, d\underline{t}\right) T_{y_j} h_2 ,$$

where $\Phi = \{\phi_j\}$, $\Psi = \{\psi_i\}$ are two partitions of unity associated with the two sets of sampling points $X = \{\underline{x}_i\}$, $Y = \{\underline{y}_j\}$; $h_1, h_2 \in C^1_\alpha(\Re^N)$,

$$(D_\Phi g) * h = \sum_j <\phi_j, g> T_{y_j} h \quad \text{and} \quad T_y h(x) = h(x-y) ,$$

they have obtained the following sampling theorem which we state without proof.

THEOREM 9.2 (Feichtinger and Gröchenig [4,5])

Let Ω, Ω_0 be two compact subsets of $\Re^N$ with $\Omega \subseteq \Omega_0$ and $g \in B^1_\alpha(\Omega_0)$ such that $\hat{g} \neq 0$ on Ω. Then there exist $\delta_1, \delta_2 > 0$ such that for any separated sets $X = \{\underline{x}_i\}$, $Y = \{\underline{y}_j\}$ of density δ_1 and δ_2 respectively, we can reconstruct any $f \in B^p_\alpha(\Omega)$ as a series of the form

$$f = \sum_j C_j(f(x_i)) T_{y_j} g ,$$

where the series converges in $C_\alpha^p(\Re^N)$, in particular, uniformly on compact sets and the coefficients C_j are functions of the sample values $f(x_i)$. The dependence of the coefficients C_j on the sample values is continuous in the sense that

$$\left\| \sum_j C_j \phi_j \right\|_{p,\alpha} \le C \|f\|_{p,\alpha} ,$$

for some constant $C > 0$.

9.2 Sampling by Using the Green's Function in Several Variables

Following some of the ideas presented in Chapters 6 and 8, we shall now derive a Kramer-type sampling theorem for integral transforms of functions supported in a simply connected bounded region in $\Re^N$ with a smooth boundary. Some of these integral transforms represent functions band-limited in the classical sense. For example, if

$$f(t) = \int_{-1}^{1} g^*(y) P_{t-1/2}(y)\, dy ,$$

for some $g^* \in L^2(-1,1)$, where $P_t(y)$ is the Legendre function, then with the aid of the Mehler-Dirichlet formula [1, p. 159]

$$P_{t-1/2}(y) = \frac{\sqrt{2}}{\pi} \int_0^{\operatorname{Arccos} y} \frac{\cos tx}{\sqrt{\cos x - y}}\, dx , \quad y \in (-1,1)$$

we can show [13] that

$$f(t) = \int_0^{\pi} g(x) \cos tx \, dx ,$$

for some $g \in L^2(0,\pi)$. Therefore, f is band-limited in the classical sense.

Unlike our approach in Chapter 8, here we shall no longer require that the region be an N-dimensional rectangle or that the underlying boundary-value problem be separable. However, like our sampling series in previous chapters, the one we are about to derive is also a Langrange-type interpolation series. The technique we shall use utilizes the Green's function of the boundary-value problem.

Let E be a simply connected, bounded region in $\mathfrak{R}^N$ with a smooth boundary ∂E and consider the Dirichlet problem:

$$\Delta\psi + (\lambda - q(\underline{x}))\psi = 0, \qquad \underline{x} = (x_1, \ldots, x_N) \in E \tag{9.2.1}$$

with

$$\psi(\underline{x}) = 0 \qquad \text{on} \qquad \partial E, \tag{9.2.2}$$

where $q(\underline{x})$ has continuous partial derivatives $\partial q(\underline{x})/\partial x_i$ on the closure $\overline{E}$ of E for $i = 1, \ldots, N$.

It is known that problem (9.2.1) and (9.2.2) has eigenvalues $\{\lambda_n\}_{n=1}^{\infty}$ which we shall assume, without loss of generality, that they are all different from zero, i.e., $\lambda_n \neq 0$ for all n. To any eigenvalue λ_n, there may correspond more than one eigenfunction, but there will be at most finitely many linearly independent ones. By standard orthogonalization process, we can choose them so that they are orthogonal over E. That is, if we denote the eigenfunctions corresponding to the eigenvalue λ_n by $\psi_{n,1}(\underline{x}), \ldots, \psi_{n,k_n}(\underline{x})$, then

$$\int_E \psi_{n,i}(\underline{x})\psi_{n,j}(\underline{x})\,d\underline{x} = 0 \qquad \text{if} \qquad i \neq j,$$

$1 \le i,\ j \le k_n$. We call k_n the multiplicity of the eigenvalue λ_n.

Since eigenfunctions belonging to different eigenvalues are orthogonal, we, thus, can arrange the eigenfunctions in a sequence of the form $\{\psi_{n,i}(\underline{x})\}$ where $n = 1, 2, \ldots, 1 \le i \le k_n$ such that

$$\int_E \psi_{n,i}(\underline{x})\psi_{m,j}(\underline{x})\,d\underline{x} = 0$$

if $n \neq m$ or $i \neq j$.

It is known that the series $\sum_{n=1}^{\infty} 1/\lambda_n$ in general is not convergent, but the series $\sum_{n=1}^{\infty} 1/\lambda_n^{\alpha}$ is, provided that $\alpha > N/2$ [10, p. 169]. Let p be the smallest integer greater than $(N/2) - 1$. Then the infinite product

$$P(\lambda) = \prod_{n=1}^{\infty}\left(1 - \frac{\lambda}{\lambda_n}\right)\exp\left[\left(\frac{\lambda}{\lambda_n}\right) + \frac{1}{2}\left(\frac{\lambda}{\lambda_n}\right)^2 + \ldots + \frac{1}{p}\left(\frac{\lambda}{\lambda_n}\right)^p\right],$$

converges for any complex number λ and defines an entire function in λ [9, p. 250].

Without loss of generality, we assume that the eigenvalues are arranged in such a way that $|\lambda_1| \le |\lambda_2| \le |\lambda_3| \le \ldots$.

Let $G(\underline{x},\underline{\xi},\lambda)$ be the Green's function of problem (9.2.1) and (9.2.2). Let $\underline{\xi}=\underline{\xi}_0$ be some fixed point in E such that $G(\underline{x},\underline{\xi}_0,\lambda)\neq 0$ for all $\underline{x},\lambda$ and define

$$\Phi(\underline{x},\lambda)=P(\lambda)G(\underline{x},\underline{\xi}_0,\lambda), \tag{9.2.3}$$

where $\underline{x}\in E$, $\lambda\in \mathbf{C}$.

Now we are able to state and prove our sampling theorem for functions that are band-limited, in a very broad sense, to the region E. These functions are integral transforms of functions with support in E.

THEOREM 9.3

Let

$$f(\lambda)=\int_E F(\underline{x})\Phi(\underline{x},\lambda)\,d\underline{x} \tag{9.2.4}$$

for some $F\in L^2(E)$. Then f is an entire function in λ that admits the following sampling series expansion

$$f(\lambda)=\sum_{n=1}^{\infty} f(\lambda_n)\,\frac{P(\lambda)}{(\lambda-\lambda_n)P'(\lambda_n)}, \tag{9.2.5}$$

where the series converges uniformly on any compact subset of the complex λ-plane if either $N=2,3$ or if $N=4,5,\ldots$ and $\sum_{n=1}^{\infty}\sum_{i=1}^{k_n}|\hat{F}_{n,i}|^{\beta}<\infty$, where

$$\hat{F}_{n,i}=\int_E F(\underline{x})\psi_{n,i}(\underline{x})\,d\underline{x} \quad \text{and} \quad 0<\beta<N/(N-2) \quad (1<\beta<2).$$

Proof. Since $G(\underline{x},\underline{\xi}_0,\lambda)$ is a meromorphic function in λ whose poles are all simple and located exactly at the points $\{\lambda_n\}_{n=1}^{\infty}$ [10, p. 106], it follows that $\Phi(\underline{x},\lambda)$ is an entire function in λ and is in $L^2(E)$ as a function of $\underline{x}$. In view of the Cauchy-Schwarz inequality, we have

$$|f(\lambda)|^2\leq\left(\int_E|F(\underline{x})|^2d\underline{x}\right)\left(\int_E|\Phi(\underline{x},\lambda)|^2d\underline{x}\right)<\infty;$$

hence $f(\lambda)$ is well defined. Because of the uniform convergence of the integral in (9.2.4) on compact subsets of the complex λ-plane, it follows that $f(\lambda)$ is indeed an entire function.

Since $F \in L^2(E)$, then

$$F(\underline{x}) = \sum_{n=1}^{\infty} \sum_{i=1}^{k_n} \hat{F}_{n,i}\ \psi_{n,i}(\underline{x}) \tag{9.2.6}$$

where

$$\hat{F}_{n,i} = \int_E F(\underline{x}) \psi_{n,i}(\underline{x}) d\underline{x}\ .$$

Similarly,

$$\Phi(\underline{x}, \lambda) = \sum_{n=1}^{\infty} \sum_{i=1}^{k_n} \hat{\Phi}_{n,i}(\lambda) \psi_{n,i}(\underline{x}),$$

where

$$\hat{\Phi}_{n,i}(\lambda) = \int_E \Phi(\underline{x}, \lambda) \psi_{n,i}(\underline{x}) d\underline{x}\ .$$

From Parseval's equality it follows that

$$f(\lambda) = \sum_{n=1}^{\infty} \sum_{i=1}^{k_n} \hat{F}_{n,i} \hat{\Phi}_{n,i}(\lambda)\,. \tag{9.2.7}$$

In view of (9.2.3) and the fact that the residue of $G(\underline{x}, \underline{\xi}_0, \lambda)$ at λ_n is

$$-\sum_{i=1}^{k_n} \psi_{n,i}(\underline{x}) \psi_{n,i}(\underline{\xi}_0),$$

it follows that

$$f(\lambda_n) = \lim_{\lambda \to \lambda_n} f(\lambda) = \lim_{\lambda \to \lambda_n} \int_E F(\underline{x}) \Phi(\underline{x}, \lambda) d\underline{x}$$

$$= -P'(\lambda_n) \int_E F(\underline{x}) \left(\sum_{i=1}^{k_n} \psi_{n,i}(\underline{x}) \psi_{n,i}(\underline{\xi}_0) \right) d\underline{x}$$

$$= -P'(\lambda_n) \sum_{i=1}^{k_n} \hat{F}_{n,i} \psi_{n,i}(\underline{\xi}_0)\,. \tag{9.2.8}$$

Moreover, from formula (14.16.1) in [10], we conclude that

$$\hat{\Phi}_{n,i}(\lambda) = P(\lambda) \frac{\psi_{n,i}(\underline{\xi}_0)}{(\lambda_n - \lambda)}\,. \tag{9.2.9}$$

Therefore, by combining (9.2.9), (9.2.8) and (9.2.7) we obtain (9.2.5).

To show that (9.2.5) converges uniformly on any compact subset K of the complex λ-plane, we choose a nonnegative real number R and a positive integer M so that

$$K \subset D_R = \{z : |z| \le R\} \quad \text{and} \quad |\lambda_n| \ge 2R \quad \text{for all} \quad n \ge M+1 .$$

For any $\lambda \in K$, $|\lambda - \lambda_n| \ge |\lambda_n| - |\lambda| \ge |\lambda_n| - R$, for any $n \ge M+1$. Thus, if we denote $\max_{\lambda \in K} |P(\lambda)|$ by A, then

$$\left(|P(\lambda)|^\alpha \sum_{n=M+1}^{\infty} \frac{1}{|\lambda - \lambda_n|^\alpha} \right) \le A^\alpha \sum_{n=M+1}^{\infty} \frac{1}{(|\lambda_n| - R)^\alpha} = C_\alpha(R) < \infty .$$

The last series is convergent because so is $\sum_{n=1}^{\infty} 1/|\lambda_n|^\alpha$, provided that $\alpha > N/2$. Moreover, $C_\alpha(R)$ is independent of $\lambda \in K$.

For $N = 2, 3$, we apply the Cauchy-Schwarz inequality to (9.2.5) to obtain

$$\left| f(\lambda) - \sum_{n=1}^{M} f(\lambda_n) \frac{P(\lambda)}{(\lambda - \lambda_n) P'(\lambda_n)} \right|^2 \le \left(\sum_{n=M+1}^{\infty} \left| \frac{f(\lambda_n)}{P'(\lambda_n)} \right|^2 \right) \left(\sum_{n=M+1}^{\infty} \left| \frac{P(\lambda)}{(\lambda - \lambda_n)} \right|^2 \right)$$

$$\le \left(\sum_{n=M+1}^{\infty} \sum_{i=1}^{k_n} |\hat{F}_{n,i}|^2 \right) \left(|P(\lambda)|^2 \sum_{n=M+1}^{\infty} \frac{1}{|\lambda - \lambda_n|^2} \right) \le C_2(R) \sum_{n=M+1}^{\infty} \sum_{i=1}^{k_n} |\hat{F}_{n,i}|^2 \to 0 ,$$

uniformly in λ as $M \to \infty$.

But for $N = 4, 5, \ldots$, we use Hölder's inequality to obtain

$$\left| f(\lambda) - \sum_{n=1}^{M} f(\lambda_n) \frac{P(\lambda)}{(\lambda - \lambda_n) P'(\lambda_n)} \right| \le C_\alpha^{1/\alpha}(R) \left(\sum_{n=M+1}^{\infty} \sum_{i=1}^{k_n} |\hat{F}_{n,i}|^\beta \right)^{1/\beta} \to 0 ,$$

uniformly in λ as $M \to \infty$, where $1/\alpha + 1/\beta = 1$. ▮

A similar sampling theorem associated with the Neumann boundary-value problem:

$$\Delta\psi + (\lambda - q(\underline{x}))\psi = 0 , \qquad \underline{x} \in E$$

with

$$\frac{\partial \psi}{\partial n} = 0 \quad \text{on} \quad \partial E ,$$

where $\partial/\partial n$ denotes the derivative along the outward normal, can also be obtained by a minor modification of the above argument.

References

1. A. Erdelyi, W. Magnus, F. Oberhettinger, and F. Tricomi, *Higher Transcendental Functions*, Vol. 1, McGraw-Hill, New York (1954).

2. H. G. Feichtinger, Discretization of convolution and reconstruction of band-limited functions from irregular sampling, Progress in Approximation Theory, Special Issue of the *J. Approx. Theory* (1991), 333-345.

3. H. G. Feichtinger and K. Gröchenig, Reconstruction of band-limited functions from irregular sampling values, manuscript (1989).

4. _______, Irregular sampling theorems and series expansions of band-limited functions, *J. Math. Anal. Appls.*, 167 (1992), 530-556.

5. _______, Iterative reconstruction of multivariate band-limited functions from irregular sampling values, *SIAM J. Math. Anal.*, 23 (1992), 244-261.

6. _______, Multidimensional irregular sampling of band-limited functions in L^p-spaces, *Multivariate Approx. Theory*, IV (Oberwolfach), *Internat. Ser. Numer. Math.*, 90, Birkhäuser, Basel (1989), 135-142.

7. K. Gröchenig, Reconstruction algorithms in irregular sampling, *Math. Comp.*, 59 (1992), 181-194.

8. _______, A new approach to irregular sampling of band-limited functions, Recent advances in Fourier analysis and its applications, *NATO Adv. Sci. Inst. Ser. C: Math. Phys. Sci.*, Kluwer Acad. Publ., Dordrecht (1990), 251-260.

9. E. Titchmarsh, *The Theory of Functions*, Oxford University Press, Oxford (1978).

10. _______, *Eigenfunction Expansions Associated with Second Order Differential Equations*, Part II, Clarendon Press, Oxford (1958).

11. A. Zayed, Sampling theorem for functions bandlimited to a disc, to appear in the *J. Complex Variables Theory Applications*.

12. _______, A new role of Green's function in interpolation and sampling theory, to appear in the *J. Math. Anal. Appls.*

13. A. Zayed, G. Hinsen and P. Butzer, On Lagrange interpolation and Kramer-type sampling theorems associated with Sturm-Liouville problems, *SIAM J. Appl. Math.*, 50, 3 (1990) 893-909.

10

FRAMES AND WAVELETS: A NEW PERSPECTIVE ON SAMPLING THEOREMS

10.0 Introduction

In the last two decades not only new results in sampling theory have been obtained, but also new approaches and ways of understanding sampling concepts have emerged. The last few years, in particular, have witnessed a great mingling of sampling theory and more recent results in mathematical analysis, such as the theory of frames and wavelets.

The chief purpose of this chapter is to report on some of the advances that these new theories have made in sampling theory. Among all the different new techniques used in sampling, the most remarkable is the theory of frames. In this regard, the elegant work of J. Benedetto [3], and J. Benedetto and W. Heller [4] plays a central role in making the viewing of sampling theorems through the theory of frames easy to fathom. Frames, like bases, are building blocks in Hilbert spaces, through them one can obtain representations for the space elements; but, unlike bases, frames are not necessarily linearly independent.

Frames for $L^2(\mathfrak{R})$ that are generated from one single function, whether by translation, modulation, dilation, or combination thereof, have been the center of intensive research activities in recent years because of their intimate connection with both the discrete wavelet and windowed Fourier transforms. In this chapter we shall be concerned with the role that these frames play in sampling theory.

When the frames are appropriately chosen in the Hilbert space B_π^2 of the classical band-limited signals, sampling theorems are obtained. The Frames that Benedetto and Heller used comprise functions that are generated from one single function by modulation and translation, i.e., functions of the form $g_{mb,na}(x) = e^{2\pi imbx}g(x-na); m,n = 0, \pm1, \pm2, \ldots$. We shall call these functions the discrete windowed Fourier transform functions or the short-time Fourier transform functions.

Benedetto and Heller's approach of using frames to derive sampling theorems is parallel to a relatively older approach by Higgins [14, 15] in which he used bases rather than frames to derive a more general form of the Paley-Wiener-Levinson sampling theorem for band-limited functions. To derive their sampling theorems, Higgins used the duality between a basis and its biorthogonal basis in the setting of reproducing-kernel Hilbert spaces, whereas Benedetto and Heller used frames and their dual frames in $L^2(\mathfrak{R})$. In view of the parallelism of these two approaches, it seems appealing from an aesthetic point of view to present the two approaches. This will be accomplished in Sections 1 and 2.

We shall begin with an introduction to reproducing-kernel Hilbert spaces, in which Higgins' construction lives. We then give a brief presentation of the theory of frames followed by some of its applications in sampling theory as developed by Benedetto and Heller.

Another recent development in mathematical analysis that has produced a tremendous impact on different areas of applied mathematics, physics and engineering, in particular on signal and image processing, is the theory of wavelets. Notwithstanding the fact that the word "wavelets" started to appear in the literature only about 10 years ago, wavelets themselves were recognized in one form or another by different scientists in various disciplines much earlier than that. For example, subband coding systems and tree structured digital filter banks, ideas that have been known to electric and communication engineers for awhile, turned out to be related to multiresolution analysis which is a mathematical structure contrived to generate orthonormal wavelet bases. Also, one of the main features of tree structured filtered banks is that they give rise to non-uniform filter bandwidths and non-uniform decimation ratios in the subbands; a feature that characterizes wavelets as well; however, non-uniform filter banks had been used in speech processing [26] long before wavelets were formally introduced. In addition, some relatively older notions in quantum physics and engineering, like the notions of coherent state representations in physics and windowed Fourier transform (short-time Fourier transform) in electrical engineering also turned out to be related to wavelets.

Simply put, wavelets are functions generated from one single function, known as the mother wavelet, by dilation and translation. They are most important when they form building blocks for some function space whether as a frame or as an orthonormal basis. Most appealing is their ability to be building blocks not only for $L^2(\Re)$, but for much more general function spaces.

From the standpoint of signal analysis, the wavelet transform is a mathematical technique that can be used to split a signal into different frequency components and then studies each component with a resolution matched to its scale, thus providing a very good frequency and spatial resolution. Unlike other techniques used to study signals in the time-frequency domain, such as the windowed Fourier transform, in the wavelet transform the analyzing wavelets have time-width adapted to their frequency: high frequency wavelets are very narrow, while low frequency wavelets are much broader. This is in contrast with the windowed Fourier transform, where the analyzing signals all have the same envelope function, but translated to the proper time location and filled with higher frequency oscillations. This adaptability property of wavelets is especially useful for certain classes of signals, e.g., voiced speech signals as the energy is concentrated at lower frequencies, while the higher frequencies contain very little energy.

The construction of the first few orthonormal wavelet bases was more or less an art rather than a procedure; it required ingenuity, special tricks and subtle computations. In fact, some constructions were attributed to miraculous cancelations! This path for constructing orthonormal wavelet bases has taken a sharp turn with the advent of multiresolution analysis.

For a mathematician, a multiresolution analysis is a mathematical structure that has been concocted to generate an orthonormal wavelet basis for $L^2(\Re)$ in a systematic manner. It is an idea that has cut across various constructions of known orthonormal wavelet bases, tying together different methods and at the end offering a unified approach and a recipe for generating an orthonormal wavelet basis. But for a signal analyst, a multiresolution analysis is a sequence of successive approximations of the class of all finite-energy signals that allows one to study a finite-energy signal at different scales and describes mathematically the increment of information needed to go from a coarse approximation to a higher resolution approximation. This makes multiresolution analysis an excellent mathematical technique in the field of image processing, where the idea of studying images at different scales and comparing the results has been very popular for many years.

The wavelet and the discrete windowed Fourier transforms, as it turns out, both arise from the representation of two topological groups acting on $L^2(\Re)$, called the affine and the Weyl-Heisenberg groups respectively, and therefore they both can be introduced in a unified way using the concept of topological group representation on a Hilbert space; see [13]. Notwithstanding the elegance of this approach, we favor another approach that may be less elegant but more direct.

We shall introduce wavelets in Section 3 and discuss conditions under which wavelets will form a frame or an orthonormal basis for $L^2(\Re)$. This will be followed by an introduction to multiresolution analyses to show how orthonormal wavelet bases of $L^2(\Re)$ can be generated. We finally conclude this chapter with a discussion on some wavelet applications in sampling theory.

10.1 Sampling in Reproducing-Kernel Hilbert Spaces

The notion of a reproducing kernel in a Hilbert space of functions and reproducing-kernel Hilbert spaces was first introduced in 1950 by N. Aronszajn [1] and since then it has become an important technique in mathematical analysis. In this section, we begin by giving some definitions and properties pertaining to Hilbert spaces, in general, and to reproducing-kernel Hilbert spaces, in particular, that will be needed to state and prove Higgins' sampling theorem. The sampling theorem and some of its generalizations will then follow.

Let $\mathcal{H}$ be a Hilbert space with inner product $\langle\cdot,\cdot\rangle$ and norm $\|x\| = \sqrt{\langle x,x\rangle}$, for $x \in \mathcal{H}$.

Definition 10.1.1. A sequence of vectors $\{x_n\}_{n=1}^{\infty}$ in a Hilbert space $\mathcal{H}$ is said to be a basis (Schauder basis) of $\mathcal{H}$ if to each $x \in \mathcal{H}$, there corresponds a unique sequence of scalars $\{c_n\}_{n=1}^{\infty}$ such that

$$x = \sum_{n=1}^{\infty} c_n x_n \,, \tag{10.1.1}$$

where the convergence is understood to be in the norm, that is

$$\left\| x - \sum_{n=1}^{N} c_n x_n \right\| \to 0 \quad \text{as} \quad N \to \infty \,.$$

A basis $\{x_n\}_{n=1}^{\infty}$ of $\mathcal{H}$ is said to be orthogonal if $\langle x_m, x_n\rangle = 0$, whenever $m \neq n$, and an orthogonal basis is said to be orthonormal if, in addition, $\langle x_n, x_n\rangle = 1$ for all n.

An orthogonal basis is complete in the sense that if $\langle x, x_n\rangle = 0$ for all n, then $x = 0$.

Every separable Hilbert space has an orthonormal basis and for orthonormal basis the expansion (10.1.1) is given by

$$x = \sum_{n=1}^{\infty} \langle x, x_n\rangle x_n \,, \tag{10.1.2}$$

with

$$\|x\|^2 = \sum_{n=1}^{\infty} |\langle x, x_n\rangle|^2 \,. \tag{10.1.3}$$

More generally, for any $x, y \in \mathcal{H}$

$$\langle x,y\rangle = \sum_{n=1}^{\infty} \langle x,x_n\rangle \overline{\langle y,x_n\rangle}. \tag{10.1.4}$$

It can be shown [31, p. 29] that every basis $\{x_n\}_{n=1}^{\infty}$ of a Hilbert space possesses a unique biorthonormal basis $\{x_n^*\}_{n=1}^{\infty}$, which means that

$$\langle x_m, x_n^*\rangle = \delta_{m,n},$$

and for every $x \in \mathcal{H}$

$$x = \sum_{n=1}^{\infty} \langle x,x_n^*\rangle x_n = \sum_{n=1}^{\infty} \langle x,x_n\rangle x_n^*.$$

If $\langle x_m, x_n^*\rangle = 0$ whenever $m \neq n$, but $\langle x_n, x_n^*\rangle$ is not necessarily equal to one, $\{x_n^*\}_{n=1}^{\infty}$ will be called a biorthogonal basis of $\{x_n\}_{n=1}^{\infty}$. In this case we have for any $x \in \mathcal{H}$

$$x = \sum_{n=1}^{\infty} (\overline{d}_n)^{-1} \langle x,x_n^*\rangle x_n = \sum_{n=1}^{\infty} (d_n)^{-1} \langle x,x_n\rangle x_n^*, \tag{10.1.5}$$

where $d_n = \langle x_n^*, x_n\rangle \neq 0$. It can be easily verified that for all n, $d_n \neq 0$.

Throughout the rest of this section $\mathcal{H}$ will denote a Hilbert space consisting of functions defined on some set X. A function $K(x,t)$, $x,t \in X$ is called a reproducing kernel of $\mathcal{H}$ if

i) For every $t, K(x,t)$ as a function of x belongs to $\mathcal{H}$.
ii) For every $x,t \in X$ and $f \in \mathcal{H}$,

$$f(t) = \langle f(x), K(x,t)\rangle_x, \tag{10.1.6}$$

where the subscript x indicates that the scalar product applies to functions of x.

Definition 10.1.2. A Hilbert space $\mathcal{H}$ is called a reproducing-kernel Hilbert space if it possesses a reproducing kernel.

It is readily seen that if a reproducing kernel exists, it is unique. As for its existence, it is known that $\mathcal{H}$ has a reproducing kernel if and only if for every $x \in X$, the evaluation functional $f(x)$, $f \in \mathcal{H}$ is continuous [1].

If $\{e_n\}_{n=1}^{\infty}$ is an orthonormal basis of $\mathcal{H}$ and if K is its reproducing kernel, then

$$K(t,x) = \sum_{n=1}^{\infty} e_n(t)\overline{e}_n(x). \tag{10.1.7.a}$$

For, if we fix $t \in X$ and set

$$\langle f, K_t \rangle = \langle f(y), K(y,t) \rangle_y = f(t),$$

then by (10.1.2) and (10.1.6)

$$K_t = \sum_{n=1}^{\infty} \langle K_t, e_n \rangle e_n = \sum_{n=1}^{\infty} \overline{e_n(t)} e_n,$$

which, in view of (10.1.6) once more and (10.1.4), yields

$$K(t,x) = \langle K_x, K_t \rangle = \sum_{n=1}^{\infty} e_n(t) \overline{e}_n(x).$$

Similarly, we can verify that if $\{e_n\}_{n=1}^{\infty}$ is a basis (not necessarily orthogonal), then the reproducing kernel may be given by

$$K(t,x) = \sum_{n=1}^{\infty} (\overline{d}_n)^{-1} e_n(t) \overline{e_n^*(x)} = \sum_{n=1}^{\infty} (d_n)^{-1} e_n^*(t) \overline{e_n(x)}, \qquad (10.1.7.b)$$

where $d_n = \langle e_n^*, e_n \rangle$ and $\{e_n^*\}_{n=1}^{\infty}$ is a biorthogonal basis for $\{e_n\}_{n=1}^{\infty}$. When $\{e_n\}_{n=1}^{\infty}$ is an orthonormal basis, $e_n = e_n^*$, $d_n = 1$, and (10.1.7.b) reduces to (10.1.7.a).

If $\lim_{n \to \infty} \|f_n - f\| = 0$ in a reproducing-kernel Hilbert space, then $\lim_{n \to \infty} f_n(x) = f(x)$ for each $x \in X$. The convergence is uniform on every subset of X in which $K(x,x)$ is uniformly bounded as can be seen from the relation

$$|f_n(y) - f(y)| = |\langle f_n(x) - f(x), K(x,y) \rangle_x| \le \|f_n - f\| \, \|K(x,y)\|_x = \|f_n - f\| \sqrt{K(y,y)},$$

where

$$\|K(x,y)\|_x^2 = \langle K(x,y), K(x,y) \rangle_x .$$

Definition 10.1.3. A basis $\{S_n\}$ of a reproducing-kernel Hilbert space $\mathcal{H}$ is called a sampling basis if there exists a sequence $\{t_n\}$ in X such that

$$f(x) = \sum_n f(t_n) S_n(x), \quad f \in \mathcal{H}, \qquad (10.1.8)$$

where the series converges in the norm and hence pointwise on x.

Lemma 10.1.1. Let $\{S_n\}$ be a basis of a reproducing-kernel Hilbert space $\mathcal{H}$. Then $\{S_n\}$ is a sampling basis if and only if its biorthogonal basis $\{S_n^*\}$ is given by

$$S_n^*(x) = d_n K(x,t_n), \tag{10.1.9}$$

where $K(x,t)$ is the reproducing kernel which can be given by

$$K(x,t) = \sum_{n=1}^{\infty} (\overline{d}_n)^{-1} S_n(x) \overline{S_n^*(t)} = \sum_{n=1}^{\infty} (d_n)^{-1} S_n^*(x) \overline{S_n(t)}, \tag{10.1.10}$$

where $d_n = \langle S_n^*, S_n \rangle$. $\{S_n^*\}$ is the biorthonormal basis of $\{S_n\}$ if and only if $S_n^*(x) = K(x,t_n)$ for all n.

Proof. Let $\{S_n\}$ be a sampling basis, then by 10.1.5

$$f(x) = \sum_n f(t_n) S_n(x) = \sum_n (\overline{d}_n)^{-1} \langle f, S_n^* \rangle S_n(x), \tag{10.1.11}$$

which, by the uniqueness of the expansion, gives

$$f(t_n) = \langle f, S_n^*/d_n \rangle = \langle f(x), K(x,t_n) \rangle_x ;$$

thus,

$$S_n^*(x) = d_n K(x,t_n) .$$

Conversely, we can easily show that if (10.1.9) holds, then so does (10.1.8). The rest of the proof follows immediately from (10.1.7.b). ▮

An immediate consequence of Lemma 10.1.1 is that an orthonormal basis $\{S_n\}$ of a reproducing-kernel Hilbert space is a sampling basis if and only if it is generated from the reproducing kernel of the space via the relation

$$S_n(x) = K(x,t_n) .$$

Also, if $\{S_n\}_{n=1}^{\infty}$ is a sampling basis, then $S_n(t_m) = \delta_{m,n}$. For, by (10.1.7.b) and (10.1.9)

$$K(x,t_m) = \frac{1}{d_m} S_m^*(x) = \sum_{n=1}^{\infty} (d_n)^{-1} \overline{S_n(t_m)} S_n^*(x) ,$$

and since $\{S_n^*\}_{n=1}^{\infty}$ is a basis, then by the uniqueness of the expansion, all the terms in the series are zeros, except for $n = m$; thus $S_n(t_m) = \delta_{m,n}$.

The space $L^2(\Re)$ is not a reproducing-kernel Hilbert space, but one of its important subspaces, namely the Paley-Wiener space $\mathbf{B}_\pi^2$, is. Its reproducing kernel is $K(x,t) = \sin\pi(x-t)/\pi(x-t)$, hence

$$f(t) = \int_{-\infty}^{\infty} f(x)\frac{\sin\pi(x-t)}{\pi(x-t)}dx\,, \quad f\in\mathbf{B}_\pi^2\,.$$

Moreover,

$$\left\{K(x,t_n) = \frac{\sin\pi(x-t_n)}{\pi(x-t_n)}, \quad t_n = n\right\}_{n=-\infty}^{\infty}$$

is an orthonormal basis of $\mathbf{B}_\pi^2$. This is a consequence of the fact that the map $T: L^2[-\pi,\pi]\rightarrow\mathbf{B}_\pi^2$ defined by

$$(Tf)(t) = \frac{1}{2\pi}\int_{-\pi}^{\pi} f(x)e^{-itx}dx\,, \quad f\in L^2[-\pi,\pi] \tag{10.1.12}$$

is an isomorphism and that $\{e^{inx}\}_{n=-\infty}^{\infty}$ is an orthonormal basis of $L^2[-\pi,\pi]$ with respect to the inner product

$$\langle f,g\rangle = \frac{1}{2\pi}\int_{-\pi}^{\pi} f(x)\overline{g(x)}dx\,, \quad f,g\in L^2[-\pi,\pi]\,.$$

A more elegant proof of the fact that $\{\sin\pi(x-n)/\pi(x-n)\}_{n=-\infty}^{\infty}$ is orthonormal will be given in Section 10.3.

Clearly, $\{\sin\pi(t-t_n)/\pi(t-t_n)\}$ is a sampling basis of $\mathbf{B}_\pi^2$ and (10.1.8) is nothing more than the Whittaker-Shannon-Kotel'nikov (WSK) sampling expansion. Moreover, (10.1.7.a) now takes the form

$$\frac{\sin\pi(x-t)}{\pi(x-t)} = \sum_{n=-\infty}^{\infty}\frac{\sin\pi(x-n)}{\pi(x-n)}\,\frac{\sin\pi(t-n)}{\pi(t-n)}\,.$$

THEOREM 10.1 (Higgins [14])

Let $\{t_n\}$ be a sequence of real numbers such that $\{e^{it_n x}\}$ is a basis for $L^2[-\pi,\pi]$. Then

i) $\{S_n(t) = \sin\pi(t-t_n)/\pi(t-t_n)\}$ is a basis for $\mathbf{B}^2_\pi$, whose biorthonormal basis $\{S^*_n(t)\}$ is a sampling basis for $\mathbf{B}^2_\pi$ with sampling expansion

$$f(t) = \sum_n f(t_n) S^*_n(t), \tag{10.1.13}$$

where the series converges pointwise on the real axis and uniformly on any compact subset thereof.

ii) Conversely, any sampling basis for $\mathbf{B}^2_\pi$ is necessarily the biorthonormal basis of some basis of $\mathbf{B}^2_\pi$ of the form $\{S_n(t) = \sin\pi(t-t_n)/\pi(t-t_n)\}$, where $\{e^{it_n x}\}$ is a basis of $L^2[-\pi,\pi]$.

iii) The only complete orthonormal sampling basis for $\mathbf{B}^2_\pi$ is $\{\sin\pi(t-n-a)/\pi(t-n-a)\}$, where a is a real number.

Proof.

i) Since $\{e^{it_n x}\}$ is a basis for $L^2[-\pi,\pi]$, and its image under the isomorphism T, given by (10.1.12), is

$$\{S_n(t) = \sin\pi(t-t_n)/\pi(t-t_n)\},$$

it follows that $\{S_n(t)\}$ is a basis for $\mathbf{B}^2_\pi$. The reproducing kernel for $\mathbf{B}^2_\pi$ is $K(x,t) = \sin\pi(x-t)/\pi(x-t)$ and since $S_n(x) = K(x,t_n)$, we have by Lemma 10.1.1 the biorthonormal basis $\{S^*_n(t)\}$ of $\{S_n(t)\}$ is a sampling basis for $\mathbf{B}^2_\pi$.

ii) The proof is similar to (i).

iii) It is easy to verify that $\{e^{it_n x}\}$ is a complete orthonormal basis for $L^2[-\pi,\pi]$ if and only if $t_n = n + a$, for some real a. ∎

Corollary 10.1.1. Let $\{t_n\}$ be a sequence of real numbers satisfying

$$|t_n - n| \le D < \frac{1}{4}, \quad n = 0, \pm 1, \pm 2, \ldots. \tag{10.1.14}$$

Then

$$\left\{S^*_n(t) = \frac{G(t)}{(t-t_n)G'(t_n)}\right\}$$

is a sampling basis for $\mathbf{B}_\pi^2$, where

$$G(t) = (t - t_0)\prod_{n=1}^{\infty}\left(1 - \frac{t}{t_n}\right)\left(1 - \frac{t}{t_{-n}}\right),$$

and for any $f \in \mathbf{B}_\pi^2$,

$$f(t) = \sum_{n=-\infty}^{\infty} f(t_n)\frac{G(t)}{(t - t_n)G'(t_n)}.$$

Proof. Under condition (10.1.14), $\{e^{it_n x}\}$ is a basis for $L^2[-\pi,\pi]$ that possesses a unique biorthonormal basis $\{h_n\}$ given by

$$\frac{G(t)}{(t - t_n)G'(t_n)} = \frac{1}{2\pi}\int_{-\pi}^{\pi} h_n(x)e^{-itx}dx\,;$$

see (3.1.7) and [19, p. 58]; hence,

$$S_n^* = \frac{G(t)}{(t - t_n)G'(t_n)},$$

and the rest of the proof follows from Theorem 10.1. ▌

Lemma 10.1.1 can be slightly generalized if, for example, we assume that the reproducing kernel $K(x,t)$ is such that $0 < \varepsilon \le K(t,t) \le A < \infty$ and $K(t_n,t_m) = 0$ if $n \neq m$ for some uniqueness set $\{t_n\}$ of $\mathcal{H}$. A set $\{t_n\}$ is called a uniqueness set of $\mathcal{H}$ if for any $f \in \mathcal{H}$, $f(t_n) = 0$ for all n implies that $f = 0$. In other words, the only function $f \in \mathcal{H}$ that can vanish at $t = t_n$ for all n is the zero function. In this case, the sampling expansion becomes

$$f(t) = \sum_n f(t_n)D_n(t), \quad \text{where} \quad D_n(t) = K(t,t_n)/K(t_n,t_n). \tag{10.1.15}$$

For, if we set

$$S_n(t) = \frac{K(t,t_n)}{\sqrt{K(t_n,t_n)}},$$

then

$$\langle S_n, S_m\rangle = \frac{1}{[K(t_n,t_n)K(t_m,t_m)]^{1/2}}K(t_m,t_n) = \delta_{m,n}.$$

Moreover,

$$\langle f, S_n \rangle = \frac{1}{\sqrt{K(t_n, t_n)}} f(t_n),$$

which, in view of the fact that $\{t_n\}$ is a uniqueness set of $\mathcal{H}$, implies that $\{S_n\}$ is complete; hence an orthonormal basis of $\mathcal{H}$. The sampling series follows immediately from (10.1.2).

Using the sampling expansion (10.1.15) and other generalizations thereof, Nashed and Walter [25] extended Higgins' theorem (Theorem 10.1) to closed subspaces of the Sobolev space H^1 that are more general than B_σ^2. These subspaces are related to the spaces H_π^1 of Chapter 2.

10.2 Frames and Sampling Theorems

10.2.A Frames

The concept of frames in a Hilbert space was first introduced in 1952 by R. Duffin and A. Schaeffer [10] in the context of non-harmonic Fourier series. Their work was motivated by their finding that, under certain restrictions on a sequence of numbers $\{\lambda_n\}$, the set of functions $\{e^{i\lambda_n t}\}$ in $L^2(-\pi, \pi)$ exhibits properties quite similar to those of an orthonormal basis like $\{e^{int}/\sqrt{2\pi}\}$. However, the situation is more complicated with the set $\{e^{i\lambda_n t}\}$ since it is in general over complete. This means that it may contain a proper subset that is itself complete; a property that is impossible for an orthonormal basis to have.

We begin this relatively long section by introducing the concept of frames in a general Hilbert space, then we focus our attention on a special class of frames in the Hilbert space $L^2(\mathfrak{R})$. This special class of frames will play a vital role in sampling theory. Each frame in this class is generated from one single function $g \in L^2(\mathfrak{R})$ by two operations, translation and modulation, each of which depends on one real parameter. Admittedly, this may appear a bit peculiar since the frame spans a space of functions of only one variable while the frame itself depends on two variables. A redundancy in the representation of the function space elements in terms of the frame is therefore expected; however, sometimes this redundancy can be advantageous as we shall see later. To study this special class of frames for certain critical values of the parameters, the Zak transform will be needed. Therefore, we shall include a short discussion of the Zak transform and some of its applications to this special class of frames.

The main theorems in this section are Theorems 10.9 and 10.11 on uniform and non-uniform sampling. Most of the material presented in this section has been conglomerated from various sources, to mention a few [3, 4, 6, 13].

Definition 10.2.1. Given a function f, we define the translation, modulation and dilation operators respectively as follows:

$$T_a f(x) = f(x-a),$$

$$E_a f(x) = e^{2\pi i a x} f(x),$$

$$D_a f(x) = \frac{1}{\sqrt{|a|}} f(x/a), \quad a \neq 0,$$

where a is a real number.

We shall adopt the notation that $E_a(x) = E_a 1(x) = e^{2\pi i a x}$. These three operators are unitary operators from $L^2(\Re)$ onto itself. Hereafter, the Fourier transform of a function $f(x)$ will be defined as

$$F(f)(\omega) = \hat{f}(\omega) = \int_{-\infty}^{\infty} f(x) e^{2\pi i w x} dx,$$

so that the inverse transform will be given by

$$F^{-1}(f)(x) = f(x) = \int_{-\infty}^{\infty} \hat{f}(\omega) e^{-2\pi i x \omega} d\omega.$$

We may also use the notation

$$\check{f}(x) = \int_{-\infty}^{\infty} f(y) e^{-2\pi i x y} dy,$$

so that $(\hat{f})^{\vee} = f$.

Using the standard inner product $\langle , \rangle$ on $L^2(\Re)$

$$\langle f, g \rangle = \int_{-\infty}^{\infty} f(x) \overline{g(x)} dx, \quad f, g \in L^2(\Re),$$

one can easily verify that

$$\langle f, T_a g \rangle = \langle T_{-a} f, g \rangle; \quad \langle f, E_a g \rangle = \langle E_{-a} f, g \rangle;$$

$$\langle f, D_a g \rangle = \langle D_{1/a} f, g \rangle.$$

Let $\mathcal{H}$ be a separable Hilbert space with inner product $\langle \cdot, \cdot \rangle$ and norm $\|f\| = \sqrt{\langle f, f \rangle}$, for any $f \in \mathcal{H}$.

Definition 10.2.2. Let $\mathcal{G} = \{g_n\}$ be a basis in $\mathcal{H}$. Then

i) $\mathcal{G}$ is called unconditional if

$$\sum c_n g_n \in \mathcal{H} \quad \text{implies that} \quad \sum |c_n| g_n \in \mathcal{H}.$$

ii) $\mathcal{G}$ is said to be bounded if there exist two nonnegative numbers A and B such that for all n

$$A \le \|g_n\| \le B .$$

iii) $\mathcal{G}$ is said to be a Riesz basis if there exist a topological isomorphism $T: \mathcal{H} \to \mathcal{H}$ and an orthonormal basis $\{u_n\}$ of $\mathcal{H}$ such that $Tg_n = u_n$ for each n.

Definition 10.2.3. Let $\mathcal{G} = \{g_n\}$ be a sequence in $\mathcal{H}$ (not necessarily a basis of $\mathcal{H}$). Then $\mathcal{G}$ is called a frame if there exist two numbers $A, B > 0$ such that for any $f \in \mathcal{H}$, we have

$$A\|f\|^2 \le \sum_n |\langle f, g_n\rangle|^2 \le B\|f\|^2 .$$

The numbers A and B are called the frame bounds. The frame is said to be tight if $A = B$ and is exact if it ceases to be a frame whenever any single element is deleted from the frame. To every frame $\mathcal{G}$ there corresponds an operator S, known as the frame operator, from $\mathcal{H}$ into itself defined by

$$Sf = \sum \langle f, g_n\rangle g_n , \quad \text{for all} \quad f \in \mathcal{H} .$$

The next theorem, which we shall state without a proof, relates some of the aforementioned definitions.

THEOREM 10.2

In a separable Hilbert space $\mathcal{H}$ the following conditions are equivalent:

a) $\{g_n\}$ is an exact frame for $\mathcal{H}$.
b) $\{g_n\}$ is a bounded unconditional basis of $\mathcal{H}$.
c) $\{g_n\}$ is a Riesz basis of $\mathcal{H}$.

The equivalence of (a) and (b) is proved in [13] and that of (a) and (c) is proved [31, pp. 188-189]. That (b) implies (c) directly is not trivial and it was first proved by Köthe [18].

If $\{g_n\}$ is an orthonormal basis of $\mathcal{H}$, then in view of the Parseval equality, we have for any $f \in \mathcal{H}$

$$\|f\|^2 = \sum_n |\langle f, g_n\rangle|^2 ,$$

which shows that an orthonormal basis is a tight exact frame with frame bounds $A = B = 1$. However, tight frames are not necessarily orthonormal bases. For example,

$$\left\{\left(\frac{1}{\sqrt{2}},\frac{1}{\sqrt{2}}\right),\left(-\frac{1}{\sqrt{2}},\frac{1}{\sqrt{2}}\right),\left(\frac{1}{\sqrt{2}},-\frac{1}{\sqrt{2}}\right),\left(-\frac{1}{\sqrt{2}},-\frac{1}{\sqrt{2}}\right)\right\}$$

is a tight frame in $\mathfrak{R}^2$ with frame bounds $A = 2 = B$, but it is not even orthonormal since these four vectors are linearly dependent.

Lemma 10.2.1. If $\{g_n\}$ is a tight frame with frame bound $A = 1$, and if $\|g_n\| = 1$ for all n, then $\{g_n\}$ is an orthonormal basis.

Proof. Since $\{g_n\}$ is a tight frame with frame bound $= 1$, then

$$\|g_k\|^2 = \sum_n |\langle g_k, g_n\rangle|^2 = \|g_k\|^4 + \sum_{n \neq k} |\langle g_k, g_n\rangle|^2 .$$

But since $\|g_k\| = 1$, it follows that $\langle g_k, g_n\rangle = 0$ for all $n \neq k$. The completeness of $\{g_n\}$ is a consequence of the fact that frames are complete. To see this, let $f \in \mathcal{H}$ be such that $\langle f, g_n\rangle = 0$ for all n. Then, in view of the relation

$$A\|f\|^2 \leq \sum_n |\langle f, g_n\rangle|^2 = 0,$$

we have $f = 0$. ∎

Therefore, we can say that for a tight frame $\{g_n\}$, the frame bound is a measurement of how close the frame to being an orthonormal basis, the closer the frame bound to 1, the closer the frame to being orthonormal, provided that $\|g_n\| = 1$ for all n. The last restriction cannot be dropped since if $\{g_n\}_{n=1}^{\infty}$ is an orthonormal basis of $\mathcal{H}$, then $\{g_1, g_2/\sqrt{2}, g_2/\sqrt{2}, g_3/\sqrt{3}, g_3/\sqrt{3}, g_3/\sqrt{3}, \ldots\}$ is a tight frame with frame bound $A = 1$, but clearly it is not an orthonormal basis.

Tightness and exactness are not related as illustrated by the following example. If $\{g_n\}$ is an orthonormal basis of $\mathcal{H}$, then $\{g_1, g_1, g_2, g_2, \ldots\}$ is a tight frame with frame bounds $A = 2 = B$, but it is not exact, whereas $\{\sqrt{2}g_1, g_2, g_3, \ldots\}$ is exact but not tight since the frame bounds are easily seen to be $A = 1$ and $B = 2$. More examples of this nature can be found in [13, p. 635].

The following theorem summarizes some of the main properties of frames and frame operators that we shall use in the sequel to derive the sampling theorems.

THEOREM 10.3

Let $\mathcal{H}$ be a separable Hilbert space and $\{g_n\}_{n=-\infty}^{\infty} \subset \mathcal{H}$ be a frame with frame bounds A and B. Then

i) The frame operator $Sf = \sum_n \langle f, g_n \rangle g_n$ is a bounded linear operator on $\mathcal{H}$ with $AI \le S \le BI$.
ii) S is invertible with $B^{-1}I \le S^{-1} \le A^{-1}I$. Moreover, S^{-1} is a positive operator; hence it is self-adjoint.
iii) $\{S^{-1}g_n\}$ is a frame with frame bounds B^{-1}, A^{-1}.
iv) Every $f \in \mathcal{H}$ can be written in the form

$$f = \sum_n \langle f, S^{-1}g_n \rangle g_n = \sum_n \langle f, g_n \rangle S^{-1} g_n \,. \tag{10.2.1}$$

v) If there exists a sequence of scalers $\{c_n\}$ such that $f = \sum_n c_n g_n$, then

$$\sum_n |c_n|^2 = \sum_n |a_n|^2 + \sum_n |a_n - c_n|^2,$$

where $a_n = \langle f, S^{-1}g_n \rangle$.

vi) In addition, if $\{g_n\}$ is an exact frame, then $\{g_n\}$ and $\{S^{-1}g_n\}$ are bi-orthonormal, i.e., $\langle g_m, S^{-1}g_n \rangle = \delta_{m,n}$.

Proof: i) Recall that in any Hilbert space $\mathcal{H}$ the norm of any $f \in \mathcal{H}$ may be given by $\|f\| = \sup_{|h|=1} |\langle f, h \rangle|$. Fix $f \in \mathcal{H}$ and set $S_N f = \sum_{n=-N}^{N} \langle f, g_n \rangle g_n$. For $0 \le M \le N$, we have by the Cauchy-Schwarz inequality

$$\begin{aligned}
\|S_N f - S_M f\|^2 &= \sup_{|h|=1} |\langle S_N f - S_M f, h \rangle|^2 = \sup_{|h|=1} \left| \sum_{M+1 \le |n| \le N} \langle f, g_n \rangle \langle g_n, h \rangle \right|^2 \\
&\le \sup_{|h|=1} \left(\sum_{M+1 \le |n| \le N} |\langle f, g_n \rangle|^2 \right) \left(\sum_{M+1 \le |n| \le N} |\langle g_n, h \rangle|^2 \right) \\
&\le \sup_{|h|=1} \left(\sum_{M+1 \le |n| \le N} |\langle f, g_n \rangle|^2 \right) (B\|h\|^2) = B \left(\sum_{M+1 \le |n| \le N} |\langle f, g_n \rangle|^2 \right) \to 0
\end{aligned}$$

as $M, N \to \infty$.

Thus, $\{S_N f\}$ is a Cauchy sequence in $\mathcal{H}$ and hence $\lim_{N \to \infty} S_N f = Sf$ is well defined. By repeating the above argument once more, we obtain

$$\|Sf\|^2 = \sup_{|h|=1} |\langle Sf, h \rangle|^2 = \sup_{|h|=1} \left| \sum_n \langle f, g_n \rangle \langle g_n, h \rangle \right|^2 \le B\left(\sum_n |\langle f, g_n \rangle|^2 \right) \le B^2 \|f\|^2,$$

which implies that $\|S\| \le B$, hence S is bounded. The relation $AI \le S \le BI$ is equivalent to $\langle AIf,f\rangle \le \langle Sf,f\rangle \le \langle BIf,f\rangle$ and the latter follows from the definition of frames.

ii) From the relation $AI \le S \le BI$ it follows that

$$\|I - B^{-1}S\| \le \left\|\left(1-\frac{A}{B}\right)I\right\| < 1\,;$$

hence $B^{-1}S$ is invertible and consequently so is S. By multiplying the relation $AI \le S \le BI$ by S^{-1} and keeping in mind that S^{-1} commutes with S and I, we obtain that $B^{-1}I \le S^{-1} \le A^{-1}I$. Since $\langle S^{-1}f,f\rangle = \langle S^{-1}f, S(S^{-1}f)\rangle \ge A\|S^{-1}f\|^2 \ge 0$, it follows that S^{-1} is positive.

iii) Since S^{-1} is positive it is self-adjoint and therefore,

$$\sum_n \langle f, S^{-1}g_n\rangle S^{-1}g_n = S^{-1}\Big(\sum_n \langle S^{-1}f, g_n\rangle g_n\Big) = S^{-1}(S(S^{-1}f)) = S^{-1}f\,,$$

which gives

$$\sum_n |\langle f, S^{-1}g_n\rangle|^2 = \langle S^{-1}f, f\rangle\,.$$

From part (ii), it now follows that

$$B^{-1}\|f\|^2 \le \sum_n |\langle f, S^{-1}g_n\rangle|^2 \le A^{-1}\|f\|^2\,,$$

hence $\{S^{-1}g_n\}$ is a frame with the prescribed frame bounds.

iv) Replacing f by $S^{-1}f$ in the relation

$$Sf = \sum_n \langle f, g_n\rangle g_n\,,$$

yields

$$f = \sum_n \langle S^{-1}f, g_n\rangle g_n = \sum \langle f, S^{-1}g_n\rangle g_n\,.$$

The other expansion is obtained similarly.

v) Substituting $f = \sum_n a_n g_n$ into the first term in the inner product $\langle f, S^{-1}f\rangle$ yields

$$\langle f, S^{-1}f\rangle = \sum_n a_n \langle g_n, S^{-1}f\rangle = \sum_n |a_n|^2\,,$$

while substituting $f = \sum_n c_n g_n$ yields

$$\langle f, S^{-1} f\rangle = \sum_n c_n \overline{a}_n ,$$

hence $\sum_n |a_n|^2 = \sum_n c_n \overline{a}_n$. Combining this with the relation

$$\sum_n |a_n|^2 + \sum_n |a_n - c_n|^2 = \sum_n |a_n|^2 + \sum_n (|a_n|^2 - a_n \overline{c}_n - \overline{a}_n c_n + |c_n|^2),$$

yields the result.

vi) Fix m and let $a_n = \langle g_m, S^{-1} g_n\rangle$. Assume that $a_m \neq 1$. It is easy to see that

$$g_m = \left(\frac{1}{1-a_m}\right) \sum_{n \neq m} a_n g_n ,$$

and therefore, in virtue of the Cauchy-Schwarz inequality, we have for any $f \in \mathcal{H}$

$$|\langle f, g_m\rangle|^2 = \frac{1}{|1-a_m|^2} \left|\sum_{n \neq m} a_n \langle f, g_n\rangle\right|^2 \leq C \sum_{n \neq m} |\langle f, g_n\rangle|^2 ,$$

where $C = (1/|1-a_m|^2) \sum_{n \neq m} |a_n|^2$. Hence

$$\sum_n |\langle f, g_n\rangle|^2 = |\langle f, g_m\rangle|^2 + \sum_{n \neq m} |\langle f, g_n\rangle|^2 \leq (1+C) \sum_{n \neq m} |\langle f, g_n\rangle|^2 ,$$

which implies that

$$\left(\frac{A}{1+C}\right) \|f\|^2 \leq \sum_{n \neq m} |\langle f, g_n\rangle|^2 \leq B \|f\|^2 .$$

But this is equivalent to saying that $\{g_n\}_{n \neq m}$ is a frame with frame bounds $(A/(1+C))$ and B, which is a contradiction since $\{g_n\}$ is exact and hence no subset of it can be a frame. Thus, if $\{g_n\}$ is exact, $a_m = 1$. To show that $a_n = 0$ if $n \neq m$, we invoke part (v). Since $g_m = \sum_n \delta_{m,n} g_n$, we set $c_n = \delta_{m,n}$ in (v) to obtain

$$1 = |a_m|^2 + \sum_{n \neq m} |a_n|^2 + |a_m - 1|^2 + \sum_{n \neq m} |a_n|^2 ,$$

but since $a_m = 1$, we immediately obtain that $a_n = 0$ for all $n \neq m$. ∎

The frame $\{S^{-1}g_n\}$ is called the dual frame of the frame $\{g_n\}$. Equation (10.2.1) is the analogue of (10.1.5) for frames and the two coincide if any of the conditions of Theorem 10.2 is satisfied and in this case $S^{-1}g_n = g_n^*$.

Part (v) shows that although the expansion of f in terms of the frame $\{g_n\}$ is not necessarily unique, using the expansion with coefficients $f_n = \langle f, S^{-1}g_n \rangle$ is the most economical in the sense that for any other coefficients $\{c_n\}$ with $f = \sum_n c_n g_n$, we have $\sum_n |c_n|^2 \geq \sum_n |f_n|^2$.

10.2.B The Discrete Windowed Fourier Transform Functions and the Gabor Frames

In what follows we shall focus our attention on frames for $L^2(\mathfrak{R})$ that are generated from one single function by translation and modulation, that is frames of the form $\{g_{m,n}(x) = E_{b_m} T_{a_n} g(x) = \exp(2\pi i b_m x) g(x - a_n)\}_{m,n \in Z}$ for some $g \in L^2(\mathfrak{R})$. When $b_m = bm$ and $a_n = an$ for some $a \neq 0 \neq b$, we shall write $g_{mb,na}(x) = e^{2\pi i b m x} g(x - na)$ instead, to exhibit the dependency on the parameters a and b.

We shall call the functions $g_{m,n}(x)$ the discrete windowed Fourier transform functions or the short-time Fourier transform functions. Sometimes, they are called the Weyl-Heisenberg coherent states [13]. Occasionally we shall call g the mother or analyzing wavelet. When the functions $\{g_{m,n}\}_{m,n=-\infty}^{\infty}$ form a frame for $L^2(\mathfrak{R})$, it is sometimes called the Gabor frame although, technically speaking, the Gabor frame is the one in which g is the Gaussian function, $g(x) = g(0)\exp(-bx^2)$, for some $b > 0$. Whether the functions $\{g_{m,n}\}_{m,n=-\infty}^{\infty}$ form a frame for $L^2(\mathfrak{R})$ and under what conditions they do form a frame or an orthonormal basis of $L^2(\mathfrak{R})$ will be the focus of the next few theorems. But before we address these questions, let us say a few words about the role that these functions play in signal analysis and how it came about.

The discrete windowed Fourier transform functions $g_{m\omega_0, nt_0}$ are obtained from one single function g by what is equivalent to a translation by amount nt_0 in the time domain and modulation by $m\omega_0$ in the frequency domain. Translating a signal by a certain amount, physically means delaying it by that amount, while modulating it by a certain amount means shifting its spectrum by that amount. Modulation is very practical in signal transmission. For example, speech signals fall in the range 0 to 4000 Hz, while music signals contain frequencies in the range 0 to 20,000 Hz. If several of these signals were transmitted over the same medium without modulation, they would interfere and produce an unintelligible signal at the receiver. By modulation, each signal can be moved to a different frequency band before transmission; thus eliminating the interference.

The reader should be well aware that determining the Fourier transform (the frequency content) of a signal $f(t)$ requires that one knows f for almost all t. This is one of the shortfalls of the Fourier transform since it precludes a very practical notion, the notion of frequencies changing with time or equivalently the notion of finding the frequency content of a signal locally in time. For example, when a musician plays a piece of music and passes from a low to a high note, the frequency changes with the time, so in order to reproduce this musical piece, one needs to know which frequency to play at any given instant of time.

One of the earliest mathematical techniques used to remedy this deficiency in the Fourier transform was introduced in 1946 by a Hungarian-British physicist and engineer, Dennis Gabor [11], who won the Nobel Prize in physics in 1971 for his invention and development of the holographic method. Gabor introduced another transform that is akin to the Fourier transform and which is now known as the Gabor transform. The Gabor transform $G(f)(\omega,t)$ of a signal f is defined by

$$G(f)(\omega,t) = \int_{-\infty}^{\infty} f(x)g(x-t)e^{2\pi i\omega x}dx ,$$

where g is the Gaussian function $g(x) = g(0)e^{-bx^2}$, for some $b > 0$.

A more general transform, known as the windowed Fourier transform or the short-time Fourier transform, can be defined similarly as

$$F_{WFT}(f)(\omega,t) = F_{STFT}(f)(\omega,t) = \int_{-\infty}^{\infty} f(x)g(x-t)e^{2\pi i\omega x}dx ,$$

where g is now an arbitrary, but nice function. g is called the window function. Clearly, when the window function is chosen as the Gaussian function, the windowed Fourier transform yields the Gabor transform.

Typically, g is chosen such that either g or its Fourier transform $\hat{g}$ has compact support. To calculate $F_{WFT}(f)(\omega,t)$ at time $t = t_0$, we first shift the window function by amount t_0, multiply the shifted window by f, then calculate the Fourier transform. If, for example, $\operatorname{supp} g = [-1,1]$ and $g = 1$ on $[-1+\varepsilon, 1-\varepsilon]$, then $f(x)g(x-t_0)$ represents a well-localized slice of f centered at t_0, and $F_{WFT}(f)(\omega,t_0)$ will give a good approximation of the frequency content of f in the time interval $[-1+t_0, 1+t_0]$. Essentially, the window captures the features of the signal f in a local region around t_0 in the time domain before obtaining the frequency domain information.

It is more often than not that the variables ω and t are discretized so that

$$F_{WFT}(f)(m\omega_0, nt_0) = F_{STFT}(f)(m\omega_0, nt_0) = \int_{-\infty}^{\infty} f(x)g(x - nt_0)e^{2\pi i m\omega_0 x}dx$$

$$= \int_{-\infty}^{\infty} f(x)g_{m\omega_0, nt_0}(x)\,dx \ ,$$

with $\omega_0, t_0 > 0$ and $m, n = 0, \pm 1, \pm 2, \ldots$.

To calculate the windowed Fourier transform of f, the Fourier transform of the product of f and g is computed, then the window is shifted in time and the Fourier transform of the product is computed again, and so on. This produces separate Fourier transforms for each location of the window center. Therefore, unlike the classical Fourier transform, the windowed Fourier transform allows us to view the signal in the time and frequency domains simultaneously.

Since signals with compact support in the time domain has infinite support in the frequency domain and vice versa, there is a tradeoff between time localization and frequency resolution. This tradeoff is made more precise by the uncertainty principle (cf. (2.2.8)), which heuristically asserts that if a signal has a narrow support in the time domain, its Fourier transform has a wide support (bandwidth) in the frequency domain. Neither the Gaussian function used in the Gabor transform nor its Fourier transform has compact support, yet the Gaussian function serves as an excellent window function in some cases. The reason is: it has another redeeming feature, namely, it is optimally localized in both the time and the frequency domains in the sense that it attains the equality in (2.2.8).

The set of discretized values of the windowed Fourier transform $\{F_{WFT}(f)(m\omega_0, nt_0)\}_{m,n=-\infty}^{\infty}$ of f consists of uniformly sampled values of the two-dimensional windowed Fourier transform $F_{WFT}(f)(\omega, t)$. The set of sample points $\{(m\omega_0, nt_0)\}_{m,n=-\infty}^{\infty}$ is called the Gabor lattice. A question of great importance is this: can we reconstruct $f(t)$ from the sample values $\{F_{WFT}(f)(m\omega_0, nt_0)\}_{m,n=-\infty}^{\infty}$ of its windowed Fourier transform? For an appropriate function g, the answer is in the affirmative, provided that $0 < \omega_0 t_0 \le 1$. When $0 < \omega_0 t_0 < 1$, the reconstruction is stable and g can have a good time and frequency localization; this is in contrast with the case when $\omega_0 t_0 = 1$, where the reconstruction is unstable and g cannot have a good time and frequency localization. When $\omega_0 t_0 > 1$, the reconstruction of f is in general impossible, no matter how we choose g. The reconstruction formula is a series of the form

$$f(t) = \sum_{m,n} (F_{WFT}(f)(m\omega_0, nt_0))\tilde{g}_{m\omega_0,nt_0}(t),$$

where $\tilde{g}_{m\omega_0,nt_0}$ are functions related to $g_{m\omega_0,nt_0}$. Another related series is

$$f(t) = \sum_{m,n} (\tilde{F}(f)(m\omega_0, nt_0))g_{m\omega_0,nt_0}(t),$$

where $\{\tilde{F}(f)(m\omega_0, nt_0)\}_{m,n=-\infty}^{\infty}$ is a set of sample values of another related transform called the dual transform.

It turns out that the question of reconstructing f from the sample values of its windowed Fourier transform is quite related to the question of whether the functions $\{g_{m\omega_0,nt_0}\}$ form a frame for $L^2(\Re)$ or not. Therefore, we will now return to our investigation of $\{g_{m\omega_0,nt_0}\}$ in the context of frames to see when they do form a frame or an orthonormal basis for $L^2(\Re)$.

First, it is easily seen that $\{E_{b_m}T_{a_n}g\}$ is a frame for $L^2(\Re)$ if and only if $\{T_{a_n}E_{b_m}g\}$ is a frame for $L^2(\Re)$. Second, since the Fourier transform is a unitary transformation from $L^2(\Re)$ onto itself, it follows that the Fourier transform of a frame for $L^2(\Re)$ is also a frame for $L^2(\Re)$ and

$$\left(E_{b_m}T_{a_n}g\right)^{\wedge} = e^{2\pi i b_m a_n}E_{a_n}T_{-b_m}\hat{g}, \tag{10.2.2}$$

and

$$\left(E_{b_m}T_{a_n}\hat{g}\right)^{\vee} = e^{2\pi i b_m a_n}E_{-a_n}T_{b_m}g. \tag{10.2.3}$$

It is also easy to see that if $a_n = n$, $b_m = m$ and $g(x) = \chi_{[0,1)}$, then $\{g_{m,n}(x) = e^{2\pi i m x}g(x-n)\}_{m,n\in Z}$ constitutes an orthonormal basis for $L^2(\Re)$. This orthonormal basis of $L^2(\Re)$ is a translation of $\{e^{2\pi i m x}\}_{m=-\infty}^{\infty}$ which is an orthonormal basis of $[0,1)$.

More generally, Duffin and Schaeffer [10] proved that if $\{b_m\}_{m\in Z}$ is a sequence of real numbers with uniform density $d > 0$, i.e., there exist constants L and δ such that

$$|b_m - \frac{m}{d}| \le L, \quad m = 0, \pm 1, \pm 2, \ldots$$

$$|b_n - b_m| \ge \delta > 0, \quad n \ne m$$

and if $0 < 2\gamma < d$, then $\{E_{b_m}\}$ is a frame for $L^2(-\gamma,\gamma)$. In particular, if $d = 1$, $b_m = m$, then $\{E_m\}$ is a frame for $L^2(-\gamma,\gamma)$ when $0 < \gamma < 1/2$ and is an orthonormal basis when $\gamma = 1/2$.

We now begin by summarizing some properties of the functions $\{g_{m,n}(x)\}_{m,n\in Z}$ that will be needed in our investigation. The first question that comes to mind is this: what kind of a function g can generate a frame of the form $\left\{g_{m,n}(x) = e^{2\pi i b_m x} g(x-a_n)\right\}_{m,n\in Z}$ for $L^2(\mathfrak{R})$? The function g, if it exists, will be called the mother wavelet or analyzing wavelet. This question is easier to answer if $b_m = mb$, $a_n = na$ for $a,b > 0$. In this case, it has been shown [13] that for $g \in L^2(\mathfrak{R})$ to generate a frame it is necessary that the function

$$G(x) = \sum_n |g(x-na)|^2$$

satisfies

$$0 < A \leq G(x) \leq B < \infty, \text{ a.e. on } \mathfrak{R} \tag{10.2.4}$$

for some constants A and B. This condition clearly implies that g must be bounded. A sufficient condition is provided in the following theorem.

THEOREM 10.4

Let $g \in L^2(\mathfrak{R})$ be such that $G(x)$ satisfies condition (10.2.4). If g has compact support, with supp $(g) \subset I \subset \mathfrak{R}$, where I is some interval of length $1/b$, then $\{E_{mb}T_{na}g\}$ is a frame for $L^2(\mathfrak{R})$ with frame bounds $b^{-1}A$ and $b^{-1}B$. Moreover, for any $f \in L^2(\mathfrak{R})$, we have

$$Sf(x) = \frac{1}{b} f(x)\, G(x), \tag{10.2.5}$$

where S is the frame operator. Moreover, S is invertible and

$$S^{-1}f(x) = b f(x)/G(x). \tag{10.2.6}$$

Proof. Let $I_n = I + na = \{x + na : x \in I\}$. For fixed n,

$$f(x)\overline{g(x-na)} \in L^2(I_n)$$

since g is bounded and $f \in L^2(\mathfrak{R})$. Let $h_n(x) = f(x)\overline{g(x-na)}$ and $e_m(x) = \sqrt{b}\, E_{mb}(x) = \sqrt{b}\, e^{2\pi i m b x}$. In view of the fact that $\{e_m(x)\}$ is an orthonormal basis of $L^2(I_n)$, we obtain

$$h_n(x) = \sum_m \langle h_n, e_m\rangle e_m(x),$$

which by Parseval's equality yields

$$\int_{I_n} |h_n(x)|^2 dx = \sum_m |\langle h_n, e_m\rangle|^2$$

$$= b \sum_m \left| \int_{I_n} f(x)\, \overline{g(x-na)}\ \overline{E_{mb}(x)} \right|^2 dx = b \sum_m |\langle f, E_{mb}T_{na}g\rangle|^2 .$$

Therefore, by the compactness of the support of g, we have

$$\sum_n \int_{I_n} |h_n(x)|^2 dx = \sum_n \int_{-\infty}^{\infty} |f(x)|^2 |g(x-na)|^2 dx = \int_{-\infty}^{\infty} |f(x)|^2 G(x)\, dx$$

$$= b \sum_{n,m} |\langle f, E_{mb}T_{na}g\rangle|^2 . \qquad (10.2.7)$$

From (10.2.7) and (10.2.4), it follows that

$$b^{-1}A\|f\|^2 \le \sum_{n,m} |\langle f, E_{mb}T_{na}g\rangle|^2 \le b^{-1}B\|f\|^2 ,$$

hence $\{E_{mb}T_{na}g\}$ is a frame for $L^2(\Re)$ with the prescribed bounds.

Next, from the definition of the frame operator S, we obtain

$$Sf(x) = \sum_{m,n} \langle f, E_{mb}T_{na}g\rangle\, e^{2\pi imbx} g(x-na) = \sum_n g(x-na) H_n(x) , \qquad (10.2.8)$$

where

$$H_n(x) = \sum_m \langle f, E_{mb}T_{na}g\rangle\, e^{2\pi imbx} . \qquad (10.2.9)$$

But it is easy to see that $H_n(x)$ is a periodic function with period $1/b$ and hence (10.2.9) is its Fourier series expansion. Therefore,

$$\int_{I_n} H_n(x)\, e^{-2\pi imbx} dx = \frac{1}{b} \int_{-\infty}^{\infty} f(x)\, \overline{g(x-na)}\, e^{-2\pi imbx} dx$$

$$= \frac{1}{b} \int_{I_n} f(x)\, \overline{g(x-na)}\, e^{-2\pi imbx} dx ,$$

since $\operatorname{supp} g \subset I$. From the completeness of $\{E_{mb}\}$, we conclude that

$$H_n(x) = \frac{1}{b} f(x)\, \overline{g(x-na)} . \qquad (10.2.10)$$

By substituting (10.2.10) into (10.2.8) we obtain

$$Sf(x) = \frac{1}{b}\sum_n f(x)\,|\,g(x-na)\,|^2 = \frac{1}{b}f(x)\,G(x)\,.$$

Finally, from Theorem 10.3, it follows that S is invertible and by replacing f by $S^{-1}f$ in (10.2.5) we obtain (10.2.6). ▮

Corollary 10.4.1. Let $g \in L^2(\mathfrak{R})$ have compact support, with $\operatorname{supp} g \subset I$, where I is some interval of length $1/b$. If $a > 1/b$, then g cannot generate a frame for $L^2(\mathfrak{R})$. In fact, $\{g_{mb,na} = E_{mb}T_{na}g\}$ is not even complete.

Proof. It is not hard to see that $\bigcup_n \operatorname{supp}(g(x-na))$ does not cover the whole real line and $G(x) = 0$ on a set of positive measure. Furthermore, any function supported in $\mathfrak{R} - \bigcup_n \operatorname{supp}(g(x-na))$ will be orthogonal to every $g_{mb,na}$. ▮

Actually, more than this can be said: it can be shown ([6, p. 107]; [7]) that if $ab > 1$, then $\{g_{mb,na}\}$ can never be complete in $L^2(\mathfrak{R})$ whether g has compact support or not.

For nonnegative integers a, b and a is even, it is easy to construct a function $f \in L^2(\mathfrak{R})$ such that f is not identically zero, but $\langle f, g_{mb,na}\rangle = 0$ for all m and n. For example, for $0 < x < 1$, set

$$f(x+k) = (-1)^k\,\overline{g(x-k-1)}\,.$$

Hence,

$$\int_{-\infty}^{\infty} |\,f(x)\,|^2\,dx = \int_{-\infty}^{\infty} |\,g(x)\,|^2\,dx\,,$$

but since $g \in L^2(\mathfrak{R})$ and it is not identically zero, then so is f. Now

$$\langle f, g_{mb,na}\rangle = \int_{-\infty}^{\infty} f(x)\,e^{-2\pi imbx}\,\overline{g(x-na)}\,dx = \sum_{k=-\infty}^{\infty}\int_k^{k+1} f(x)\,\overline{g(x-na)}\,e^{-2\pi imbx}dx$$

$$= \sum_{k=-\infty}^{\infty}\int_0^1 f(x+k)\,\overline{g(x+k-na)}\,e^{-2\pi imbx}dx$$

$$= \int_0^1 e^{-2\pi imbx}h_n(x)\,dx\,,$$

where $h_n(x) = \sum_{k=-\infty}^{\infty} (-1)^k \overline{g(x-k-1)}\ \overline{g(x+k-na)}$. Upon replacing k by $na-k-1$ in the above summation, we obtain

$$h_n(x) = (-1) \sum_{k=-\infty}^{\infty} (-1)^k \overline{g(x+k-na)}\ \overline{g(x-k-1)},$$

so $h_n(x) = -h_n(x)$, which implies that $\langle f, g_{mb,na}\rangle = 0$.

For general positive numbers a and b such that ab is a rational number greater than 1, I. Daubechies [7], using properties of the Zak transform, constructed a nonzero function $f \in L^2(\mathfrak{R})$ so that for any $g \in L^2(\mathfrak{R})$, $\langle f, g_{mb,na}\rangle = 0$ for all m and n. If ab is an irrational number greater than 1, the existence of such an f has been proved by Rieffel [27], but no explicit representation for it is yet known.

Corollary 10.4.2. Let g be a continuous function with compact support such that $\operatorname{supp}(g) = I$ for some interval I of length L. Then $\{E_{mb}T_{na}g\}$ is a frame for $L^2(\mathfrak{R})$ for any $0 < a < L$ and $0 < b \le 1/L$.

Proof. This will follow from Theorem 10.4 if we can show that G satisfies (10.2.4). Since g has compact support, it follows that for any fixed $x \in \mathfrak{R}$, the series defining G actually contains only a finite number of non-zero terms. The number of these non-zero terms is at most $[1/ab]+1$ and since g is bounded, $G(x)$ is bounded above. Let J be a subinterval of I with the same center but with length a. Then for any $x \in \mathfrak{R}$, there is an $n \in Z$ such that $x - na \in J$; hence $\inf_{x\in\mathbf{R}} G(x) \ge \inf_{x \in J} |g(x)|^2 = A > 0$; hence $G(x) \ge A > 0$ for all x. ▌

If g does not have compact support, it is still possible for $\{E_{mb}T_{na}g\}$ to be a frame for $L^2(\mathfrak{R})$ as illustrated in the following theorem whose proof can be found in [13].

THEOREM 10.5

Let $g \in L^2(\mathfrak{R})$ and $a > 0$ be such that $G(x)$ satisfies (10.2.4) and

$$\lim_{b\to 0} \sum_{k=0} \beta(k/b) = 0, \quad \text{where}$$

$$\beta(s) = \text{ess.} \sup_{x\in\mathbf{R}} \left|\sum_n g(x-na)\overline{g(x-s-na)}\right|.$$

Then there exists $b_0 > 0$ such that $\{E_{mb}T_{na}g\}$ is a frame for $L^2(\mathfrak{R})$ for each $0 < b < b_0$.

Now we turn our attention to other frames related to the frame $\{E_{mb}T_{na}g\}$. It is interesting to note that the dual frame $\{S^{-1}E_{mb}T_{na}g\}$ of the frame $\{E_{mb}T_{na}g\}$ is also generated by one single function, namely,

$S^{-1}E_{mb}T_{na}g = E_{mb}T_{na}\tilde{g}$, where $\tilde{g}(x) = S^{-1}g(x)$, whether g has compact support or not. To see this, first observe that for any integer k

$$S(T_{ka}f) = \sum_{m,n}\langle T_{ka}f, g_{mb,na}\rangle g_{mb,na} = \sum_{m,n} e^{-2\pi imkba}\langle f, g_{mb,(n-k)a}\rangle g_{mb,na}$$

$$= \sum_{m,n}\langle f, g_{mb,(n-k)a}\rangle e^{2\pi imb(x-ka)}g(x-na)$$

$$= \sum_{m,n}\langle f, g_{mb,na}\rangle e^{2\pi imb(x-ka)}g(x-ka-na)$$

$$= \sum_{m,n}\langle f, g_{mb,na}\rangle T_{ka}g_{mb,na}(x) = T_{ka}Sf .$$

Similarly, we can verify that $S(E_{kb}f) = E_{kb}Sf$. Since S commutes with both T_{ka} and E_{kb}, then so does S^{-1}. Therefore,

$$S^{-1}(E_{mb}T_{na}g) = E_{mb}T_{na}\tilde{g} , \tag{10.2.11}$$

or

$$S^{-1}(g_{mb,na}) = \tilde{g}_{mb,na} .$$

10.2.C The Zak Transform

Some of the techniques that have been used to study the frames $\{g_{mb,na}\}$, when $0 < ab < 1$, cease to work in the critical case when $ab = 1$; a new approach is needed. It turns out that the Zak transform is an excellent alternative technique to use in this critical case.

The Zak transform of a signal f can be considered as a generalization of its discrete Fourier transform in which an infinite sequence of samples of f in the form $\{f(ta+ka)\}_{k=-\infty}^{\infty}$, $a > 0$ are used. It can also be considered as a mixed time-frequency representation of f.

The Zak transform was discovered by several people in different fields and it was called different names depending on the field in which it was discovered. However, it seems that its first appearance in the literature was in the work of I. M. Gel'fand on "eigenfunction expansions for an equation with periodic coefficients" [12] and it was called the Gel'fand mapping in the Russian literature. It is also said [28, p. 110] that some properties of the Zak transform were even known to Gauss. Almost 17 years after the publication of Gel'fand's work, the Zak transform was rediscovered independently and was called the $k-q$ transform by J. Zak in his work on solid state physics [32]. Although the Gel'fand-Zak transform seems to be a more appropriate name for this transform, as suggested by A. Janssen [16], there is now a general consent among scientists to call it the Zak transform since Zak was indeed the first to systematically study this transform and recognize its usefulness in a more general setting. Among the different areas where the Zak transform has been used successfully is signal analysis. A very good review article on this topic can be found in [16]; see also [17].

Definition 10.2.4. The Zak transform (Zf) of a function f is defined by

$$(Zf)(t,\omega) = \sqrt{a} \sum_{k=-\infty}^{\infty} f(ta + ka) e^{2\pi i k\omega},$$

for t, ω real and $a > 0$.

The Zak transform exists if, for example, $\sum_k |f(at + ka)| < \infty$, for all t, in particular if $|f(t)| \leq C(1 + |t|)^{-(1+\varepsilon)}$, $\varepsilon > 0$. The following two properties are easy to verify

$$(Zf)(t+1,\omega) = e^{-2\pi i\omega}(Zf)(t,\omega)$$

and

$$(Zf)(t,\omega+1) = (Zf)(t,\omega).$$

This implies that (Zf) is completely determined by its values on the unit square $Q = [0,1) \times [0,1)$. We define

$$L^2(Q) = \left\{F \mid F : Q \to C \text{ such that } \int_0^1 \int_0^1 |F(t,\omega)|^2 \, dt \, d\omega < \infty\right\}.$$

This is a Hilbert space with inner product

$$\langle f, G\rangle = \int_0^1 \int_0^1 F(t,\omega) \overline{G(t,\omega)} \, dt \, d\omega,$$

and norm

$$\|F\| = \left(\int_0^1 \int_0^1 |F(t,\omega)|^2 \, dt \, d\omega\right)^{1/2}.$$

The set $\{E_{(m,n)} = E_{(m,n)}(t,\omega) = e^{2\pi i m t} e^{2\pi i n \omega}\}$ constitutes an orthonormal basis of $L^2(Q)$.

THEOREM 10.6

The Zak transform is a unitary map of $L^2(\Re)$ onto $L^2(Q)$.

Proof. Let $f \in L^2(\Re)$ and set $(Zf)(t,\omega) = \sum_k F_k(t,\omega)$, where $F_k(t,\omega) = \sqrt{a}\, f(ta+ka)e^{2\pi ik\omega}$, then

$$\|F_k\|^2 = a\int_0^1\int_0^1 |f(ta+ka)|^2\,dt\,d\omega = \int_0^a |f(t+ka)|^2\,dt < \infty,$$

hence $F_k \in L^2(Q)$. Moreover,

$$\langle F_k, F_m\rangle = a\int_0^1\int_0^1 f(ta+ka)\overline{f(ta+ma)}\,e^{2\pi i(k-m)\omega}dt\,d\omega = 0,$$

which implies that

$$\|Zf\|^2 = \|\sum_k F_k\|^2 = \sum_k \|F_k\|^2 = \sum_k \int_0^a |f(t+ka)|^2\,dt$$

$$= \sum_k \int_{ka}^{(k+1)a} |f(t)|^2\,dt = \int_{-\infty}^{\infty} |f(t)|^2\,dt = \|f\|^2.$$

This shows that the Zak transform is a well-defined, linear and norm-preserving map from $L^2(\Re)$ into $L^2(Q)$.

Now let

$$e_{m,n}(x) = \frac{1}{\sqrt{a}}T_{na}E_{m/a}\chi_{[0,a)} = \frac{1}{\sqrt{a}}e^{2\pi im(x-na)/a}\chi_{[0,a)}(x-na)$$

$$= \frac{1}{\sqrt{a}}e^{(2\pi imx)/a}\chi_{[na,(n+1)a)}(x).$$

$\{e_{m,n}(x)\}_{m,n=-\infty}^{\infty}$ constitutes an orthonormal basis of $L^2(\Re)$. Taking the Zak transform of $e_{m,n}(x)$ yields

$$(Ze_{m,n})(t,\omega) = \sum_k e^{2\pi ik\omega}e^{2\pi im(at+ak)/a}\chi_{[na,(n+1)a)}(at+ak)$$

$$= e^{2\pi imt}\sum_k e^{2\pi ik\omega}\chi_{[n-k,n-k+1)}(t).$$

Since the only nonzero term in the series is the term that corresponds to $k = n$, we immediately obtain that $(Ze_{m,n})(t,\omega) = e^{2\pi imt}e^{2\pi in\omega}$. Therefore, the Zak transform maps the orthonormal basis $\{e_{m,n}(x)\}$ of $L^2(\mathfrak{R})$ onto the orthonormal basis $\{E_{m,n}(t,\omega)\}$ of $L^2(Q)$ and this completes the proof. ▮

The inverse Zak transform is easily seen to be

$$(Z^{-1}F)(x) = \frac{1}{\sqrt{a}} \int_0^1 F(t/a, \omega)\, d\omega .$$

Also, one can easily verify that if $ab = 1$, then

$$\left.\begin{aligned} &(Z(E_{mb}T_{na}g))(t,\omega) = E_{m,n}(t,\omega)(Zg)(t,\omega), \\ &\text{or} \\ &(Zg_{mb,na})(t,\omega) = E_{m,n}(t,\omega)(Zg)(t,\omega) \end{aligned}\right\} \qquad (10.2.12)$$

THEOREM 10.7

Given $a, b > 0$ with $ab = 1$ and $g \in L^2(\mathfrak{R})$, then the following are equivalent:

i) there exist $A, B > 0$ such that $0 < A \le |Zg|^2 \le B < \infty$,

ii) $\{g_{mb,na}(x) = e^{2\pi imbx}g(x - na)\}$ is a frame for $L^2(\mathfrak{R})$ with frame bounds A and B,

iii) $\{g_{mb,na}(x)\}$ is an exact frame for $L^2(\mathfrak{R})$ with frame bounds A, B, and for any $f \in L^2(\mathfrak{R})$, $f = \Sigma\, c_{mn}\, g_{mb,na}$, where

$$c_{m,n} = \int_0^1\int_0^1 \frac{(Zf)(t,\omega)}{(Zg)(t,\omega)} e^{-2\pi imt} e^{-2\pi in\omega} dt\, d\omega . \qquad (10.2.13)$$

Proof. (i) $\Rightarrow$ (ii) Since the Zak transform is a unitary map from $L^2(\mathfrak{R})$ onto $L^2(Q)$, it suffices to show that $\{(Zg_{mb,na})(t,\omega)\}$ is a frame for $L^2(Q)$. Given $F \in L^2(Q)$, we have $F(\overline{Zg}) \in L^2(Q)$ because Zg is bounded. But in view of (10.2.12)

$$\langle F, Zg_{mb,na}\rangle = \langle F, E_{m,n}Zg\rangle = \langle F\,\overline{Zg}, E_{m,n}\rangle ,$$

and since $\{E_{m,n}\}$ is an orthonormal basis for $L^2(Q)$, Parseval's equality implies that

$$\sum_{m,n} |\langle F, Zg_{m,n}\rangle|^2 = \|F\,\overline{Zg}\|^2 .$$

From the inequalities $A\|F\|^2 \le \|F\,\overline{Zg}\|^2 \le B\|F\|^2$, we obtain

$$A\|F\|^2 \le \sum_{m,n} |\langle F, Zg_{mb,na}\rangle|^2 \le B\|F\|^2;$$

hence $\{Zg_{mb,na}\}$ is a frame for $L^2(Q)$.

(ii) ⇒ (i) If (ii) holds, then $\{E_{m,n}Zg\}$ is a frame for $L^2(Q)$ with bounds A, B and hence for any $F \in L^2(Q)$

$$A\|F\|^2 \le \sum_{m,n} |\langle F, E_{m,n}Zg\rangle|^2 \le B\|F\|^2. \tag{10.2.14}$$

But as shown above

$$\sum_{m,n} |\langle F, E_{m,n}Zg\rangle|^2 = \sum_{m,n} |\langle F\,\overline{Zg}, E_{m,n}\rangle|^2 = \|F\,\overline{Zg}\|^2, \tag{10.2.15}$$

and by combining (10.2.14) and (10.2.15) we obtain

$$A\|F\|^2 \le \|F\,\overline{Zg}\|^2 \le B\|F\|^2,$$

for all $F \in L^2(Q)$, which implies (i).

(ii) ⇒ (iii) If (ii) holds, then $\{E_{m,n}Zg\}$ is a frame for $L^2(Q)$. But from (i) Zg is bounded from above and below by positive numbers B and A; hence the mapping $W: L^2(Q) \to L^2(Q)$, defined by $W(F) = F(Zg)$ for any $F \in L^2(Q)$, is a topological isomorphism that maps the orthonormal basis $\{E_{m,n}(t,\omega)\}$ onto $\{(Zg_{mb,na})(t,\omega)\}$. Therefore, $\{(Zg_{mb,na})(t,\omega)\}$ is a Riesz basis for $L^2(Q)$; hence so is $\{g_{mb,na}(x)\}$ for $L^2(\mathfrak{R})$ and by Theorem 10.2 $\{g_{mb,na}(x)\}$ is an exact frame for $L^2(\mathfrak{R})$.

(iii) ⇒ (ii) Is trivial; we only need to show (10.2.13). First, we show that

$$Z(Sf) = (Zf)|Zg|^2, \tag{10.2.16}$$

where S is the frame operator associated with the frame $\{g_{mb,na}(x)\}$. Since $\{E_{m,n}\}$ is an orthonormal basis for $L^2(Q)$, we have

$$Z(Sf) = \sum_{m,n}\langle f, g_{mb,na}\rangle(Zg_{mb,na}) = (Zg)\sum_{m,n}\langle f, g_{mb,na}\rangle E_{m,n} = (Zg)\sum_{m,n}\langle Zf, Zg_{mb,na}\rangle E_{m,n}$$

$$= (Zg)\sum_{m,n}\langle Zf, ZgE_{m,n}\rangle E_{m,n} = (Zg)\sum_{m,n}\langle (Zf)(\overline{Zg}), E_{m,n}\rangle E_{m,n} = (Zf)|Zg|^2,$$

where

$$\langle f, g_{mb,na}\rangle = \langle Zf, Zg_{mb,na}\rangle = \langle Zf, E_{m,n}Zg\rangle$$

$$= \int_0^1 \int_0^1 (Zf)(t,\omega)\overline{(Zg)}(t,\omega)e^{-2\pi imt}e^{-2\pi in\omega}dt\, d\omega .$$

Replacing f by $S^{-1}f$ in (10.2.16) yields

$$Z(S^{-1}f) = \frac{(Zf)}{|Zg|^2}. \tag{10.2.17}$$

In particular, if $f = g$, we have

$$Z(\tilde{g}) = \frac{1}{\overline{(Zg)}} \tag{10.2.18}$$

where $\tilde{g} = S^{-1}g$. Hence, by combining (10.2.1), (10.2.11), and (10.2.18) we obtain

$$c_{m,n} = \langle f, S^{-1}g_{mb,na}\rangle = \langle f, \tilde{g}_{m,n}\rangle = \langle Zf, Z\tilde{g}_{mb,na}\rangle$$

$$= \langle Zf, E_{m,n}(Z\tilde{g})\rangle = \left\langle Zf, E_{m,n}\frac{1}{\overline{(Zg)}}\right\rangle = \left\langle \frac{Zf}{Zg}, E_{m,n}\right\rangle,$$

which is (10.2.13). ▮

The relation

$$\langle Zg_{kb,la}, Zg_{mb,na}\rangle = \int_0^1 \int_0^1 E_{k,l}(t,\omega)\overline{E}_{m,n}(t,\omega)\,|(Zg)(t,\omega)|^2\, dt\, d\omega ,$$

implies that the set $\{Zg_{mb,na}\}$, and consequently $\{g_{mb,na}\}$, is orthonormal if and only if $|Zg| = 1$, a.e., and hence by (i) and Lemma 10.2.1, it is an orthonormal basis if and only if $|Zg| = 1$, a.e.

Similarly, from the relation

$$\langle f, g_{mb,na}\rangle = \langle Zf, Zg_{mb,na}\rangle = \langle (Zf)\overline{(Zg)}, E_{m,n}\rangle$$

it follows that $\{g_{mb,na}\}$ is complete if and only if $(Zg) \neq 0$, a.e.

It is not hard to show [13] that if $g \in L^2(\mathfrak{R})$ is such that Zg is a continuous function, then Zg must have a zero in Q; hence, $|Zg|$ cannot be bounded below away from zero. Consequently, such a function g cannot generate a frame for $L^2(\mathfrak{R})$ when $ab = 1$; however, if $Zg = 0$ only on a set of measure zero, then the system of functions $\{g_{mb,na}\}$ is complete though it may not be a frame. An example of such a system of functions is the one generated by the Gaussian function $g(x) = e^{-\gamma x^2}$ because the Zak transform of g is continuous and has a single zero in Q. Nevertheless, the Gaussian function is known to generate a frame when $0 < ab < 1$, see [6, p. 84].

On the other hand, for $a = b = 1$ and $g(x) = \chi_{[-1/2, 1/2)}$, the Zak transform of g satisfies $|Zg| = 1$ on Q; hence g generates an orthonormal basis for $L^2(\Re)$, but when $0 < a, b < 1$, g generates only a frame as seen from Theorem 10.4. Since the Fourier transform of a frame of $L^2(\Re)$ is also a frame for $L^2(\Re)$, it follows from (10.2.2) that $\hat{g}(\omega) = \sin \pi\omega / \pi\omega$ also generates a frame of $L^2(\Re)$. This is an example of a frame that is generated by a non-compactly supported function. Other examples can be found in [13 and 6].

The above discussion indicates that nice functions cannot generate a frame when $ab = 1$. This is indeed the case. In fact, if g is a nice function in the sense that it is either smooth or it decays very fast, it cannot generate a frame when $ab = 1$. This is stated more precisely in the following theorem of R. Balian [2] and F. Low [20].

THEOREM 10.8

Let $g \in L^2(\Re)$ and $a, b > 0$ with $ab = 1$. If g generates a frame for $L^2(\Re)$, then either $xg(x)$ or $\omega\hat{g}(\omega)$ is not in $L^2(\Re)$.

To sum it all up, for $g \in L^2(\Re)$ we have:

i) If $ab > 1$, it is impossible for $\{g_{mb,na}\}$ to be a frame for $L^2(\Re)$.
ii) If $ab = 1$, then $\{g_{mb,na}\}$ can be a frame for $L^2(\Re)$, but g cannot be a nice function.
iii) If $0 < ab < 1$, then $\{g_{mb,na}\}$ can be a frame for $L^2(\Re)$ with a nice mother wavelet g. In fact, g can be infinitely differentiable with compact support.

For example, let $\phi \in C^\infty(\Re)$ be such that $\text{supp}\,\phi = I$ for some interval I of length $1/b$. By Corollary 10.4.2 $\{\phi_{mb,na}(x)\}$ is a frame for $L^2(\Re)$ if $a > 0$, $ab < 1$, and consequently $G(x) = \sum_n |\phi(x - na)|^2$ is bounded above and below by some positive constants. If we set $g(x) = \sqrt{b}\,\phi(x)/\sqrt{G(x)}$, then $\text{supp}\, g \subseteq I$, and $G_1(x) = \sum_n |g(x - na)|^2 = b$. Thus, in view of Corollary 10.4.2, $\{g_{mb,na}(x)\}$ is a tight frame with frame bounds $A = B = 1$; see equation (10.2.7).

10.2.D Sampling Theorems by Using Frames

Having introduced frames and some of their properties, we are now able to show how frames can be used to derive sampling theorems.

THEOREM 10.9 (Benedetto and Heller [4])
Let $\Omega, \sigma > 0$ with $0 < \Omega < 1/2\sigma$, and let $g \in S(\Re)$ (Schwartz space of functions) be such $\operatorname{supp} \hat{g} \subset [-1/(2\Omega), 1/(2\Omega)]$, $\hat{g} = 1$ on $[-\sigma, \sigma]$ and $\hat{g} > 0$ on $(-1/(2\Omega), -\sigma] \cup [\sigma, 1/(2\Omega))$. Set

$$G(\omega) = \sum_m |\hat{g}(\omega - mb)|^2$$

and

$$\phi(t) = \left(\frac{\hat{g}}{G}\right)^{\vee}(t),$$

where $\sigma + (1/2\Omega) \le b < 1/\Omega$. Then for any σ-band-limited function f, we have

$$f(t) = \Omega \sum_n f(n\Omega)\phi(t - n\Omega) \quad \text{in} \quad L^2(\Re). \tag{10.2.19}$$

Proof. Our choice of b guarantees that G is bounded above and below away from zero by some constants B and A such that $0 < A \le G(\omega) \le B < \infty$. Since $\hat{g}$ has compact support and $\operatorname{supp} \hat{g}$ is contained in an interval of length $1/\Omega$, it follows from Theorem 10.4 that $\{E_{na}T_{mb}\hat{g}\}$ is a frame for $L^2(\Re)$, where $a = \Omega$. Therefore,

$$\begin{aligned} \hat{f} &= \sum_{m,n} \langle \hat{f}, E_{na}T_{mb}\hat{g}\rangle S^{-1}(E_{na}T_{mb}\hat{g}) \\ &= \sum_{m,n} \langle \hat{f}, E_{na}T_{mb}\hat{g}\rangle E_{na}T_{mb}S^{-1}(\hat{g}), \end{aligned}$$

in view of (10.2.11). But from (10.2.6) we have

$$S^{-1}(\hat{g}) = \Omega\frac{\hat{g}}{G},$$

hence

$$\begin{aligned} \hat{f} &= \Omega \sum_{m,n} \langle \hat{f}, E_{n\Omega}T_{mb}\hat{g}\rangle E_{n\Omega}T_{mb}\left(\frac{\hat{g}}{G}\right) \\ &= \Omega \sum_{m,n} \langle \hat{f}, E_{n\Omega}T_{mb}\hat{g}\rangle E_{n\Omega}T_{mb}\hat{\phi}. \end{aligned} \tag{10.2.20}$$

Now we calculate the coefficients $\langle \hat{f}, E_{n\Omega}T_{mb}\hat{g}\rangle$

$$\langle \hat{f}, E_{n\Omega}T_{mb}\hat{g}\rangle = \int_{-\infty}^{\infty} \hat{f}(\omega)e^{-2\pi i n\Omega\omega}\overline{\hat{g}(\omega - mb)}\,d\omega.$$

Since $\hat{g}$ is supported in $[-1/2\Omega, 1/2\Omega]$, by our choice of b, the above integral is zero for all m, except when $m = 0$, and in this case since $\hat{g} = 1$ on $[-\sigma, \sigma]$,

$$\langle \hat{f}, E_{n\Omega}\hat{g}\rangle = \int_{-\sigma}^{\sigma} \hat{f}(\omega)e^{-2\pi i n\Omega\omega}d\omega = f(n\Omega). \tag{10.2.21}$$

Therefore, by combining (10.2.20) and (20.2.21) we obtain

$$\hat{f} = \Omega \sum_n f(n\Omega) E_{n\Omega} \hat{\phi},$$

and with the aid of (10.2.3) and the inverse Fourier transform we further obtain

$$f = \Omega \sum_n f(n\Omega) T_{n\Omega} \phi,$$

which is (10.2.19). ∎

To obtain a sampling theorem for irregular samples by using frames, we need additional properties of frames that we state in the following theorem:

THEOREM 10.10

Let g be a σ-band-limited function and assume that $\{a_n\}, \{b_m\}$ are real numbers for which $\{E_{a_n}(x)\}$ is a frame for $L^2[-\sigma, \sigma]$. Assume that there exist $A, B > 0$ such that

$$0 < A \le G(\omega) = \sum_m |\hat{g}(\omega - b_m)|^2 \le B < \infty .$$

Then $\{E_{a_n} T_{b_m} \hat{g}\}$ is a frame for $L^2(\mathfrak{R})$, in addition, it is a tight frame if and only if $\{E_{a_n}(x)\}$ is a tight frame for $L^2[-\sigma, \sigma]$ and G is a constant a.e. on $\mathfrak{R}$. The frame operator for $\{E_{a_n} T_{b_m} \hat{g}\}$ can be given by

$$S\hat{h} = \sum_m T_{b_m} \hat{g} S_m\left(\hat{h}\, T_{b_m} \overline{\hat{g}}\right), \tag{10.2.22}$$

where S_m is the frame operator for $\{T_{b_m} E_{a_n}\}$ that is given by

$$S_m \hat{k} = \sum_n \left\langle \hat{k}, T_{b_m} E_{a_n} \right\rangle_{I_m} T_{b_m} E_{a_n}, \tag{10.2.23}$$

for $\hat{h} \in L^2(\mathfrak{R}), \hat{k} \in L^2(I_m)$, and $\langle f, g \rangle_{I_m} = \int_{I_m} f(x) \overline{g(x)}\, dx.$

Proof. Let $I = [-\sigma, \sigma]$ and set $I_m = I + b_m = \{x + b_m : x \in I\}$. For fixed m, $\{T_{b_m} E_{a_n}(x)\}$ is a frame for I_m with the same frame bounds for $\{E_{a_n}(x)\}$, say A_I and B_I. Let $f \in L^2(\mathfrak{R})$, then $\hat{h}_m = \hat{f}\, T_{b_m} \overline{\hat{g}} \in L^2(I_m)$ since $\hat{g}$ is supported in $[-\sigma, \sigma]$ and bounded thereon. Clearly, $\hat{h}_m$ has compact support and supp $\hat{h}_m \subset I_m$; hence,

$$A_I \|\hat{h}_m\|^2_{I_m} \le \sum_n |\left\langle \hat{h}_m, T_{b_m} E_{a_n} \right\rangle_{I_m}|^2 \le B_I \|\hat{h}_m\|^2_{I_m},$$

and by summing over m, we obtain

$$A_I \sum_m \|\hat{f} T_{b_m} \hat{g}\|^2_{I_m} \le \sum_{m,n} |\langle \hat{f} T_{b_m} \overline{\hat{g}}, T_{b_m} E_{a_n} \rangle_{I_m}|^2 \le B_I \sum_m \|\hat{f} T_{b_m} \hat{g}\|^2_{I_m}. \tag{10.2.24}$$

As shown in the proof of Theorem 10.4; cf. (10.2.7)

$$\sum_m \|\hat{f} T_{b_m} \hat{g}\|^2_{I_m} = \sum_m \int_{I_m} |\hat{f}(\omega)|^2 |\hat{g}(\omega - b_m)|^2 d\omega = \int_{-\infty}^{\infty} |\hat{f}(\omega)|^2 G(\omega)\, d\omega, \tag{10.2.25}$$

and

$$\langle \hat{f} T_{b_m} \overline{\hat{g}}, T_{b_m} E_{a_n} \rangle_{I_m} = \langle \hat{f}, T_{b_m}(E_{a_n} \hat{g}) \rangle_{I_m} =$$

$$e^{2\pi i a_n b_m} \int_{-\infty}^{\infty} \hat{f}(\omega) e^{-2\pi i a_n \omega} \overline{\hat{g}(\omega - b_m)}\, d\omega = e^{2\pi i a_n b_m} \langle \hat{f}, E_{a_n} T_{b_m} \hat{g} \rangle, \tag{10.2.26}$$

since $\operatorname{supp} \hat{g} = [-\sigma, \sigma]$. By substituting (10.2.25) and (10.2.26) into (10.2.24), we obtain

$$A_I A \|f\|^2 \le \sum_{m,n} |\langle \hat{f}, E_{a_n} T_{b_m} \hat{g} \rangle|^2 \le B_I B \|f\|^2. \tag{10.2.27}$$

Thus, $\{E_{a_n} T_{b_m} \hat{g}\}$ is a frame for $L^2(\mathfrak{R})$. The assertion about the tightness of the frame follows from (10.2.27).

Finally,

$$\begin{aligned} S\hat{h} &= \sum_{m,n} \langle \hat{h}, E_{a_n} T_{b_m} \hat{g} \rangle E_{a_n} T_{b_m} \hat{g} \\ &= \sum_m T_{b_m} \hat{g} \left(\sum_n \langle \hat{h}, E_{a_n} T_{b_m} \hat{g} \rangle E_{a_n} \right) \\ &= \sum_m T_{b_m} \hat{g} \left(\sum_n \langle \hat{h}\, T_{b_m} \overline{\hat{g}}, E_{a_n} \rangle_{I_m} E_{a_n} \right), \end{aligned}$$

but

$$\begin{aligned} \langle \hat{h}\, T_{b_m} \overline{\hat{g}}, E_{a_n} \rangle_{I_m} E_{a_n}(\omega) &= \left(\int_{I_m} \hat{h}(\gamma) \overline{\hat{g}(\gamma - b_m)} e^{-2\pi i a_n \gamma} d\gamma \right) e^{2\pi i a_n \omega} \\ &= \left(\int_{I_m} \hat{h}(\gamma) \overline{\hat{g}(\gamma - b_m)} e^{-2\pi i a_n (\gamma - b_m)} d\gamma \right) e^{2\pi i a_n (\omega - b_m)} \\ &= \langle \hat{h}\, T_{b_m} \overline{\hat{g}}, T_{b_m} E_{a_n} \rangle_{I_m} T_{b_m} E_{a_n}(\omega). \end{aligned}$$

Therefore,

$$S\hat{h} = \sum_m T_{b_m}\hat{g}\left(\sum_n\left\langle \hat{h}\, T_{b_m}\bar{\hat{g}}, T_{b_m}E_{a_n}\right\rangle_{I_m} T_{b_m}E_{a_n}\right)$$

$$= \sum_m T_{b_m}\hat{g}\left(S_m\left(\hat{h}\, T_{b_m}\bar{\hat{g}}\right)\right).$$

▮

Corollary 10.10.1. Under the hypothesis of Theorem 10.10 if $a_n = na$ and $a = 1/2\sigma$, then for any $f \in L^2(\mathfrak{R})$

$$S^{-1}\hat{f} = \frac{1}{2\sigma}\frac{\hat{f}}{G}. \tag{10.2.28}$$

Moreover,

$$S^{-1}(E_{na}T_{b_m}\hat{g}) = \frac{1}{2\sigma}\frac{E_{na}T_{b_m}\hat{g}}{G}, \tag{10.2.29}$$

and

$$E_{na}T_{b_m}S^{-1}\hat{g} = \frac{1}{2\sigma}\frac{E_{na}T_{b_m}\hat{g}}{T_{b_m}G}. \tag{10.2.30}$$

Proof. For the given values of a_n and a, $\{(1/\sqrt{2\sigma})E_{na}(x)\}$ is an orthonormal basis, hence a frame, for $L^2[-\sigma,\sigma]$. Since $\{E_{na}T_{b_m}\hat{g}\}$ is a frame for $L^2(\mathfrak{R})$ and $\hat{f}\,T_{b_m}\bar{\hat{g}} \in L^2(I_m)$, we have

$$S\hat{f} = \sum_{m,n}\left\langle \hat{f}, E_{na}T_{b_m}\hat{g}\right\rangle E_{na}T_{b_m}\hat{g}$$

$$= \sum_m T_{b_m}\hat{g}\left(\sum_n\left\langle \hat{f}, E_{na}T_{b_m}\hat{g}\right\rangle E_{na}\right)$$

$$= \sum_m T_{b_m}\hat{g}\left(\sum_n\left\langle \hat{f}\,T_{b_m}\bar{\hat{g}}, E_{na}\right\rangle E_{na}\right)$$

$$= 2\sigma\hat{f}\sum_m (T_{b_m}\hat{g})\left(T_{b_m}\bar{\hat{g}}\right) = 2\sigma\hat{f}G. \tag{10.2.31}$$

Replacing $\hat{f}$ by $S^{-1}\hat{f}$ in (10.2.31) yields (10.2.28), and (10.2.29), (10.2.30) follow immediately from (10.2.28). ▮

When $b_m = mb$, the operator S^{-1} will commute with $E_{n_a}T_{m_b}$ as seen from (10.2.11).

THEOREM 10.11 (Benedetto and Heller [4])
Let $\sigma > 0$, and $\{a_n\}$ be a sequence of real numbers such that $\{E_{a_n}(\omega)\}$ is a frame for $L^2[-\sigma,\sigma]$ with frame operator S_0 and let $\hat{g}(\omega) = (1/\sqrt{2\sigma})\chi_{[-\sigma,\sigma)}$. Then for any σ-band-limited function f, we have

$$f = \sqrt{2\sigma}\sum_n f(a_n)\phi_n, \qquad (10.2.32)$$

where

$$\phi_n(t) = (K_n(\omega))^{\vee}, \quad K_n = \sum_m \langle E_{a_n}\hat{g}, h_m\rangle_{I_0} h_m,$$

$$h_m = S_0^{-1}(E_{a_m}) \quad \text{and} \quad I_0 = [-\sigma,\sigma].$$

In particular, if $\{E_{a_n}(\omega)\}$ is an exact frame, then

$$f = \sum_n f(a_n)\Psi_n,$$

where

$$\Psi_n(t) = \int_{-\sigma}^{\sigma} h_n(\omega)e^{-2\pi i t\omega}d\omega.$$

Proof. Set $b_m = 2\sigma m$. Since

$$G(\omega) = \sum_m |\hat{g}(\omega - 2\sigma m)|^2 = \frac{1}{2\sigma},$$

by Theorem 10.10 $\{E_{a_n}T_{b_m}\hat{g}\}$ is a frame for $L^2(\mathfrak{R})$. Let us denote its frame operator by S. Thus, for any $\hat{f} \in L^2(\mathfrak{R})$ we have

$$\hat{f} = \sum_{m,n}\langle \hat{f}, E_{a_n}T_{b_m}\hat{g}\rangle S^{-1}(E_{a_n}T_{b_m}\hat{g}),$$

and

$$S\hat{f} = \sum_{m,n}\langle \hat{f}, E_{a_n}T_{b_m}\hat{g}\rangle E_{a_n}T_{b_m}\hat{g}.$$

Since f is band-limited to $[-\sigma,\sigma]$, it follows, as in the proof of Theorem 10.9, (cf. (10.2.21)) that

$$\langle \hat{f}, E_{a_n}T_{b_m}\hat{g}\rangle = \begin{cases} \frac{1}{\sqrt{2\sigma}}f(a_n) & \text{if} \quad m = 0 \\ 0 & \text{if} \quad m \neq 0 \end{cases}.$$

Therefore,

$$\hat{f} = \frac{1}{\sqrt{2\sigma}} \sum_n f(a_n) S^{-1}(E_{a_n} \hat{g}), \tag{10.2.33}$$

and

$$S\hat{f} = \frac{1}{\sqrt{2\sigma}} \sum_n f(a_n) E_{a_n} \hat{g} = \frac{1}{\sqrt{2\sigma}} \sum_n \langle \hat{f}, E_{a_n} \rangle_{I_0} E_{a_n} \hat{g}. \tag{10.2.34}$$

Since the frame operator of the frame $\{E_{a_n}\}$ is S_0 (which can also be obtained from (10.2.23) by putting $m = 0$), then

$$S_0^{-1}\hat{f} = \sum_n \langle \hat{f}, S_0^{-1} E_{a_n} \rangle_{I_0} S_0^{-1} E_{a_n} = \sum_n \langle \hat{f}, h_n \rangle_{I_0} h_n, \tag{10.2.35}$$

and

$$S_0 \hat{f} = \sum_n \langle \hat{f}, E_{a_n} \rangle_{I_0} E_{a_n}. \tag{10.2.36}$$

By combining (10.2.34) and (10.2.36), we obtain

$$S\hat{f} = \frac{1}{2\sigma} S_0 \hat{f},$$

which in turn gives

$$S^{-1}\hat{f} = (2\sigma) S_0^{-1}(\hat{f}) \quad \text{in} \quad L^2[-\sigma, \sigma]. \tag{10.2.37}$$

From (10.2.35) and (10.2.37), we obtain

$$S^{-1}(E_{a_n}\hat{g}) = (2\sigma) S_0^{-1}(E_{a_n}\hat{g}) = (2\sigma) \sum_m \langle E_{a_n}\hat{g}, h_m \rangle_{I_0} h_m,$$

and upon substituting this into (10.2.33) and taking the inverse Fourier transform, we obtain (10.2.32).

If $\{E_{a_n}(\omega)\}$ is exact, then from Theorem 10.3 $S_0^{-1}(E_{a_n}) = h_n$ is the unique biorthonormal sequence associated with $\{E_{a_n}(\omega)\}$, and this implies that

$$K_n = \frac{1}{\sqrt{2\sigma}} h_n,$$

and

$$\phi_n(t) = \frac{1}{\sqrt{2\sigma}} \int_{-\sigma}^{\sigma} h_n(\omega) e^{-2\pi i t \omega} d\omega = \frac{1}{\sqrt{2\sigma}} \Psi_n(t).$$

∎

Corollary 10.11.1. Let $\sigma > 0$ and $\{a_n\}$ satisfy

$$\sup_n |n - 2\sigma a_n| < \frac{1}{4}, \tag{10.2.38}$$

then

$$f(t) = \sum_n f(a_n)\Psi_n(t),$$

where

$$\Psi_n(t) = \frac{G(t)}{G'(a_n)(t - a_n)}, \tag{10.2.39}$$

and

$$G(t) = (t - a_0)\prod_{n=1}^{\infty}\left(1 - \frac{t}{a_n}\right)\left(1 - \frac{t}{a_{-n}}\right). \tag{10.2.40}$$

Proof. Condition (10.2.38) ensures that $\{E_{a_n}\}$ is an exact frame in $L^2(-\sigma, \sigma)$ (cf. [31, p. 42]) and hence it has a unique biorthonormal sequence $\{h_n\}$. That Ψ_n is given by (10.2.39) and (10.2.40) can be found in [19, p. 58]; see also (3.1.7). ▮

In the next section we shall study different kinds of frames for $L^2(\Re)$, but again with more emphasis on their role in sampling theory.

10.3 Wavelet Analysis and Sampling

10.3.A Wavelets

Wavelets, like the discrete windowed Fourier transform functions $\{g_{mb,na}(x) = e^{2\pi i m b x} g(x - na)\}_{m,n=-\infty}^{\infty}$ introduced in the preceding section, are functions generated from one single function, called the mother wavelet or the analyzing wavelet. But unlike the discrete windowed Fourier transform functions, they are generated from the mother wavelet by dilation and translation instead of modulation and translation. More precisely, if ψ is the mother wavelet, then the functions

$$\psi_{m,n}(x) = a^{-m/2}\psi(a^{-m}x - nb),$$

are called wavelets, where $m, n = 0, \pm 1, \pm 2, \ldots, a > 0,\ b \in \Re$. Since m and n take positive and negative values, there is no loss of generality in restricting a and b so that $a \geq 1$, $b \geq 0$, and since the cases where either $a = 1$ or $b = 0$

are degenerate, they will also be excluded. Therefore, hereafter it will be always assumed that $a > 1$ and $b > 0$.

Wavelets, like the discrete windowed Fourier transform functions, are most interesting when they form building blocks for the space $L^2(\Re)$ whether as a frame or as an orthonormal basis. Interestingly enough, it turns out that wavelets form an unconditional basis not only for $L^2(\Re)$, but for many other function spaces including $L^p(\Re)$, $1 < p < \infty$; a feature that the system $\{g_{mb,na}(x)\}_{m,n=-\infty}^{\infty}$ does not seem to have.

Let $\psi \in L^2(\Re)$ and define

$$\psi_{m,n}(x) = a^{-m/2}\psi(a^{-m}x - nb),$$

for $m,n \in \mathbf{Z}$ and $a > 1$, $b > 0$.

Although it is more appropriate to use a notation that indicates the dependency of $\psi_{m,n}$ on a and b, say for example $\psi_{m,n}^{a,b}(x)$, for the sake of consistency with the conventional notation used in the literature, we shall refrain from doing so; however, if such dependency on a and b is to be displayed, we may use the notation introduced in Section 10.2 to write

$$\psi_{m,n}(x) = D_{a^m} T_{nb}\psi(x).$$

The constant $a^{-m/2}$ is a normalization constant so that

$$\int_{-\infty}^{\infty} |\psi_{m,n}(x)|^2\, dx = \int_{-\infty}^{\infty} |\psi(x)|^2\, dx.$$

If the support of ψ is $[c,d]$, then the support of $\psi_{m,n}(x)$ is $[a^m c + nb, a^m d + nb]$. But whether ψ has compact support or not the shape of the graph of $\psi_{m,n}$ is a scaled dilated version of the graph of ψ and translated by amount nb. This is one of the main differences between the discrete windowed Fourier transform functions and wavelets. For the discrete windowed Fourier transform functions, $\{g_{mb,na}(x)\}$, all the functions $g_{mb,na}(x)$ have the same envelope as the original function g but translated to the proper time location and filled in with high frequency oscillations. But for the wavelets, $\{\psi_{m,n}\}$, the shape of the graph of each $\psi_{m,n}$ is a scaled dilated version of the graph of ψ and each graph is also translated to the proper time location. However, the shape of the graph of each $\psi_{m,n}$ is adapted to the wavelet frequency that is represented by the integer m. For positive m, the graph is wider and flatter while for negative m the graph is narrower and sharper. In other words, high frequency wavelets are narrower and sharper while low frequency wavelets are broader and flatter. This adaptability of the shape of the wavelets to their frequencies makes the wavelet transform better than the discrete windowed Fourier transform in analyzing signals because it has the so-called "zoom in" property, that is, it can zoom in on very short-lived high frequency phenomena.

In order to represent finite-energy signals by wavelets, the system of wavelets has to be at least a frame or preferably an orthonormal basis for $L^2(\Re)$. For this to hold, certain conditions on ψ, a and b must be imposed.

For given a and b, a necessary condition for ψ to be a mother wavelet is given in the following theorem whose proof can be found in [6].

THEOREM 10.12 (Daubechies [6, p. 63])

If $\{\psi_{m,n}(x)\}$ is a frame for $L^2(\Re)$ with frame bounds A, B, then

$$(b \ln a)A \le \int_0^\infty \frac{|\hat{\psi}(\omega)|^2}{\omega} d\omega \le (b \ln a)B \qquad (10.3.1.a)$$

and

$$(b \ln a)A \le \int_{-\infty}^0 \frac{|\hat{\psi}(\omega)|^2}{|\omega|} d\omega \le (b \ln a)B \,. \qquad (10.3.1.b)$$

Because of the singularity of the integrand at $\omega = 0$, in order for (10.3.1.a) and (10.3.1.b) to hold $\hat{\psi}(0)$ must be zero, i.e.,

$$\hat{\psi}(0) = \int_{-\infty}^{\infty} \psi(x)\, dx = 0 \,. \qquad (10.3.2)$$

This condition is called the admissibility condition of the mother wavelet ψ. For $\{\psi_{m,n}(x)\}$ to be a tight frame ($A = B$), the following relation must hold

$$A = \frac{1}{b \ln a} \int_0^\infty \frac{|\hat{\psi}(\omega)|^2}{\omega} d\omega = \frac{1}{b \ln a} \int_{-\infty}^0 \frac{|\hat{\psi}(\omega)|^2}{|\omega|} d\omega = B \,.$$

In particular, for $\{\psi_{m,n}(x)\}$ to be an orthonormal basis, we must have

$$(b \ln a) = \int_0^\infty \frac{|\hat{\psi}(\omega)|^2}{\omega} d\omega = \int_{-\infty}^0 \frac{|\hat{\psi}(\omega)|^2}{|\omega|} d\omega \,.$$

Let

$$H_+^2(\Re) = \{f \in L^2(\Re) : \operatorname{supp} \hat{f} \subset [0,\infty)\}$$

$$H_-^2(\Re) = \{f \in L^2(\Re) : \operatorname{supp} \hat{f} \subset (-\infty,0]\}\,.$$

It is easy to see that $H_\pm^2(\Re)$ is a closed subspace of $L^2(\Re)$ such that

$$L^2(\Re) = H_+^2(\Re) \oplus H_-^2(\Re)\,,$$

with

$$\|f\|_{H_+^2} = \left(\int_0^\infty |\hat{f}(\omega)|^2\, d\omega\right)^{1/2}, \quad \|f\|_{H_-^2} = \left(\int_{-\infty}^0 |\hat{f}(\omega)|^2\, d\omega\right)^{1/2}.$$

If $f \in L^2(\Re)$ is real-valued, then $\hat{f}(\omega) = \overline{\hat{f}(-\omega)}$, so the values of $\hat{f}$ on the positive or the negative ω-axis completely determine $\hat{f}$, and hence f, Moreover,

$$\|f\|_{L^2}^2 = \int_{-\infty}^{\infty} |f(x)|^2\, dx = \int_{-\infty}^{\infty} |\hat{f}(\omega)|^2\, d\omega = 2\int_0^{\infty} |\hat{f}(\omega)|^2\, d\omega = 2\int_{-\infty}^{0} |\hat{f}(\omega)|^2\, d\omega$$

$$= 2\|f\|_{H_+^2}^2 = 2\|f\|_{H_-^2}^2\,.$$

The following theorems give sufficient conditions for $\{\psi_{m,n}(x)\}$ to be a frame for $L^2(\Re)$ for a given a and $\psi \in L^2(\Re)$ that satisfies the admissibility condition. It will be shown that for a given a and $\psi \in L^2(\Re)$ with $\int_{-\infty}^{\infty} \psi(x)\,dx = 0$, there exists $b_0 > 0$ such that for any $0 < b < b_0$, $\{D_{a^m} T_{nb} \psi(x)\}$ is a frame for $L^2(\Re)$. This will be proved in the case where $\hat{\psi}$ has compact support, but the general case will be stated without a proof.

THEOREM 10.13

Let $\psi_1, \psi_2 \in L^2(\Re)$ be such that $\operatorname{supp} \hat{\psi}_1 \subset [l, L]$ and $\operatorname{supp} \hat{\psi}_2 \subset [-L, -l]$, where $0 \le l < L < \infty$. Furthermore, let there exist $A, B > 0$ such that

$$0 < A \le \sum_{m=-\infty}^{\infty} |\hat{\psi}_1(a^m\omega)|^2 \le B < \infty, \quad \text{a.e. in } [0,\infty) \qquad (10.3.3.a)$$

and

$$0 < A \le \sum_{m=-\infty}^{\infty} |\hat{\psi}_2(a^m\omega)|^2 \le B < \infty, \quad \text{a.e. in } (-\infty,0]. \qquad (10.3.3.b)$$

Then $\{D_{a^m} T_{nb}\psi_1, D_{a^m} T_{nb}\psi_2\}$ is a frame for $L^2(\mathfrak{R})$ with frame bounds $b^{-1}A$ and $b^{-1}B$, where b is any number such that $0 < b \le 1/(L - l)$.

Proof. First, let us observe that the Fourier transform of the wavelet $\{\psi_{m,n}(x)\}$ can be written as

$$\hat{\psi}_{m,n}(\omega) = \left(D_{a^m} T_{nb}\psi\right)^\wedge(\omega) = a^{-m/2}\int_{-\infty}^{\infty} \psi(a^{-m}x - nb)\, e^{2\pi i x\omega}\, dx$$

$$= a^{m/2} e^{2\pi i n b a^m \omega}\, \hat{\psi}(a^m\omega) = D_{a^{-m}}(E_{nb}\hat{\psi}(\omega)). \quad (10.3.4)$$

Therefore, for any $f \in L^2(\mathfrak{R})$

$$\left\langle f, D_{a^m} T_{nb}\,\psi\right\rangle = \left\langle \hat{f}, \left(D_{a^m} T_{nb}\,\psi\right)^\wedge\right\rangle = \left\langle \hat{f}, D_{a^{-m}} E_{nb}\,\hat{\psi}\right\rangle,$$

but

$$\left\langle \hat{f}, \left(D_{a^m} T_{nb}\,\psi\right)^\wedge\right\rangle = \int_{-\infty}^{\infty} \hat{f}(\omega)\, a^{m/2} e^{-2\pi i n b a^m \omega}\, \overline{\hat{\psi}(a^m\omega)}\, d\omega$$

$$= a^{-m/2}\int_{-\infty}^{\infty} \hat{f}\left(\frac{\omega}{a^m}\right) e^{-2\pi i n b\omega}\, \overline{\hat{\psi}(\omega)}\, d\omega$$

$$= \int_{-\infty}^{\infty} \left(D_{a^m}\hat{f}\right) \overline{\hat{\psi}}(\omega)\, e^{-2\pi i n b\omega}\, d\omega = \left\langle \left(D_{a^m}\hat{f}\,\right)\overline{\hat{\psi}}, E_{nb}\right\rangle.$$

Hence

$$\left\langle f, D_{a^m} T_{nb}\psi\right\rangle = \left\langle \left(D_{a^m}\hat{f}\right)\overline{\hat{\psi}}, E_{nb}\right\rangle. \quad (10.3.5)$$

Now since $\operatorname{supp}\hat{\psi}_1 \subset [l, L] \subset \left[l, l + \frac{1}{b}\right] = I$, $(D_{a^m}\hat{f}) \cdot \overline{\hat{\psi}_1} \in L^2(I)$, and since $\{\sqrt{b}\, E_{nb}\}_{n=-\infty}^{\infty}$ is an orthonormal basis of $L^2(I)$, it follows that

$$\sum_{n=-\infty}^{\infty} \left|\left\langle \left(D_{a^m}\hat{f}\right) \cdot \overline{\hat{\psi}_1}, E_{nb}\right\rangle_I\right|^2 = \frac{1}{b}\int_I \left|D_{a^m}\hat{f}(\omega)\right|^2 |\hat{\psi}_1(\omega)|^2 d\omega$$

$$= \frac{1}{b}\int_0^{\infty} \left|D_{a^m}\hat{f}(\omega)\right|^2 |\hat{\psi}_1(\omega)|^2 d\omega = \frac{1}{b}\int_0^{\infty} |\hat{f}(\omega)|^2\, |\hat{\psi}_1(a^m\omega)|^2 d\omega. \quad (10.3.6)$$

Therefore, by combining (10.3.5) and (10.3.6), we obtain

$$\sum_{m,n=-\infty}^{\infty}\left|\left\langle f,D_{a^m}T_{nb}\psi_1\right\rangle\right|^2=\sum_{m,n=-\infty}^{\infty}\left|\left\langle\left(D_{a^m}\hat{f}\right)\cdot\overline{\hat{\psi}}_1,E_{nb}\right\rangle_l\right|^2$$

$$=\frac{1}{b}\int_0^{\infty}|\hat{f}(\omega)|^2\left(\sum_{m=-\infty}^{\infty}|\hat{\psi}_1(a^m\omega)|^2\right)d\omega\,,\qquad(10.3.7)$$

which, in view of (10.3.3.a), yields

$$b^{-1}A\|f\|^2_{H^2_+}\le\sum_{m,n=-\infty}^{\infty}\left|\left\langle f,D_{a^m}T_{nb}\psi_1\right\rangle\right|^2\le b^{-1}B\|f\|^2_{H^2_+}.$$

Hence, $\{D_{a^m}T_{nb}\psi_1\}$ is a frame for H^2_+. Similarly, we can show that $\{D_{a^m}T_{nb}\psi_2\}$ is a frame for H^2_- and since L^2 is the direct sum of H^2_+ and H^2_- the proof is now complete. ∎

Conditions (10.3.3.a) and (10.3.3.b) are satisfied if, for example, we assume further that $\hat{\psi}_1$, $\hat{\psi}_2$ are continuous and do not vanish on (l,L), $(-L,-l)$ respectively, and that $1<a<L/l$. To show for example that (10.3.3.a) is satisfied, it suffices to show that for any $\omega\in\mathfrak{R}^+$, there are only finitely many integers n for which $l\le a^n\omega\le L$ and since $\hat{\psi}_1$ is continuous and does not vanish on (l,L), it will follow that the series in (10.3.3.a) is bounded above for all $\omega\in\mathfrak{R}^+$ and bounded below away from zero, except possibly for countably many points. If $l<\omega<L$, then there are at most n_0 non-zero terms in (10.3.3.a), where n_0 is the smallest integer for which $L/l<a^{n_0}$. If $0<\omega<l$, then there is at least one integer n for which $l\le a^n\omega$. Let n_0 be the smallest such an integer. Similarly, since $a\omega<L$, let m_0 be the greatest integer such that $a^{m_0}\omega\le L$. We claim that $n_0\le m_0$. For, otherwise $n_0>m_0$, which implies that $a^{m_0}<l/\omega$. Hence $a^{m_0+1}<L/\omega$, which contradicts the definition of m_0. Therefore, for any ω such that $0<\omega<l$, the series (10.3.3.a) contains at most m_0-n_0+1 non-zero terms.

It is, of course, more desirable to have just one single function ψ generating the frame for $L^2(\mathfrak{R})$ and a natural candidate for that would be the function $\psi=\psi_1+\psi_2$, assuming that ψ_1, ψ_2 satisfy the hypothesis of Theorem 10.13. But this is not true in general. For example, if $a=2$, $b=1$, $\hat{\psi}_1(\omega)=\chi_{[1,2)}(\omega)$ and $\hat{\psi}_2(\omega)=\chi_{(-2,-1]}(\omega)$, then in view of (10.3.4) and the fact that $\langle\psi_{m,n},\psi_{k,l}\rangle=\langle\hat{\psi}_{m,n},\hat{\psi}_{k,l}\rangle$, it is easy to show that $\{D_{2^m}T_n\psi_1,D_{2^m}T_n\psi_2\}$ is an orthonormal basis for $L^2(\mathfrak{R})$, but there is at least one function $f\in L^2(\mathfrak{R})$ with $f\ne 0$ such that $\langle\psi_{m,n},f\rangle=0$ for all m, n, where $\psi=\psi_1+\psi_2$; hence $\{\psi_{m,n}\}$ is not complete and thereby cannot be a frame. One such function f is defined by $\hat{f}(\omega)=\chi_{(-2,-1]}(\omega)-\chi_{[1,2)}(\omega)$.

The next theorem shows that this deficiency can be corrected if b is chosen appropriately.

THEOREM 10.14

Let ψ_1, ψ_2 satisfy the hypothesis of Theorem 10.13. If $b < 1/(2L)$, then $\psi = \psi_1 + \psi_2$ generates a frame for $L^2(\mathfrak{R})$.

Proof. Since $\operatorname{supp}\psi \subset [-L, L] \subset [-1/2b, 1/2b] = I$, then for any $f \in L^2(\mathfrak{R})$, $(D_{a^m}\hat{f}) \cdot \overline{\hat{\psi}} \in L^2(I)$ and it follows as in the proof of Theorem 10.13 (see (10.3.6), (10.3.7)) that

$$\sum_{m,n=-\infty}^{\infty} \left|\left\langle f, D_{a^m} T_{nb} \psi\right\rangle\right|^2 = \sum_{m,n=-\infty}^{\infty} \left|\left\langle (D_{a^m}\hat{f}) \cdot \overline{\hat{\psi}}, E_{nb}\right\rangle\right|^2 = \frac{1}{b} \sum_{m=-\infty}^{\infty} \int_{-\infty}^{\infty} |\hat{f}(\omega)|^2 |\hat{\psi}(a^m\omega)|^2 \, d\omega$$

$$= \frac{1}{b}\left(\int_0^{\infty} |\hat{f}(\omega)|^2 \left(\sum_{m=-\infty}^{\infty} |\hat{\psi}_1(a^m\omega)|^2 \right) d\omega + \int_{-\infty}^{0} |\hat{f}(\omega)|^2 \left(\sum_{m=-\infty}^{\infty} |\hat{\psi}_2(a^m\omega)|^2 \right) d\omega \right),$$

which, together with (10.3.3.a) and (10.3.3.b), yields

$$\frac{1}{b} A \|f\|^2 \le \sum_{m,n=-\infty}^{\infty} \left|\left\langle f, D_{a^m} T_{nb} \psi\right\rangle\right|^2 \le \frac{1}{b} B \|f\|^2 .$$

∎

If the support of $\hat{\psi}$ is not compact, the conclusion of Theorem 10.13 still holds. That is, for a given $\psi \in L^2(\mathfrak{R})$ that satisfies the admissibility condition and $a > 1$, there exists $b_0 > 0$ such that $\{D_{a^m} T_{nb} \psi\}$ is a frame for $L^2(\mathfrak{R})$ for any $0 < b < b_0$; see [6, p. 69] for the proof.

THEOREM 10.15

Let $\psi \in L^2(\mathfrak{R})$, $\int_{-\infty}^{\infty} \psi(x)\, dx = 0$, and $a > 1$. If

$$\inf_{1 \le |\omega| \le a} \sum_{m=-\infty}^{\infty} |\hat{\psi}(a^m\omega)|^2 > 0,$$

$$\sup_{1 \le |\omega| \le a} \sum_{m=-\infty}^{\infty} |\hat{\psi}(a^m\omega)|^2 < \infty,$$

and

$$\beta(s) = \sup_{\omega} \sum_{m=-\infty}^{\infty} |\hat{\psi}(a^m\omega)|\, |\hat{\psi}(a^m\omega + s)|$$

decays at least as fast as $(1+|s|)^{-(1+\varepsilon)}$, $\varepsilon > 0$ as $|s| \to \infty$, then there exists $b_0 > 0$ such that $\{D_{a^m} T_{nb} \psi\}$ is a frame for $L^2(\mathfrak{R})$ for any b such that $0 < b < b_0$ and the frame bounds can be given by

$$A = \frac{1}{b}\left\{ \inf_{1 \le |\omega| \le a} \sum_{m=-\infty}^{\infty} |\hat{\psi}(a^m \omega)|^2 - \sum_{\substack{k=-\infty \\ k \ne 0}}^{\infty} \left[\beta\left(\frac{k}{b}\right)\beta\left(-\frac{k}{b}\right)\right]^{1/2} \right\},$$

$$B = \frac{1}{b}\left\{ \sup_{1 \le |\omega| \le a} \sum_{m=-\infty}^{\infty} |\hat{\psi}(a^m \omega)|^2 + \sum_{\substack{k=-\infty \\ k \ne 0}}^{\infty} \left[\beta\left(\frac{k}{b}\right)\beta\left(-\frac{k}{b}\right)\right]^{1/2} \right\}.$$

In particular, the conclusion is valid if there exist $A, B > 0$ such that

$$0 < A \le \sum_{m=-\infty}^{\infty} |\hat{\psi}(a^m \omega)|^2 \le B < \infty, \quad \text{a.e. in } \mathfrak{R}.$$

and

$$\lim_{b \to 0} \sum_{k \ne 0} \left(\beta\left(\frac{k}{b}\right)\beta\left(-\frac{k}{b}\right)\right)^{1/2} = 0.$$

Before we proceed any further, let us give some examples of wavelets.

Example 1. The simplest and also the oldest system of orthonormal wavelets is the Haar orthonormal basis $\{\psi_{m,n}(x) = D_{2^m} T_n \psi(x)\}$ for $L^2(\mathfrak{R})$ (see [6, p. 11] for details) for which

$$\psi(x) = \begin{cases} -1, & 0 \le x < 1/2 \\ 1, & 1/2 < x < 1 \\ 0 & \text{otherwise} \end{cases}.$$

That ψ is indeed a mother wavelet is easier to show by using multiresolution analysis; therefore, we will postpone the proof until we discuss multiresolution analyses in Section 10.3.C.

The wavelets of this system are not continuous and their Fourier transforms decay like $|\omega|^{-1}$ as $|\omega| \to \infty$. It is worth noting that the discrete windowed Fourier transform functions

$$g_{m,n}(x) = e^{2\pi i m x} g(x-n),$$

where $g(x) = \chi_{[0,1)}(x)$, is also an orthonormal basis for $L^2(\mathfrak{R})$. However, whereas the Haar system is an unconditional basis for $L^p(\mathfrak{R})$, $1 \le p < \infty$, the system $\{g_{m,n}(x)\}$ is not, unless $p = 2$.

***Example* 2.** Let

$$\hat{\psi}(\omega) = \begin{cases} 1 & , \quad \frac{1}{2} \le |\omega| \le 1 \\ 0 & , \quad \text{otherwise} \end{cases} ;$$

hence $\psi(x) = (\sin 2\pi x - \sin \pi x)/(\pi x)$. To show that $\{\psi_{m,n}(x) = D_{2^m} T_n \psi(x)\}$ is an orthonormal basis for $L^2(\Re)$, we invoke Lemma 10.2.1 to conclude that it is sufficient to show that $\|\psi_{m,n}\| = 1$ for all m, n and that $\{\psi_{m,n}(x)\}$ is a tight frame with frame bound $= 1$. First,

$$\|\psi_{m,n}\| = \|\psi\| = \|\hat{\psi}\| = 1 .$$

Second, with the aid of (10.3.4), we obtain

$$\sum_{m,n} |\langle f, \psi_{m,n}\rangle|^2 = \sum_{m,n} |\langle \hat{f}, \hat{\psi}_{m,n}\rangle|^2$$

$$= \sum_{m,n} 2^m \left| \int_{2^{-m-1} \le |\omega| \le 2^{-m}} \hat{f}(\omega) e^{2^{m+1}\pi i n \omega} d\omega \right|^2 = \sum_{m,n} 2^{-m} \left| \int_{1/2 \le |\omega| \le 1} \hat{f}\left(\frac{\omega}{2^m}\right) e^{2\pi i n \omega} d\omega \right|^2$$

$$= \sum_{m,n} 2^{-m} \left| \int_{1/2}^{1} \hat{f}\left(\frac{\omega}{2^m}\right) e^{2\pi i n \omega} d\omega + \int_{-1}^{-1/2} \hat{f}\left(\frac{\omega}{2^m}\right) e^{2\pi i n \omega} d\omega \right|^2$$

$$= \sum_{m} 2^{-m} \left(\sum_{n} \left| \int_{0}^{1} e^{2\pi i n \omega} \left[\hat{f}\left(\frac{\omega}{2^m}\right) \chi_{[1/2,1]}(\omega) + \hat{f}\left(\frac{\omega - 1}{2^m}\right) \chi_{[0,1/2]}(\omega) \right] d\omega \right|^2 \right)$$

$$= \sum_{m} 2^{-m} \int_{0}^{1} \left[\hat{f}\left(\frac{\omega}{2^m}\right) \chi_{[1/2,1]}(\omega) + \hat{f}\left(\frac{\omega - 1}{2^m}\right) \chi_{[0,1/2]}(\omega) \right]^2 d\omega$$

$$= \sum_{m} \left(\int_{1/2^{m+1}}^{1/2^m} |\hat{f}(\omega)|^2 d\omega + \int_{-1/2^m}^{-1/2^{m+1}} |\hat{f}(\omega)|^2 d\omega \right) = \int_{0}^{\infty} |\hat{f}(\omega)|^2 d\omega + \int_{-\infty}^{0} |\hat{f}(\omega)|^2 d\omega$$

$$= \int_{-\infty}^{\infty} |\hat{f}(\omega)|^2 d\omega = \int_{-\infty}^{\infty} |f(x)|^2 dx = \|f\|^2 ,$$

which proves that $\{\psi_{m,n}(x)\}$ is a tight frame with frame bound $= 1$.

The next example was first introduced in [9].

***Example* 3.** Let ν be a C^k or C^∞-function such that

$$\nu(x) = \begin{cases} 0 & , \quad x \le 0 \\ 1 & , \quad x \ge 1 \end{cases} \qquad (10.3.8)$$

For $a > 1$, $b > 0$, define

$$\hat{\psi}_1(\omega) = \frac{1}{\sqrt{\ln a}} \begin{cases} 0 & , \quad \omega < l \quad \text{or} \quad \omega \geq a^2 l \\ \sin\left[\frac{\pi}{2}\nu\left(\frac{\omega - l}{l(a-1)}\right)\right] & , \quad l \leq \omega \leq al \\ \cos\left[\frac{\pi}{2}\nu\left(\frac{\omega - al}{al(a-1)}\right)\right] & , \quad al < \omega \leq a^2 l , \end{cases}$$

and $\hat{\psi}_2(\omega) = \overline{\hat{\psi}_1(-\omega)}$, where $l = [b(a^2 - 1)]^{-1}$.

Clearly, $\text{supp}\,\hat{\psi}_1 \subset [l, L]$, where $L = a^2 l$, and $\hat{\psi}_1$ is as smooth as ν. Moreover, for any $\omega > 0$, the series $\sum_{m=-\infty}^{\infty} |\hat{\psi}_1(a^m \omega)|^2$ contains only two non-zero terms whose sum is $1/\ln a$ for almost all ω. For, if $0 < \omega < l$, we choose $m_0 > 0$ to be the smallest integer such that $l/\omega \leq a^{m_0}$. It is easy to see that $l \leq a^{m_0}\omega \leq al$, $al \leq a^{m_0+1}\omega \leq L$ and $L \leq a^{m_0+2}\omega$. Hence

$$\sum_{m=-\infty}^{\infty} |\hat{\psi}_1(a^m \omega)|^2 = |\hat{\psi}_1(a^{m_0}\omega)|^2 + |\hat{\psi}_1(a^{m_0+1}\omega)|^2 = \frac{1}{\ln a} .$$

The proof for $\omega \geq l$ is similar. Therefore, by Theorem 10.13, it follows that $\{D_{a^m} T_{nb} \psi_1, D_{a^m} T_{nb} \psi_2\}$ is a tight frame for $L^2(\mathfrak{R})$ with a frame bound $1/(b \ln a)$; however, it is not a basis as can be seen from part (vi) of Theorem 10.3. The normalization constant $(1/\sqrt{\ln a})$ in the definition of $\hat{\psi}_1(\omega)$ has been chosen so that

$$\int_0^{\infty} \frac{|\hat{\psi}_1(\omega)|^2}{\omega} d\omega = 1 .$$

An example of a C^1-function ν satisfying (10.3.8) is

$$\nu(x) = \begin{cases} 0 & , \quad x \leq 0 \\ \sin^2 \frac{\pi}{2} x & , \quad 0 \leq x \leq 1 \\ 1 & , \quad x \geq 1 . \end{cases}$$

The next example has an interesting development. Recall from Section 10.2.C that in the critical case when $ab = 1$, the discrete windowed Fourier transform functions may form a basis for $L^2(\mathfrak{R})$, but when they do, they are either non-smooth or do not decay rapidly. Guided by this result, Y. Meyer [23] (see [8, p. 117]) tried to show that a similar result held for wavelets by proving the impossibility of constructing smooth wavelets with fast decay, but surprisingly he found that this was not indeed the case and he was able to construct the following orthonormal smooth wavelets with fast decay.

***Example* 4.** Let ψ be defined by

$$\hat{\psi}(\omega) = \begin{cases} e^{i\omega/2}\sin\left[\frac{\pi}{2}\nu(3|\omega|-1)\right] & \text{if } \frac{1}{3} \le |\omega| \le \frac{2}{3} \\ e^{i\omega/2}\cos\left[\frac{\pi}{2}\nu\left(\frac{3}{2}|\omega|-1\right)\right] & \text{if } \frac{2}{3} \le |\omega| \le \frac{4}{3} \\ 0 & \text{otherwise} \end{cases},$$

where ν is a C^∞-function satisfying (10.3.8) with the additional property

$$\nu(x) + \nu(1-x) = 1 .$$

Since $\hat{\psi}$ is C^∞ and has compact support, its inverse Fourier transform ψ is in the Schwartz space $S(\Re)$, i.e., ψ is C^∞ and $\psi(x) = O(1/|x|^N)$ as $|x| \to \infty$ for any positive integer N.

According to Lemma 10.2.1, in order for $\{\psi_{m,n}(x) = D_{2^m}T_n\psi(x)\}$ to be an orthonormal basis for $L^2(\Re)$, it is sufficient that $\|\psi_{m,n}\| = \|\psi\| = 1$ and that $\{\psi_{m,n}(x)\}$ be a tight frame with frame bound = 1. That $\|\psi\| = 1$ is not difficult to show. For,

$$\frac{1}{2}\int_{-\infty}^{\infty}|\hat{\psi}(\omega)|^2 d\omega = \int_0^{\infty}|\hat{\psi}(\omega)|^2 d\omega = \int_{1/3}^{2/3}\sin^2\left[\frac{\pi}{2}\nu(3\omega-1)\right]d\omega + \int_{2/3}^{4/3}\cos^2\left[\frac{\pi}{2}\nu\left(\frac{3\omega}{2}-1\right)\right]d\omega$$

$$= \frac{1}{3}\int_0^1 \sin^2\left[\frac{\pi}{2}\nu(s)\right]ds + \frac{2}{3}\int_0^1\cos^2\left[\frac{\pi}{2}\nu(s)\right]ds = \frac{1}{3}\left[1 + \int_0^1\cos^2\left[\frac{\pi}{2}\nu(s)\right]ds\right]$$

$$= \frac{1}{3}\left[1 + \int_0^{1/2}\cos^2\left[\frac{\pi}{2}\nu(s)\right]ds + \int_0^{1/2}\cos^2\left[\frac{\pi}{2}\nu\left(s+\frac{1}{2}\right)\right]ds\right].$$

But since $\nu\left(s+\frac{1}{2}\right) + \nu\left(\frac{1}{2}-s\right) = 1$, then

$$\int_0^{1/2}\cos^2\left[\frac{\pi}{2}\nu(s)\right]ds + \int_0^{1/2}\cos^2\left[\frac{\pi}{2}\nu\left(s+\frac{1}{2}\right)\right]ds$$

$$= \int_0^{1/2}\left(\cos^2\left[\frac{\pi}{2}\nu(s)\right] + \cos^2\left[\frac{\pi}{2}\left(1-\nu\left(\frac{1}{2}-s\right)\right)\right]\right)ds$$

$$= \int_0^{1/2}\left(\cos^2\left[\frac{\pi}{2}\nu(s)\right] + \sin^2\left[\frac{\pi}{2}\nu\left(\frac{1}{2}-s\right)\right]\right)ds = \int_0^{1/2}\left(\cos^2\left[\frac{\pi}{2}\nu(s)\right] + \sin^2\left[\frac{\pi}{2}\nu(s)\right]\right)ds$$

$$= \frac{1}{2}.$$

Therefore

$$\int_{-\infty}^{\infty} |\psi(x)|^2 dx = \int_{-\infty}^{\infty} |\hat{\psi}(\omega)|^2 d\omega = 1 .$$

However, showing that $\{\psi_{m,n}(x)\}$ is a tight frame with frame bound = 1, is more difficult and requires a more refined estimate of the frame bounds given in Theorem 10.15. These refined frame bound estimates are known as Tchamitchian's frame bound estimates [29]. For the details of the proof, we refer the reader to [6, p. 118]; see also [24].

10.3.B The Wavelet Frame Operator

Let the system of wavelet $\{D_{a^m} T_{nb} \psi\}$ be a frame for $L^2(\Re)$. As before, we define the associated frame operator S by

$$Sf = \sum_{m,n=-\infty}^{\infty} \langle f, D_{a^m} T_{nb} \psi \rangle D_{a^m} T_{nb} \psi , \quad f \in L^2(\Re) . \tag{10.3.9}$$

Since

$$\langle f, D_{a^k} T_{nb} \psi \rangle = a^{-k/2} \int_{-\infty}^{\infty} f(x) \psi(a^{-k}x - nb)\, dx = a^{k/2} \int_{-\infty}^{\infty} f(a^k x) \psi(x - nb)\, dx$$

$$= \langle D_{a^{-k}} f, T_{nb} \psi \rangle ,$$

it follows that

$$D_{a^k}(Sf) = \sum_{m,n=-\infty}^{\infty} \langle f, D_{a^m} T_{nb} \psi \rangle D_{a^{k+m}} T_{nb} \psi$$

$$= \sum_{m,n=-\infty}^{\infty} \langle f, D_{a^{m-k}} T_{nb} \psi \rangle D_{a^m} T_{nb} \psi$$

$$= \sum_{m,n=-\infty}^{\infty} \langle D_{a^k} f, D_{a^m} T_{nb} \psi \rangle D_{a^m} T_{nb} \psi = S(D_{a^k} f) ,$$

and, hence

$$S^{-1}(D_{a^k} f) = D_{a^k}(S^{-1} f) . \tag{10.3.10}$$

Unlike the discrete windowed Fourier transform frame operator, the wavelet frame operator does not commute with the translation operator T_{nb}.

Therefore, the dual frame of a wavelet frame is not generated by one single function. In fact, we have

$$\begin{aligned}\tilde{\psi}_{m,n} &= S^{-1}\psi_{m,n}(x) = S^{-1}\left(D_{a^m}T_{nb}\psi\right) = D_{a^m}(S^{-1}T_{nb}\psi)\\ &= D_{a^m}\tilde{\psi}_{0,n} = a^{-m/2}\tilde{\psi}_{0,n}(a^{-m}x)\end{aligned}$$

where $\tilde{\psi}_{0,n}(x) = S^{-1}(\psi(x-nb))$.

THEOREM 10.16

i) Let ψ, a, b be as in Theorem 10.13 and ψ satisfy the same hypothesis as ψ_1 in the same theorem. Then $Sf = (\hat{f}\cdot G)^{\vee}$ and $S^{-1}f = (\hat{f}/G)^{\vee}$ for $f \in H_+^2(\Re)$, where

$$G(\omega) = \begin{cases} 0 & \text{if} \quad \omega < 0 \\ \dfrac{1}{b}\sum\limits_{m=-\infty}^{\infty} |\hat{\psi}(a^m\omega)|^2 & \text{if} \quad \omega > 0 \end{cases}.$$

ii) If ψ_1, ψ_2, a, b satisfy the hypothesis of Theorem 10.13. Then

$$Sf = (\hat{f}\cdot G)^{\vee} \quad \text{and} \quad S^{-1}f = (\hat{f}/G)^{\vee} \quad \text{for} \quad f \in L^2(\Re),$$

where

$$G(\omega) = \begin{cases} \dfrac{1}{b}\sum\limits_{m=-\infty}^{\infty} |\hat{\psi}_1(a^m\omega)|^2 & , \quad \omega > 0 \\ \dfrac{1}{b}\sum\limits_{m=-\infty}^{\infty} |\hat{\psi}_2(a^m\omega)|^2 & , \quad \omega < 0 \end{cases}.$$

iii) If ψ_1, ψ_2, a, b satisfy the hypothesis of Theorem 10.14. Then

$$Sf = (\hat{f}\cdot G)^{\vee} \quad \text{and} \quad S^{-1}f = (\hat{f}/G)^{\vee} \quad \text{for} \quad f \in L^2(\Re),$$

where

$$G(\omega) = \frac{1}{b}\sum_{m=-\infty}^{\infty} |(\hat{\psi}_1 + \hat{\psi}_2)(a^m\omega)|^2.$$

Proof. We only prove (i) since the proofs of the others are similar.

By taking the Fourier transform of (10.3.9), we obtain in view of (10.3.4) and (10.3.5),

$$(Sf)^{\wedge}(\omega) = \sum_{m,n=-\infty}^{\infty} \left\langle f, D_{a^m} T_{nb} \psi \right\rangle D_{a^{-m}}(E_{nb}\hat{\psi}(\omega))$$

$$= \sum_{m,n=-\infty}^{\infty} \left\langle \left(D_{a^m}\hat{f}\right)\overline{\hat{\psi}}, E_{nb}\right\rangle D_{a^{-m}}(E_{nb}\hat{\psi}(\omega))$$

$$= \sum_{m=-\infty}^{\infty} D_{a^{-m}}\left(\left[\sum_{n=-\infty}^{\infty}\left\langle \left(D_{a^m}\hat{f}\right)\overline{\hat{\psi}}, E_{nb}\right\rangle E_{nb}\right]\hat{\psi}(\omega)\right).$$

But since $\left(D_{a^m}\hat{f}\right)\overline{\hat{\psi}} \in L^2(I)$, where I is any interval of length $1/b$ and $\{\sqrt{b}\,E_{nb}\}_{n=-\infty}^{\infty}$ is an orthonormal basis for $L^2(I)$ it follows that

$$(Sf)^{\wedge}(\omega) = \frac{1}{b}\sum_{m=-\infty}^{\infty} D_{a^{-m}}\left(\left(D_{a^m}\hat{f}\right)|\hat{\psi}(\omega)|^2\right) = \hat{f}(\omega)\frac{1}{b}\sum_{m=-\infty}^{\infty} a^{-m/2} D_{a^{-m}}|\hat{\psi}(\omega)|^2 = \hat{f}(\omega)G(\omega),$$

hence $Sf = (\hat{f}\cdot G)^{\vee}$. Here we have used the fact that

$$D_{a^m}(fg) = a^{m/2}\left(D_{a^m}f\right)\left(D_{a^m}g\right).$$

∎

The construction of wavelets delineated in the aforementioned examples undoubtedly required a fair amount of ingenuity. In the construction of Meyer's wavelets, in particular, some miraculous cancellations in calculating the frame bounds made the construction possible, yet also made it appear as a fluke. Fortunately, nowadays constructing orthonormal wavelet bases is more of a procedure than an art. This procedure is better explained in the language of multiresolution analysis.

10.3.C Multiresolution Analysis

The concept of multiresolution analysis of a Hilbert space of functions was formulated by S. Mallat and Y. Meyer in 1986 and it was eloquently presented by Mallat in [21].

Definition 10.3.1. A multiresolution analysis for $L^2(\Re)$ consists of a sequence of closed subspaces $\{V_n\}_{n=-\infty}^{\infty}$ of $L^2(\Re)$ and a function $\phi \in V_0$ such that

1) $\ldots \subset V_2 \subset V_1 \subset V_0 \subset V_{-1} \subset V_{-2} \subset \ldots$

2) $\bigcup_{i=-\infty}^{\infty} V_i$ is dense in $L^2(\Re)$

3) $\bigcap_{i=-\infty}^{\infty} V_i = \{0\}$

4) $V_{n+1} = D_2 V_n = \{f(2^{-1}x) : f(x) \in V_n\}$

5) $\{T_n \phi(x) = \phi(x-n)\}$ is an orthonormal basis for V_0.

It follows from (4) and (5) that

i) $f(x) \in V_0$ if and only if $f(2^{-m}x) \in V_m$,

ii) V_0 is invariant under translation by an integer, i.e.,

$$T_k V_0 = \{f(x-k) : f(x) \in V_0\} = V_0, \quad k = 0, \pm 1, \pm 2, \ldots$$

hence

iii) $V_m = D_{2^m} V_0 = D_{2^m} T_k V_0 = \{f(2^{-m}x - k) : f(x) \in V_0\}$ and

iv) for each fixed m, $\{\phi_{m,n}(x) = 2^{-m/2}\phi(2^{-m}x - n)\}_{n=-\infty}^{\infty}$ is an orthonormal basis for V_m.

An example of a multiresolution analysis is the following:

$$V_0 = \{f \in L^2(\Re) : f \text{ is constant on each interval of the form } [m, m+1), m \in Z\}, \; V_n = D_{2^n} V_0 \text{ and } \phi(x) = \chi_{[0,1)}.$$

Mallat's definition of a multiresolution analysis is different yet equivalent to the one above. His definition does not stipulate the existence of the function ϕ and hence does not include condition (5), but it includes two more conditions on the subspaces $\{V_n\}_{n=-\infty}^{\infty}$, namely $f(x) \in V_n$ implies that $f(x - 2^{-n}k) \in V_n$ for all $n \in \mathbf{Z}$ and that there exists an isomorphism from V_0 onto $l^2 = \left\{c : c = \{c_n\}_{n=-\infty}^{\infty}, \sum_{n=-\infty}^{\infty} |c_n|^2 < \infty\right\}$ which commutes with the action of $\mathbf{Z}$. These two conditions guarantee the existence of ϕ and the validity of (5).

The function ϕ is called the "scaling function" of the multiresolution analysis.

The importance of a multiresolution analysis lies in the simple fact that it enables us to generate an orthonormal wavelet basis for $L^2(\Re)$. To show this, let us first observe that since V_n is contained in V_{n-1}, we can define W_n to be the orthogonal complement of V_n in V_{n-1}, i.e.,

$$V_{n-1} = V_n \oplus W_n, \quad \text{for all } n, \tag{10.3.11}$$

which, by induction, yields

$$V_n = V_{n+k} \oplus \bigoplus_{i=1}^{k} W_{n+i},$$

where all the subspaces are orthogonal and

$$\bigoplus_{i=1}^{n} W_i = W_1 \oplus W_2 \oplus \ldots \oplus W_n .$$

In view of (10.3.11), and conditions (2), (3) of Definition 10.3.1, it is not hard to see that

$$L^2(\mathfrak{R}) = \bigoplus_{j=-\infty}^{\infty} W_j . \qquad (10.3.12)$$

Since $W_{n+1} \subset V_n$, it follows from condition (4) of Definition 10.3.1 and (10.3.11) that

$$f(x) \in W_{n+1} \text{ if and only if } f(2^{n+1}x) \in V_{-1} - V_0 = W_0,$$

or equivalently

$$f(x) \in W_0 \text{ if and only if } f(2^{-n}x) \in W_n . \qquad (10.3.13)$$

This means that condition (4) of Definition 10.3.1 also holds for the subspaces $\{W_n\}_{n=-\infty}^{\infty}$.

Therefore, if we can show that there exists a function $\psi \in W_0$ such that $\{\psi(x-n)\}_{n=-\infty}^{\infty}$ is a basis for W_0, it will follow from (10.3.12) and (10.3.13) that $\left\{ \psi_{m,n}(x) = D_{2^m} T_n \psi = 2^{-m/2}\psi(2^{-m}x - n) \right\}_{m,n=-\infty}^{\infty}$ is an orthonormal basis for $L^2(\mathfrak{R})$. Proving the existence of such an ψ is the crux of Theorem 10.17. In fact, Theorem 10.17 is more than an existence-type theorem, it actually provides a procedure for constructing such an ψ. But before we prove this theorem we need the following lemmas.

Lemma 10.3.1. Let $\phi \in L^2(\mathfrak{R})$. The following two conditions are equivalent:

i) $\{\phi(x-n)\}$ is orthonormal.

ii) $\sum_{n=-\infty}^{\infty} |\hat{\phi}(\omega+n)|^2 = 1$ a.e.

Proof. Let $\phi_{0,n}(x) = \phi(x-n)$. Then, $\hat{\phi}_{0,n}(\omega) = e^{2\pi i\omega n}\hat{\phi}(\omega)$. If we set $n = m - l$, we then have

$$\langle \phi_{0,l}, \phi_{0,m}\rangle = \langle \phi_{0,0}, \phi_{0,m-l}\rangle = \langle \hat{\phi}_{0,0}, \hat{\phi}_{0,n}\rangle = \int_{-\infty}^{\infty} e^{-2\pi i n\omega}\,|\hat{\phi}(\omega)|^2\,d\omega$$

$$= \sum_{k=-\infty}^{\infty} \int_{k}^{k+1} e^{-2\pi i n\omega}\,|\hat{\phi}(\omega)|^2\,d\omega = \int_0^1 e^{-2\pi i n\omega} \sum_{k=-\infty}^{\infty} |\hat{\phi}(\omega+k)|^2\,d\omega,$$

which implies that $\langle \phi_{0,l}, \phi_{0,m}\rangle = \delta_{l,m}$ if and only if

$$\sum_{k=-\infty}^{\infty} |\hat{\phi}(\omega+k)|^2 = 1 \qquad \text{a.e.}$$

∎

Lemma 10.3.2. The scaling function ϕ of the multiresolution analysis satisfies the following conditions

i) $$\sum_{k=-\infty}^{\infty} |\hat{\phi}(\omega+k)|^2 = 1 \qquad \text{a.e.} \tag{10.3.14}$$

ii) $$\hat{\phi}(\omega) = M\left(\frac{\omega}{2}\right)\hat{\phi}\left(\frac{\omega}{2}\right), \tag{10.3.15}$$

where $M(\omega)$ is a periodic function with period 1 that belongs to $L^2[0,1]$ and satisfies

$$|M(\omega)|^2 + |M(\omega + 1/2)|^2 = 1 \qquad \text{a.e.} \tag{10.3.16}$$

Proof. i) is a consequence of condition (5) of Definition 10.3.1 and Lemma 10.3.1.

ii) Since $\phi \in V_0 \subset V_{-1}$ and $\{\phi_{-1,n}(x) = 2^{1/2}\phi(2x-n)\}_{n=-\infty}^{\infty}$ is an orthonormal basis for V_{-1}, we have

$$\phi(x) = \sqrt{2}\sum_{n=-\infty}^{\infty} \alpha_n \phi(2x-n) \quad \text{with} \quad \sum_{n=-\infty}^{\infty} |\alpha_n|^2 < \infty, \tag{10.3.17}$$

where $\alpha_n = \langle \phi, \phi_{-1,n}\rangle$. By taking the Fourier transform of both sides of (10.3.17), we obtain (10.3.15), that is

$$\hat{\phi}(\omega) = M\left(\frac{\omega}{2}\right)\hat{\phi}\left(\frac{\omega}{2}\right),$$

where

$$M(\omega)=\frac{1}{\sqrt{2}}\sum_{n=-\infty}^{\infty}\alpha_n e^{2i\pi n\omega}. \qquad (10.3.18)$$

Clearly, $M(\omega)$ is a periodic function with period 1 that belongs to $L^2[0,1]$. By replacing ω by 2ω in (10.3.15) and combining (10.3.14), (10.3.15), we obtain

$$1=\sum_{k=-\infty}^{\infty}\left|M\left(\omega+\frac{k}{2}\right)\right|^2\left|\hat{\phi}\left(\omega+\frac{k}{2}\right)\right|^2=\sum_{\substack{k=2m\\m=-\infty}}^{\infty}|M(\omega+m)|^2|\hat{\phi}(\omega+m)|^2$$

$$+\sum_{\substack{k=2m+1\\m=-\infty}}^{\infty}\left|M\left(\omega+m+\frac{1}{2}\right)\right|^2\left|\hat{\phi}\left(\omega+m+\frac{1}{2}\right)\right|^2, \qquad (10.3.19)$$

and from (10.3.14) once more, along with the periodicity of M, we further obtain

$$|M(\omega)|^2+\left|M\left(\omega+\frac{1}{2}\right)\right|^2=1 \qquad \text{a.e.}$$

∎

Lemma 10.3.3. The Fourier transform $\hat{f}$ of any $f\in W_0$ can be factored out as

$$\hat{f}(\omega)=\lambda_f(\omega)\hat{\psi}(\omega), \qquad (10.3.20)$$

where λ_f is a periodic function with period 1 and $\hat{\psi}$ is independent of f. Moreover, $\lambda_f\in L^2[0,1]$ and

$$\|f\|_{L^2}=\|\lambda_f\|_{L^2(I)}, \qquad (10.3.21)$$

where $I=[0,1]$.

Proof. Let $f\in W_0$; hence $f\in V_{-1}$ and $f\perp V_0$. Since $f\in V_{-1}$, we have as in (10.3.17)

$$f(x)=\sqrt{2}\sum_{n=-\infty}^{\infty}f_n\phi(2x-n), \quad \text{with} \quad \sum_{n=-\infty}^{\infty}|f_n|^2<\infty, \qquad (10.3.22)$$

where $f_n = \langle f, \phi_{-1,n}\rangle$. By repeating the same argument used to derive (10.3.15), we obtain

$$\hat{f}(\omega) = M_f\left(\frac{\omega}{2}\right)\hat{\phi}\left(\frac{\omega}{2}\right), \tag{10.3.23}$$

in which

$$M_f(\omega) = \frac{1}{\sqrt{2}}\sum_{n=-\infty}^{\infty} f_n e^{2\pi i n\omega}. \tag{10.3.24}$$

$M_f(\omega)$ is readily seen to be a periodic function with period 1 that belongs to $L^2[0,1]$ since $\sum_{n=-\infty}^{\infty} |f_n|^2 < \infty$. And since $f \perp V_0$, we have as in Lemma 10.3.1

$$0 = \langle f, \phi_{0,n}\rangle = \langle \hat{f}, \hat{\phi}_{0,n}\rangle = \int_{-\infty}^{\infty} \hat{f}(\omega)\overline{\hat{\phi}(\omega)}\, e^{-2\pi i n\omega} d\omega$$

$$= \int_0^1 e^{-2\pi i n\omega}\left(\sum_{k=-\infty}^{\infty} \hat{f}(\omega+k)\overline{\hat{\phi}(\omega+k)}\right) d\omega,$$

hence

$$\sum_{k=-\infty}^{\infty} \hat{f}(\omega+k)\overline{\hat{\phi}(\omega+k)} = 0 \qquad \text{a.e.}, \tag{10.3.25}$$

where the series converges absolutely and in $L^1[0,1]$ since $\hat{f}, \hat{\phi} \in L^2(\Re)$. Substituting (10.3.15) and (10.3.23) into (10.3.25) and using (10.3.14), together with the same procedure as in (10.3.19), we obtain

$$M_f(\omega)\overline{M(\omega)} + M_f\left(\omega+\frac{1}{2}\right)\overline{M\left(\omega+\frac{1}{2}\right)} = 0 \qquad \text{a.e.} \tag{10.3.26}$$

In virtue of (10.3.16), $M(\omega)$ and $M\left(\omega+\frac{1}{2}\right)$ cannot vanish simultaneously on a set of positive measure; thus we can set

$$M_f(\omega) = K_f(\omega)\overline{M\left(\omega+\frac{1}{2}\right)}, \tag{10.3.27}$$

where

$$K_f(\omega) + K_f\left(\omega+\frac{1}{2}\right) = 0 \qquad \text{a.e.},$$

which in turn leads to

$$K_f(\omega) = e^{2\pi i\omega}\lambda_f(2\omega), \tag{10.3.28}$$

where

$$\lambda_f(2\omega) = \lambda_f(2\omega + 1)\,. \tag{10.3.29}$$

Thus, putting together (10.3.27) and (10.3.28) yields

$$M_f(\omega) = e^{2\pi i\omega}\lambda_f(2\omega)\,\overline{M\left(\omega + \frac{1}{2}\right)}\,. \tag{10.3.30}$$

By substituting this into (10.3.23), we obtain

$$\hat{f}(\omega) = e^{i\pi\omega}\,\overline{M\left(\frac{\omega}{2} + \frac{1}{2}\right)}\,\hat{\phi}\left(\frac{\omega}{2}\right)\lambda_f(\omega)\,, \tag{10.3.31}$$

and since the only term on the right-hand side of (10.3.31) that depends on f is $\lambda_f(\omega)$, we may put

$$\hat{f}(\omega) = \lambda_f(\omega)\,\hat{\psi}(\omega)\,, \tag{10.3.32}$$

where

$$\hat{\psi}(\omega) = e^{i\pi\omega}\,\overline{M\left(\frac{\omega}{2} + \frac{1}{2}\right)}\,\hat{\phi}\left(\frac{\omega}{2}\right). \tag{10.3.33}$$

Finally, from (10.3.30) and (10.3.16)

$$\begin{aligned}
\|M_f\|^2_{L^2(I)} &= \int_0^1 |\lambda_f(2\omega)|^2 \left|M\left(\omega + \frac{1}{2}\right)\right|^2 d\omega \\
&= \int_0^{\frac{1}{2}} |\lambda_f(2\omega)|^2 \left|M\left(\omega + \frac{1}{2}\right)\right|^2 d\omega + \int_{\frac{1}{2}}^1 |\lambda_f(2\omega)|^2 \left|M\left(\omega + \frac{1}{2}\right)\right|^2 d\omega \\
&= \int_0^{\frac{1}{2}} |\lambda_f(2\omega)|^2 \left(\left|M\left(\omega + \frac{1}{2}\right)\right|^2 + |M(\omega)|^2\right) d\omega = \int_0^{\frac{1}{2}} |\lambda_f(2\omega)|^2 d\omega \\
&= \frac{1}{2}\int_0^1 |\lambda_f(\omega)|^2 d\omega = \frac{1}{2}\|\lambda_f\|^2_{L^2(I)}\,,
\end{aligned}$$

which implies that λ_f is in $L^2[0,1]$ since M_f is. On the other hand, by (10.3.24) and (10.3.22) we have

$$\|M_f\|^2_{L^2(I)} = \frac{1}{2}\sum_{n=-\infty}^{\infty}|f_n|^2 = \frac{1}{2}\|f\|^2_{L^2(I)},$$

hence $\|f\|_{L^2} = \|\lambda_f\|_{L^2(I)}$. ∎

We can now show how a multiresolution analysis generates an orthonormal wavelet basis. In fact, we shall show that the function ψ, whose Fourier transform is defined in Lemma 10.3.3, is the mother wavelet; see equation (10.3.33).

THEOREM 10.17

For a given multiresolution analysis there exists an orthonormal wavelet basis $\{\psi_{m,n}(x) = 2^{-m/2}\psi(2^{-m}x - n)\}_{m,n=-\infty}^{\infty}$ for $L^2(\mathfrak{R})$.

Proof. As indicated in the paragraph preceding Lemma 10.3.1, it suffices to show that there exists a function $\psi \in W_0$ such that $\{\psi(x-n)\}_{n=-\infty}^{\infty}$ is an orthonormal basis for W_0. We set $\psi_{0,0}(x) = \psi(x)$ and $\psi_{0,n}(x) = \psi(x-n)$, where ψ is as defined by (10.3.33).

Let $f \in W_0$, then by Lemma 10.3.3

$$\hat{f}(\omega) = \lambda_f(\omega)\hat{\psi}(\omega), \tag{10.3.34}$$

where λ_f is a periodic function with period 1 that belongs to $L^2[0,1]$, which consequently can be written as

$$\lambda_f(\omega) = \sum_{n=-\infty}^{\infty}\hat{\lambda}_f(n)e^{2\pi i n\omega}, \tag{10.3.35}$$

with $\hat{\lambda}_f(n) = \langle\lambda_f(\omega), e^{2\pi i n\omega}\rangle$ and $\|\lambda_f\|^2_{L^2[0,1]} = \sum_{n=-\infty}^{\infty}|\hat{\lambda}_f(n)|^2 < \infty$.

By substituting (10.3.35) into (10.3.34) and then taking the inverse Fourier transform of (10.3.34), we obtain

$$f(x) = \sum_{n=-\infty}^{\infty}\hat{\lambda}_f(n)\psi(x-n). \tag{10.3.36}$$

To show that $\{\psi(x-n)\}_{n=-\infty}^{\infty}$ is a basis for W_0, we need to show that (i) $\psi \in W_0$; (ii) $\{\psi(x-n)\}_{n=-\infty}^{\infty}$ are orthonormal; and (iii) the series (10.3.36) converges and $\|f\|^2 = \sum_{n=-\infty}^{\infty}|\hat{\lambda}_f(n)|^2$.

We start with (iii) since it is the easiest. (iii) follows from (10.3.21) and (10.3.35). To prove (ii), it suffices, by Lemma 10.3.1, to show that

$$\sum_{k=-\infty}^{\infty} |\hat{\psi}(\omega+k)|^2 = 1 \qquad \text{a.e.}$$

But in view of (10.3.33), (10.3.16), (10.3.14) and the periodicity of $M(\omega)$,

$$\sum_{k=-\infty}^{\infty} |\hat{\psi}(\omega+k)|^2 = \sum_{k=-\infty}^{\infty} \left|M\left(\frac{\omega}{2}+\frac{k}{2}+\frac{1}{2}\right)\right|^2 \left|\hat{\phi}\left(\frac{\omega}{2}+\frac{k}{2}\right)\right|^2$$

$$= \sum_{m=-\infty}^{\infty} \left|M\left(\frac{\omega}{2}+m+\frac{1}{2}\right)\right|^2 \left|\hat{\phi}\left(\frac{\omega}{2}+m\right)\right|^2 + \sum_{m=-\infty}^{\infty} \left|M\left(\frac{\omega}{2}+m+1\right)\right|^2 \left|\hat{\phi}\left(\frac{\omega}{2}+m+1\right)\right|^2$$

$$= \left|M\left(\frac{\omega}{2}+\frac{1}{2}\right)\right|^2 \sum_{m=-\infty}^{\infty} \left|\hat{\phi}\left(\frac{\omega}{2}+m\right)\right|^2 + \left|M\left(\frac{\omega}{2}\right)\right|^2 \sum_{m=-\infty}^{\infty} \left|\hat{\phi}\left(\frac{\omega}{2}+m+1\right)\right|^2$$

$$= \left|M\left(\frac{\omega}{2}+\frac{1}{2}\right)\right|^2 + \left|M\left(\frac{\omega}{2}\right)\right|^2 = 1 .$$

Thus,

$$\langle \psi_{0,0}, \psi_{0,n} \rangle = \int_{-\infty}^{\infty} \psi(x)\,\psi(x-n)\,dx = \int_0^1 e^{-2\pi i n\omega} d\omega = \delta_{0,n} . \qquad (10.3.37)$$

As for (i), recall from (10.3.33) that $\hat{\psi}$ can be written in the form

$$\hat{\psi}(\omega) = G(\omega)\,\hat{\phi}\left(\frac{\omega}{2}\right), \qquad (10.3.38)$$

where $G(\omega) = e^{i\pi\omega}\overline{M\left(\frac{\omega}{2}+\frac{1}{2}\right)}$ is a periodic function in $L^2[0,2]$ with period 2. Hence, we may write

$$G(\omega) = \frac{1}{\sqrt{2}} \sum_{n=-\infty}^{\infty} g_n e^{\pi i n\omega} \quad \text{with} \quad \|G\|^2 = \sum_{n=-\infty}^{\infty} |g_n|^2 < \infty ,$$

and by substituting this into (10.3.38) and taking the inverse Fourier transform, we obtain

$$\psi(x) = \sqrt{2} \sum_{n=-\infty}^{\infty} g_n\,\phi(2x-n) ,$$

hence $\psi \in V_{-1}$. Finally, we show that $\psi \perp V_0$ and to this end, it suffices to show that $\langle \psi, \phi_{0,n} \rangle = 0$ for all n, where $\phi_{0,n}(x) = \phi(x-n)$. As in the proof of Lemma 10.3.1, we have

$$\langle \psi, \phi_{0,n} \rangle = \langle \hat{\psi}, \hat{\phi}_{0,n} \rangle = \int_0^1 e^{-2\pi i n \omega} \left(\sum_{k=-\infty}^{\infty} \hat{\psi}(\omega + k) \overline{\hat{\phi}(\omega + k)} \right) d\omega . \qquad (10.3.39)$$

Using (10.3.33), (10.3.15) and the periodicity of $M(\omega)$ yields

$$\sum_{k=-\infty}^{\infty} \hat{\psi}(\omega + k) \overline{\hat{\phi}(\omega + k)} = \sum_{k=-\infty}^{\infty} e^{i\pi(\omega + k)} \overline{M\left(\frac{\omega}{2} + \frac{k}{2} + \frac{1}{2}\right)} \; \overline{M\left(\frac{\omega}{2} + \frac{k}{2}\right)} \left| \hat{\phi}\left(\frac{\omega}{2} + \frac{k}{2}\right) \right|^2$$

$$= \sum_{\substack{m=-\infty \\ k=2m}}^{\infty} e^{i\pi\omega} \overline{M\left(\frac{\omega}{2} + m + \frac{1}{2}\right)} \; \overline{M\left(\frac{\omega}{2} + m\right)} \left| \hat{\phi}\left(\frac{\omega}{2} + m\right) \right|^2$$

$$- \sum_{\substack{m=-\infty \\ k=2m+1}}^{\infty} e^{i\pi\omega} \overline{M\left(\frac{\omega}{2} + m\right)} \; \overline{M\left(\frac{\omega}{2} + m + \frac{1}{2}\right)} \left| \hat{\phi}\left(\frac{\omega}{2} + m + \frac{1}{2}\right) \right|^2$$

$$= e^{i\pi\omega} \overline{M\left(\frac{\omega}{2}\right)} \; \overline{M\left(\frac{\omega}{2} + \frac{1}{2}\right)} \left(\sum_{m=-\infty}^{\infty} \left| \hat{\phi}\left(\frac{\omega}{2} + m\right) \right|^2 - \sum_{m=-\infty}^{\infty} \left| \hat{\phi}\left(\frac{\omega}{2} + \frac{1}{2} + m\right) \right|^2 \right) = 0,$$

where the last equality follows from (10.3.14). Therefore, (10.3.39) now yields

$$\langle \psi, \phi_{0,n} \rangle = 0 \quad \text{for all } n .$$

∎

Corollary 10.17.1. The mother wavelet ψ may be given explicitly by

$$\psi(x) = \sum_{n=-\infty}^{\infty} (-1)^{1-n} \alpha_{1-n} \phi_{-1,n}(x), \qquad (10.3.40)$$

where the coefficients α_n are given by

$$\phi(x) = \sum_{n=-\infty}^{\infty} \alpha_n \phi_{-1,n}(x),$$

with

$$\alpha_n = \langle \phi, \phi_{-1,n} \rangle \quad \text{and} \quad \sum_{n=-\infty}^{\infty} |\alpha_n|^2 < \infty .$$

Proof. In view of (10.3.33) and (10.3.18),

$$\hat{\psi}(\omega) = e^{i\pi\omega}\left(\frac{1}{\sqrt{2}}\sum_{n=-\infty}^{\infty} \alpha_n\, e^{-2\pi i n(\omega+1)/2}\right)\hat{\phi}\left(\frac{\omega}{2}\right),$$

which, by taking the inverse Fourier transform, is equivalent to

$$\psi(x) = \sqrt{2}\sum_{n=-\infty}^{\infty}(-1)^{1-n}\alpha_{1-n}\phi(2x-n) = \sum_{n=-\infty}^{\infty}(-1)^{1-n}\alpha_{1-n}\phi_{-1,n}(x).$$

∎

The mother wavelet ψ associated with a given multiresolution analysis is not unique as can be seen from the solution of (10.3.26). For example, we may take $K_f(\omega) = e^{2\pi i\omega}e^{2\pi i p(\omega)}\lambda_f(2\omega)$, where $p(\omega)$ is a real-valued function with $p\left(\omega+\frac{1}{2}\right) = p(\omega)$. This will lead to a new mother wavelet $\tilde{\psi}$ that satisfies

$$\hat{\tilde{\psi}}(\omega) = e^{2\pi i p(\omega/2)}\hat{\psi}(\omega).$$

In particular if $p(\omega) = p/2$, where p is an odd integer, then

$$\tilde{\psi}(x) = \sum_{n=-\infty}^{\infty}(-1)^n\alpha_{1-n}\phi_{-1,n}(x). \tag{10.3.41}$$

It is worth noting that the converse of Theorem 10.17 is not true, in the sense that there exists an orthonormal wavelet basis for $L^2(\mathfrak{R})$ that is not generated from any multiresolution analysis; see [6, p. 136].

Corollary 10.17.1 gives a recipe for reconstructing the mother wavelet ψ; hence, constructing an orthonormal wavelet basis for $L^2(\mathfrak{R})$. Implementing the procedure described in Corollary 10.17.1 is rather easy if the series (10.3.17) defining $\phi(x)$ in terms of $\sqrt{2}\,\phi(2x-n)$ has only a finite number of non-zero terms. For example, if we take $\phi(x) = \chi_{[0,1)}(x)$, then it is easy to see that

$$\alpha_n = \langle\phi, \phi_{-1,n}\rangle = \sqrt{2}\int_0^1 \phi(x)\phi(2x-n)\,dx = \begin{cases}\dfrac{1}{\sqrt{2}}, & n = 0, 1\\ 0\ , & \text{otherwise};\end{cases}$$

hence

$$\psi(x) = \frac{1}{\sqrt{2}}(\phi_{-1,1}(x) - \phi_{-1,0}(x)) = \begin{cases}-1, & 0 \le x < \dfrac{1}{2}\\ 1, & \dfrac{1}{2} \le x < 1\\ 0\ , & \text{otherwise},\end{cases}$$

which is the mother wavelet for the Haar orthonormal basis. True but more difficult to prove is that Meyer's wavelets can also be generated from a multiresolution analysis; see [6, p. 137] for details. For other related work on multiresolution analysis, we refer the reader to [22 and 23]; see also [8].

The assumption that $\{\phi_{0,n}(x) = \phi(x-n)\}_{n=-\infty}^{\infty}$ is an orthonormal basis for V_0 played a vital role in the construction of an orthonormal wavelet basis for $L^2(\mathfrak{R})$ in Theorem 10.17. Nevertheless, it turns out that the conclusion of that theorem is still true under a less restrictive condition. If we require that $\{\phi(x-n)\}_{n=-\infty}^{\infty}$ to be only a Riesz basis for V_0 and not necessarily an orthonormal basis, we still can construct an orthonormal wavelet basis for $L^2(\mathfrak{R})$.

First, let us recall that $\{\phi(x-n)\}_{n=-\infty}^{\infty}$ is a Riesz basis for V_0 if for any sequence $c = \{c_n\}_{n=-\infty}^{\infty}$ such that

$$\|c\|^2 = \sum_{n=-\infty}^{\infty} |c_n|^2 < \infty,$$

we have

$$A\|c\|^2 \le \left\| \sum_{k=-\infty}^{\infty} c_k \phi(\cdot - k) \right\|^2 \le B\|c\|^2, \tag{10.3.42}$$

where $A, B > 0$ are independent of c. Now

$$\begin{aligned} \left\| \sum_{k=-\infty}^{\infty} c_k \phi(\cdot - k) \right\|^2 &= \int_{-\infty}^{\infty} \left| \left(\sum_{k=-\infty}^{\infty} c_k e^{2\pi i k \omega} \right) \hat{\phi}(\omega) \right|^2 d\omega \\ &= \int_0^1 \left| \sum_{k=-\infty}^{\infty} c_k e^{2\pi i k \omega} \right|^2 \Phi(\omega)\, d\omega, \end{aligned} \tag{10.3.43}$$

where

$$\Phi(\omega) = \sum_{m=-\infty}^{\infty} |\hat{\phi}(\omega + m)|^2.$$

But since

$$\int_0^1 \left| \sum_{k=-\infty}^{\infty} c_k e^{2\pi i h \omega} \right|^2 d\omega = \sum_{k=-\infty}^{\infty} |c_k|^2 = \|c\|^2,$$

it follows from (10.3.43) that (10.3.42) is equivalent to

$$0 < A \le \Phi(\omega) \le B < \infty \qquad \text{a.e.}$$

Therefore, we can define

$$\hat{\tilde{\phi}}(\omega) = \frac{\hat{\phi}(\omega)}{\sqrt{\Phi(\omega)}}, \tag{10.3.44}$$

which yields that

$$\sum_{k=-\infty}^{\infty} |\hat{\tilde{\phi}}(\omega + k)|^2 = 1 .$$

As seen in Lemma 10.3.1, this last equation implies that $\{\tilde{\phi}(x-n)\}_{n=-\infty}^{\infty}$ is an orthonormal set. Let $\tilde{V}_0$ be the vector space spanned by $\{\tilde{\phi}(x-n)\}_{n=-\infty}^{\infty}$. Hence, for any $f \in \tilde{V}_0$,

$$f(x) = \sum_{n=-\infty}^{\infty} \tilde{f}_n \tilde{\phi}(x-n),$$

with

$$\tilde{f}_n = \langle f, \tilde{\phi}_{0,n} \rangle \quad \text{and} \quad \sum_{n=-\infty}^{\infty} |\tilde{f}_n|^2 < \infty,$$

which is equivalent to $\hat{f}(\omega) = \tilde{v}(\omega)\hat{\tilde{\phi}}(\omega)$, where $\tilde{v}(\omega)$ is a periodic function with period 1 and $\tilde{v} \in L^2[0,1]$. But since $\Phi(\omega)$ is also periodic with period 1, we have from (10.3.44), $\hat{f}(\omega) = v(\omega)\hat{\phi}(\omega)$ for some periodic function $v(\omega)$ with period 1 and $v \in L^2[0,1]$. This is equivalent to

$$f(x) = \sum_{n=-\infty}^{\infty} f_n \phi(x-n),$$

with $f_n = \langle f, \phi_{0,n} \rangle$ and $\sum_{n=-\infty}^{\infty} |f_n|^2 < \infty$; thus $f \in V_0$ (the span of $\{\phi(x-n)\}_{n=-\infty}^{\infty}$). Since the argument is reversible, we have that $\tilde{V}_0 = V_0$ and thereby $\tilde{V}_j = V_j$ for $j = 0, \pm 1, \pm 2, \ldots$. That is the multiresolution analysis is preserved under the orthogonalization process given by (10.3.44).

The mother wavelet $\tilde{\psi}$ associated with the scaling function $\tilde{\phi}$ is given, as in (10.3.33), by

$$\hat{\tilde{\psi}}(\omega) = e^{i\pi\omega}\, \overline{\tilde{M}\left(\frac{\omega}{2} + \frac{1}{2}\right)}\, \hat{\tilde{\phi}}\left(\frac{\omega}{2}\right), \tag{10.3.45}$$

where $\tilde{M}$ is related to $\tilde{\phi}$ in the same fashion as M is related to ϕ. Moreover, since

$$\hat{\tilde{\phi}}(\omega) = \tilde{M}\left(\frac{\omega}{2}\right)\hat{\tilde{\phi}}\left(\frac{\omega}{2}\right),$$

and

$$\hat{\phi}(\omega) = M\left(\frac{\omega}{2}\right)\hat{\phi}\left(\frac{\omega}{2}\right),$$

then in view of (10.3.44)

$$\tilde{M}\left(\frac{\omega}{2}\right) = \sqrt{\frac{\Phi\left(\frac{\omega}{2}\right)}{\Phi(\omega)}}\, M\left(\frac{\omega}{2}\right); \tag{10.3.46}$$

hence by combining (10.3.45) and (10.3.46), we obtain

$$\tilde{\psi}(\omega) = e^{i\pi\omega}\sqrt{\frac{\Phi\left(\frac{\omega}{2}+\frac{1}{2}\right)}{\Phi(\omega+1)\,\Phi\left(\frac{\omega}{2}\right)}}\, M\left(\frac{\omega}{2}+\frac{1}{2}\right)\hat{\phi}\left(\frac{\omega}{2}\right). \tag{10.3.47}$$

Definition 10.3.2.

i) A function $f(x)$ is said to be regular if there exists a nonnegative integer N such that f itself and its first N derivatives are rapidly decreasing, i.e., for all $p = 0, 1, 2, \dots$, and $0 \le n \le N$

$$\| (1 + |x|)^p f^{(n)}(x) \|_\infty < \infty .$$

The largest such N is called the degree of regularity of f and f is said to be N-regular.

ii) A multiresolution analysis is said to be N-regular if its scaling function ϕ is N-regular.

If $N = \infty$, then f is in the Schwartz space of rapidly decreasing functions. In a regular multiresolution analysis, the regularity of ϕ will, of course, be inherited by any function f in V_j. If ϕ is N-regular and $\{\phi(x-n)\}_{n=-\infty}^{\infty}$ is a Riesz basis for V_0, then the function $\tilde{\phi}$ obtained in (10.3.44) is also N-regular. To this end, let us first note that by using the Poisson summation formula we can verify that

$$\Phi(\omega) = \sum_{m=-\infty}^{\infty} |\hat{\phi}(\omega+m)|^2 = \sum_{n=-\infty}^{\infty} c(n)\, e^{2\pi i n\omega},$$

where

$$c(n) = \int_0^1 \Phi(\omega)\, e^{-2\pi i n\omega}\, d\omega = \int_{-\infty}^{\infty} \overline{\phi}(x)\,\phi(x+n)\, dx .$$

In view of the fact that $\phi(x)$ is regular, we can easily show that $\{c(n)\}_{n=-\infty}^{\infty}$ is a rapidly decreasing sequence; thus $\Phi(\omega)$ is a C^{∞}-function. Since $\{\phi(x-n)\}_{n=-\infty}^{\infty}$ is a Riesz basis for V_0, there exist two constants A and B such that $0 < A \le \Phi(\omega) \le B < \infty$, and therefore $1/\sqrt{\Phi(\omega)}$ is a C^{∞}-function. From the observations that $\phi(x)$ is rapidly decreasing if and only if $\hat{\phi}(\omega)$ is infinitely differentiable and that the Fourier transform of $\phi^{(k)}(x)$ is $(-2\pi i\,\omega)^k\hat{\phi}(\omega)$, it now follows from (10.3.44) that $\tilde{\phi}(x)$ is N-regular if $\phi(x)$ is. Moreover, if we set

$$\frac{1}{\sqrt{\Phi(\omega)}} = \sum_{n=-\infty}^{\infty} \alpha_n e^{2\pi i n \omega},$$

where $\{\alpha_n\}_{n=-\infty}^{\infty}$ is a rapidly decreasing sequence, we obtain

$$\tilde{\phi}(x) = \sum_{n=-\infty}^{\infty} \alpha_n \phi(x-n).$$

Now suppose we do not have a multiresolution analysis and all we have is just a function ϕ that satisfies

$$\phi(x) = \sum_{n=-\infty}^{\infty} c_n \phi(2x-n), \qquad (10.3.48)$$

with $\sum_{n=-\infty}^{\infty} |c_n|^2 < \infty$ and

$$0 < A \le \sum_{m=-\infty}^{\infty} |\hat{\phi}(\omega+m)|^2 \le B < \infty \qquad \text{a.e.} \qquad (10.3.49)$$

For any integer m, let us define

$$V_m = \left\{ f : f(x) = \sum_{n=-\infty}^{\infty} f_n \phi_{m,n}(x),\ \sum_{n=-\infty}^{\infty} |f_n|^2 < \infty,\ \text{where } \phi_{m,n}(x) = 2^{-m/2}\phi(2^{-m}x-n) \right\}.$$

Condition (10.3.49) is sufficient to ensure that $\{\phi(x-n)\}_{n=-\infty}^{\infty}$ is a Riesz basis for V_0 and by (10.3.48) we conclude that $\{\phi_{m,n}(x)\}$ is a Riesz basis for V_m. It can be easily shown that $\{V_m\}_{m=-\infty}^{\infty}$ satisfies conditions (1) and (4) of definition 10.3.1, and by applying the orthogonalization process used in (10.3.44), we obtain a function $\tilde{\phi}$ that satisfies condition (5) of the same definition. With some extra work it can be shown that condition (3) is also satisfied ([6, p. 141], but to satisfy condition (2) and obtain a multiresolution analysis, further restrictions on ϕ must be imposed. It can be shown [6, p. 142] that $\hat{\phi}(\omega)$ being bounded for all ω and continuous near $\omega = 0$, with $\hat{\phi}(0) \neq 0$

is sufficient for condition (2) to hold. As a special case, if $\phi \in L^1$ and $\int_{-\infty}^{\infty} \phi(x)\,dx \neq 0$, then we have a multiresolution analysis. The condition that $\hat{\phi}$ be continuous near $\omega = 0$ is sufficient but not necessary. However, if $\hat{\phi}$ is assumed to be bounded and continuous near $\omega = 0$, then the condition $\hat{\phi}(0) \neq 0$ is necessary.

If, in addition to (10.3.48) and (10.3.49), ϕ is assumed to be regular with $\int_{-\infty}^{\infty} \phi(x)\,dx \neq 0$, then it can be easily shown that condition (2) is satisfied and that the multiresolution analysis is regular.

Example 1. Consider the piecewise linear spline

$$\phi(x) = \begin{cases} 1 - |x|, & 0 \le |x| \le 1 \\ 0, & \text{otherwise}. \end{cases}$$

It is not hard to show that ϕ satisfies (10.3.48), namely,

$$\phi(x) = \frac{1}{2}\phi(2x+1) + \phi(2x) + \frac{1}{2}\phi(2x-1); \tag{10.3.50}$$

moreover,

$$\phi(x) = (g * g)(x) = \int_{-\infty}^{\infty} g(x-t)g(t)\,dt,$$

where

$$g(x) = \chi_{\left[-\frac{1}{2},\frac{1}{2}\right]}(x);$$

hence

$$\hat{\phi}(\omega) = \left(\frac{\sin \pi\omega}{\pi\omega}\right)^2. \tag{10.3.51}$$

To calculate $\sum_{m=-\infty}^{\infty} |\hat{\phi}(\omega+m)|^2$, we make use of the relation

$$\csc^2 \pi\omega = \frac{1}{\pi^2} \sum_{m=-\infty}^{\infty} \frac{1}{(\omega+m)^2}.$$

Differentiating this relation twice yields

$$\frac{\pi^2}{3}(1+2\cos^2\pi\omega)=\frac{\sin^4\pi\omega}{\pi^2}\sum_{m=-\infty}^{\infty}\frac{1}{(\omega+m)^4}. \qquad (10.3.52)$$

Therefore,

$$\Phi(\omega)=\sum_{m=-\infty}^{\infty}|\hat{\phi}(\omega+m)|^2=\sum_{m=-\infty}^{\infty}\left(\frac{\sin\pi(\omega+m)}{\pi(\omega+m)}\right)^4$$

$$=\left(\frac{\sin\pi\omega}{\pi}\right)^4\sum_{m=-\infty}^{\infty}\frac{1}{(\omega+m)^4}=\frac{1}{3}(1+2\cos^2\pi\omega). \qquad (10.3.53)$$

Both (10.3.48) and (10.3.49) are satisfied, and clearly $\phi\in L^1$ with $\int_{-\infty}^{\infty}\phi(x)\,dx=1\neq 0$; hence we have a multiresolution analysis with $\{\tilde{\phi}(x-n)\}_{n=-\infty}^{\infty}$ as an orthonormal basis for V_0, where $\tilde{\phi}(x)$ is defined by

$$\hat{\tilde{\phi}}(\omega)=\frac{\sqrt{3}\sin^2\pi\omega}{\pi^2\omega^2\sqrt{1+2\cos^2\pi\omega}}.$$

Here we have used (10.3.44), (10.3.51) and (10.3.53). From the relation

$$\hat{\phi}(\omega)=M\left(\frac{\omega}{2}\right)\hat{\phi}\left(\frac{\omega}{2}\right),$$

and (10.3.51), we obtain

$$M\left(\frac{\omega}{2}\right)=\cos^2\left(\frac{\pi\omega}{2}\right).$$

Combining this, (10.3.53) and (10.3.46) gives

$$\tilde{M}\left(\frac{\omega}{2}\right)=\sqrt{\frac{(1+2\cos^2(\pi\omega/2))}{(1+2\cos^2\pi\omega)}}\cos^2(\pi\omega/2),$$

and hence by (10.3.47)

$$\hat{\psi}(\omega)=e^{i\pi\omega}\sqrt{\frac{1+2\sin^2(\pi\omega/2)}{(1+2\cos^2\pi\omega)(1+2\cos^2(\pi\omega/2))}}\;\frac{4\sqrt{3}\sin^4(\pi\omega/2)}{\pi^2\omega^2}.$$

More general examples using B-splines of degree N ($N\geq 1$) can be found in [5]. Examples of orthonormal wavelet bases generated by compactly supported mother wavelets, such as Daubechies' wavelets, can be found in [5 and 8].

10.3.D Wavelets and Sampling

Multiresolution analyses provide a new way of looking at the WSK sampling theorem. Recall that if $f \in B_\sigma^2$, then

$$f(t) = \sum_{n=-\infty}^{\infty} f(t_n) \frac{\sin \sigma(t - t_n)}{\sigma(t - t_n)}, \quad t_n = \frac{n\pi}{\sigma}.$$

If we set $\sigma = \pi$ and

$$\phi(t) = \frac{\sin \pi t}{\pi t}; \quad \text{hence} \quad \hat{\phi}(\omega) = \chi_{[-1/2, 1/2)},$$

we then have

$$f(t) = \sum_{n=-\infty}^{\infty} f(n) \phi(t - n),$$

which is reminiscent of (10.3.36).

Although by now the reader is well aware that $\{\phi(t-n)\}_{n=-\infty}^{\infty}$ is an orthonormal sequence, we shall now give a new and elegant proof of this fact. It is obvious that $\sum_{m=-\infty}^{\infty} |\hat{\phi}(\omega + m)|^2 = 1$; hence by Lemma 10.3.1, $\{\phi(t-n)\}_{n=-\infty}^{\infty}$ is orthonormal! This is the proof that we alluded to in Section 10.1; see the paragraph following equation (10.1.12).

Let us as usual set $\phi_{m,n}(t) = 2^{-m/2}\phi(2^{-m}t - n)$ and note that

$$\hat{\phi}_{m,n}(\omega) = 2^{m/2} e^{2^{m+1}\pi i \omega n} \hat{\phi}(2^m \omega) = 2^{m/2} e^{2^{m+1}\pi i \omega n} \chi_{[-1/2^{m+1}, 1/2^{m+1})}(\omega).$$

Since $\hat{\phi}(\omega) = \chi_{[-\frac{1}{2}, \frac{1}{2})}(\omega)$ and $\hat{\phi}\left(\frac{\omega}{2}\right) = \chi_{[-1,1)}(\omega)$, it is easy to see that equation (10.3.15) is satisfied with $M\left(\frac{\omega}{2}\right)$ being $\chi_{[-\frac{1}{2}, \frac{1}{2})}(\omega)$ extended as a periodic function with period 2. Therefore, we can verify that

$$M\left(\frac{\omega}{2}\right) = \sum_{n=-\infty}^{\infty} \frac{\sin(n\pi/2)}{(n\pi)} e^{\pi i n \omega},$$

which leads to

$$\phi(t) = \sum_{n=-\infty}^{\infty} \frac{\sin(n\pi/2)}{(n\pi/2)} \phi(2t - n).$$

It is now easy to see that conditions (10.3.48) and (10.3.49) are satisfied, and $\hat{\phi}(\omega)$ is bounded and continuous near $\omega = 0$ with $\hat{\phi}(0) \neq 0$; hence a multiresolution analysis for $L^2(\Re)$ can be constructed as follows. Let

$$V_0 = \left\{ f \in L^2(\Re) : f(t) = \sum_{n=-\infty}^{\infty} f_n \phi(t-n),\ \sum_{n=-\infty}^{\infty} |f_n|^2 < \infty \right\},$$

and more generally

$$V_m = \left\{ f \in L^2(\Re) : f(t) = \sum_{n=-\infty}^{\infty} f_n \phi_{m,n}(t),\ \sum_{n=-\infty}^{\infty} |f_n|^2 < \infty \right\}.$$

Since our definition of the Fourier transform in this chapter as

$$\hat{f}(\omega) = \int_{-\infty}^{\infty} f(t)\, e^{2\pi i t \omega} dt ,$$

is different from the one we have used in the rest of the book, our definition of the band-width of a function is also different from the one we have used before; however, the two definitions differ only by a scaling factor. With this in mind, we can verify that V_m comprises functions band limited to $[-1/2^{m+1}, 1/2^{m+1}]$ and that

$$\ldots \subset V_1 \subset V_0 \subset V_{-1} \subset \ldots ,$$

with

$$\bigcap_{n=-\infty}^{\infty} V_m = \{0\}.$$

If $f \in L^2(\Re)$, then we can find $g \in L^2(\Re)$ such that $\hat{g}$ has support in $[-1/2^{m+1}, 1/2^{m+1}]$ for some m and $\|\hat{f} - \hat{g}\| < \varepsilon$; thus $\|f - g\| < \varepsilon$, implying that $\bigcup_{m=-\infty}^{\infty} V_m$ is dense in $L^2(\Re)$. Therefore, $\{V_m\}_{m=-\infty}^{\infty}$ and $\phi(t)$ form a multiresolution analysis in which the WSK sampling theorem provides sampling expansions for the elements of each subspace V_m. A mother wavelet associated with this multiresolution analysis is sometimes called the Shannon wavelet.

In virtue of (10.3.40) or (10.3.41), a mother wavelet ψ can be given by

$$\psi(t) = \sum_{n=-\infty}^{\infty} \frac{(-1)^{n+1} \cos(n\pi/2)}{(n-1)\pi} \phi(2t - n).$$

We can also show directly that

$$\psi(t) = \frac{(\sin 2\pi t - \sin \pi t)}{(\pi t)}$$

is a mother wavelet. For, ψ is clearly in V_{-1} and moreover $\psi(t-n)$ is orthogonal to V_0 for all n. To show this, it suffices to show that $\langle \phi_{0,m}, \psi_{0,n} \rangle = 0$ for all m, n. But this is an easy consequence of the fact that $\hat{\psi}(\omega) = \chi_{[-1,1]}(\omega) - \chi_{[-1/2,1/2]} = \chi_J(\omega)$, where

$$J = \left[-1, -\frac{1}{2}\right) \cup \left(\frac{1}{2}, 1\right],$$

because

$$\langle \phi_{0,m}, \psi_{0,n} \rangle = \langle \hat{\phi}_{0,m}, \hat{\psi}_{0,n} \rangle = \langle e^{2\pi i m \omega} \chi_{[-1/2,1/2]}, e^{2\pi i n \omega} \chi_J \rangle = 0 .$$

Thus, $\psi_{0,n} \in W_1$ for all n. Furthermore, $\{\psi_{0,n}\}_{n=-\infty}^{\infty}$ is an orthonormal basis for W_1 since

$$\langle \psi_{0,n}, \psi_{0,m} \rangle = \langle e^{2\pi i n \omega} \psi_J, e^{2\pi i m \omega} \psi_J \rangle = \delta_{m,n} .$$

Since for each fixed m, the space V_m consists of functions band-limited to $[-1/2^{m+1}, 1/2^{m+1}]$ or equivalently it consists of finite-energy signals with certain frequency content that depends on m, we may call V_m the mth frequency level of the multiresolution analysis.

We have just seen that there exists a multiresolution analysis associated with the WSK sampling theorem in which the scaling function ϕ of the multiresolution analysis is the sampling function of the WSK theorem and, in addition, each frequency level of the multiresolution analysis has a sampling function that is essentially a dilated version of ϕ, namely $\phi_{m,0}(t)$. It is worth noting that G. Walter [30] has shown some sort of a converse to this. That is, given a multiresolution analysis with scaling function ϕ satisfying some extra conditions, one can construct a function $S(t)$ that depends on ϕ so that $S_{m,0}(t)$ is the sampling function for the mth frequency level of the multiresolution analysis.

It is interesting to note that any multiresolution analysis gives rise to a sampling function that is self-similar in the sense that it satisfies (10.3.48). For, if ϕ is the scaling function of the multiresolution analysis, then define

$$S(x) = \int_{-\infty}^{\infty} \phi(t)\, \overline{\phi(t-x)}\, dt = (\phi(t) * \overline{\phi}(-t))(x) .$$

Because of the orthonormality of $\{\phi(x-k)\}_{k=-\infty}^{\infty}$, it follows that $S(x)$ is a sampling function, i.e., $S(k)=\delta_{k,0}$. Moreover, S inherits its self-similarity and regularity from ϕ. However, not every self-similar sampling function arises this way because, in view of the above definition of S, $\hat{S}(\omega)=|\hat{\phi}(\omega)|^2$, but this would imply that the Fourier transform of any sampling function is positive, which, of course, is not true. On the other hand, a sampling function S with positive Fourier transform generates an orthonormal family of functions of the form $\{\phi(x-k)\}_{k=-\infty}^{\infty}$ as we shall show shortly. In fact, sampling functions can be characterized in terms of their Fourier transforms as seen from the following theorem.

THEOREM 10.18

Let S be a sampling function such that its Fourier transform $\hat{S}(\omega)$ is in $L^1(\mathfrak{R})$. Then

$$\sum_{k=-\infty}^{\infty}\hat{S}(\omega+k)=1 \qquad \text{a.e.}$$

Proof. Let

$$S_1(\omega)=\sum_{k=-\infty}^{\infty}\hat{S}(\omega+k).$$

Since

$$\int_0^1|S_1(\omega)|\,d\omega\le\sum_{k=-\infty}^{\infty}\int_0^1|\hat{S}(\omega+k)|\,d\omega=\sum_{k=-\infty}^{\infty}\int_k^{k+1}|\hat{S}(\omega)|\,d\omega$$

$$=\int_{-\infty}^{\infty}|\hat{S}(\omega)|\,d\omega<\infty,$$

it follows that S_1 is defined almost everywhere and, in addition, it belongs to $L^1[0,1]$. In fact, we have implicitly shown that the series defining S_1 converges to it absolutely almost everywhere. Moreover, it is readily seen that S_1 is a periodic function with period 1.

For $n=0,\pm1,\pm2,\ldots$, we have

$$\delta_{n,0}=S(n)=\int_{-\infty}^{\infty}\hat{S}(\omega)e^{-2\pi in\omega}d\omega=\sum_{k=-\infty}^{\infty}\int_k^{k+1}\hat{S}(\omega)e^{-2\pi in\omega}d\omega=\int_0^1 S_1(\omega)e^{-2\pi in\omega}d\omega,$$

which, in view of the uniqueness of the Fourier coefficients of functions in $L^1[0,2\pi]$, implies that

$$S_1(\omega)=\sum_{k=-\infty}^{\infty}\hat{S}(\omega+k)=1 \qquad \text{a.e.}$$

∎

From Theorem 10.18 and Lemma 10.3.1, it follows that in order for a sampling function $S(x)$ to have the property that $\{S(x-n)\}_{n=-\infty}^{\infty}$ is an orthonormal set, it is sufficient that

$$S_1(\omega) = \sum_{k=-\infty}^{\infty} \hat{S}(\omega+k) = 1 \qquad \text{a.e.} \qquad (10.3.54)$$

and

$$S_2(\omega) = \sum_{k=-\infty}^{\infty} |\hat{S}(\omega+k)|^2 = 1 \qquad \text{a.e.} \qquad (10.3.55)$$

The simplest solution for (10.3.54) and (10.3.55) is $\hat{S}(\omega) = \chi_{[-1/2,1/2]}(\omega)$ which gives $S(x) = \sin \pi x / \pi x$. Another solution is

$$\hat{S}(\omega) = \chi_{[-1,1]}(\omega) - \chi_{[-1/2,1/2]}(\omega),$$

which leads to the Shannon wavelet

$$S(x) = \frac{\sin 2\pi x - \sin \pi x}{\pi x}.$$

Therefore, the Shannon wavelet is not only a mother wavelet that generates an orthonormal basis for $L^2(\Re)$, but also a sampling function. Solving (10.3.54) and (10.3.55) in general is not easy, unless we impose further restrictions on $\hat{S}$, such as requiring that $\hat{S}$ have compact support. If $\operatorname{supp} \hat{S} = [-1,1]$, then for any fixed ω, except for a set of measure zero, each of the series (10.3.54) and (10.3.55) will contain two terms only, and will be reduced to

$$\hat{S}(\omega - [\omega]) + \hat{S}(\omega - [\omega] - 1) = 1,$$

and

$$|\hat{S}(\omega - [\omega])|^2 + |\hat{S}(\omega - [\omega] - 1)|^2 = 1,$$

respectively, where $[\omega]$ is the greatest integer that is less than or equal to ω. Solving these two equations simultaneously is not difficult, but leads to complex-valued solutions. To obtain a real-valued sampling function $S(x)$, we require that $\hat{S}(\omega) = \overline{\hat{S}(-\omega)}$.

We now show that a sampling function S with positive Fourier transform generates an orthonormal family of functions of the form $\{\phi(x-k)\}_{k=-\infty}^{\infty}$. Let S be a sampling function such that $\hat{S} \in L^1(\Re)$ and $\hat{S}(\omega) \geq 0$. Define $\hat{\phi}(\omega) = \sqrt{\hat{S}(\omega)}$. It follows from Theorem 10.18 that

$$\sum_{k=-\infty}^{\infty} |\hat{\phi}(\omega+k)|^2 = 1 \qquad \text{a.e.};$$

hence, by Lemma 10.3.1, $\{\phi(x-k)\}_{k=-\infty}^{\infty}$ is an orthonormal family.

It is clear that the sampling function $\phi(t) = \sin \pi t/(\pi t)$ of the WSK sampling theorem plays a vital role in the construction of the associated multiresolution analysis. Now we show that this role can be played by other nice sampling functions. First, let us recall that $\phi(t) = \sin \pi t/(\pi t)$ satisfies the following three conditions:

a) $$\phi(k) = \begin{cases} 0 & \text{if} \quad k \neq 0 \\ 1 & \text{if} \quad k = 0 \, , \end{cases} \qquad \text{(sampling property)}$$

b) $$\phi(t) = \sum_{k=-\infty}^{\infty} \phi(k/2)\, \phi(2t-k) \, , \quad \text{and} \qquad \text{(self-similarity property)}$$

c) $$\| (1+|t|)^p \, \phi^{(1)}(t) \|_\infty < \infty \, , \quad p = 0, 1 \, . \qquad \text{(growth condition)}$$

So, let us assume that there exists a function g that satisfies the following conditions:

a′) $$g(k) = \begin{cases} 0 & \text{if} \quad k \neq 0 \\ 1 & \text{if} \quad k = 0 \, , \end{cases} \tag{10.3.56}$$

b′) $$g(t) = \sum_{k=-\infty}^{\infty} \alpha_k \, g(2t-k) \, , \tag{10.3.57}$$

for some sequence $\{\alpha_k\}_{k=-\infty}^{\infty}$ with $\sum_{k=-\infty}^{\infty} |\alpha_k|^2 < \infty$, and

c′) g is regular.

Because of (10.3.56), equation (10.3.57) can be written in the form

$$g(t) = \sum_{k=-\infty}^{\infty} g(k/2)\, g(2t-k) \, .$$

In order for a function g satisfying conditions (a') and (b') to generate a multiresolution analysis, it is sufficient to assume that $\hat{g}$ satisfies (10.3.49) and is bounded, continuous near $\omega = 0$ with $\hat{g}(0) \neq 0$. If, in addition, g satisfies condition (c'), the multiresolution analysis will be regular.

An example of such a function g that satisfies conditions (a'), (b') and generates a multiresolution analysis is

$$g(x) = \begin{cases} 1 - |x| & , \quad 0 \le |x| \le 1 \\ 0 & , \quad \text{otherwise} \end{cases}$$

discussed in Example 1, Section 10.3.C.

Another example which was found independently by P. Lemarie and G. Battle (see [21]) is this: let $g(x)$ be the unique cubic spline with nodes at the integers that satisfies

$$g(k) = \begin{cases} 0 & \text{if } k \neq 0 \\ 1 & \text{if } k = 0 \, . \end{cases}$$

It is known that the Fourier transform of g is given by

$$\hat{g}(\omega) = \left(\frac{\sin \pi\omega}{\pi\omega}\right)^4 \left(1 - \frac{2}{3}\sin^2 \pi\omega\right)^{-1} .$$

Clearly, $\hat{g}$ is bounded and continuous near $\omega = 0$, with $\hat{g}(0) \neq 0$. That g is self-similar is equivalent to saying that $\hat{g}(2\omega)/\hat{g}(\omega)$ is periodic with period 1, but this is evident from the definition of $\hat{g}(\omega)$. We can easily verify that

$$\Phi(\omega) = \sum_{k=-\infty}^{\infty} |\hat{g}(\omega + k)|^2 = \left(\frac{\sin \pi\omega}{\pi}\right)^8 \frac{\sigma_8(\omega)}{\left(1 - \frac{2}{3}\sin^2 \pi\omega\right)^2} ,$$

where

$$\sigma_8(\omega) = \sum_{k=-\infty}^{\infty} \frac{1}{(\omega + k)^8} .$$

Hence, by (10.3.44)

$$\hat{\phi}(\omega) = \frac{\hat{g}(\omega)}{\sqrt{\Phi(\omega)}} = \frac{1}{\omega^4 \sqrt{\sigma_8(\omega)}} \quad \text{and} \quad M(\omega) = \frac{\hat{\phi}(2\omega)}{\hat{\phi}(\omega)} = \sqrt{\frac{\sigma_8(\omega)}{2^8 \sigma_8(2\omega)}} .$$

To calculate $\sigma_8(\omega)$, differentiate (10.3.52) 4 times.

References

1. N. Aronszajn, Theory of reproducing kernels, *Trans. Amer. Math. Soc.*, 68 (1950), 337-404.
2. R. Balian, Un principe d'incertitude fort en théorie du signal on en mécanique quantique, *C. R. Acad. Sci. Paris*, 292 (1981), 1357-1362.
3. J. Benedetto, Irregular sampling and frames, *Wavelets - A Tutorial in Theory and Applications*, C. Chui, Ed., Academic Press, New York (1991) 1-63.
4. J. Benedetto and W. Heller, Irregular sampling and the theory of frames, Part I, *Note Matematica*, X, Suppl. 1 (1990), 103-125.
5. C. Chui, *An Introduction to Wavelets*, Academic Press, New York (1992).
6. I. Daubechies, *Ten Lectures on Wavelets*, CBMS-NSF Regional Conference Series in Applied Mathematics, SIAM Publ., Philadelphia (1992).
7. ______, The wavelet transform, time-frequency localization and signal analysis, *IEEE Trans. Inform. Theory*, 36 (1990), 961-1005.
8. ______, Orthonormal bases of comactly supported wavelets, *Comm. Pure Appl. Math.* 41 (1988), 909-996.
9. I. Daubechies, A. Grossman and Y. Meyer, Painless nonorthogonal expansions, *J. Math. Phys.*, 27 (1986), 1271-1283.
10. R. Duffin and A. Schaeffer, A class of nonharmonic Fourier series, *Trans. Amer. Math. Soc.*, 72 (1952), 341-366.
11. D. Gabor, Theory of communications, *J. Inst. Elec. Eng. (London)*, 93 (1946), 429-457.
12. I. Gel' fand, Eigenfunction expansions for an equation with periodic coefficients, *Dokl. Akad. Nauk. SSSR* 76 (1950), 1117-1120.
13. C. Heil and D. Walnut, Continuous and discrete wavelet transforms, *SIAM Rev.*, 31 (1989), 628-666.
14. R. Higgins, A sampling theorem for irregularly spaced sample points, *IEEE Trans. Inform. Theory*, IT-22 (1976), 621-622.
15. ______, An interpolation series associated with the Bessel-Hankel transform, *J. London Math. Soc.*, 5 (1972), 707-714.

16. A. Janssen, The Zak transform: a signal transform for sampled time-continuous signals, *Phillips J. Res.*, 43 (1988), 23-69.

17. ______, Gabor representation of generalized functions, *J. Math. Anal. Appl.* 80 (1981), 377-394.

18. G. Köthe, Das Trägheitsgesetz der quadratischen Formen im Hilbertschen Raum, *Math. Z.* 41 (1936), 137-152.

19. N. Levinson, *Gaps and Density Theorems*, Amer. Math. Soc. Colloq. Publ. Ser. Vol. 26, Providence, RI (1940).

20. F. Low, Complete sets of wave packets, *A Passion for Physics-Essayes in Honor of Geoffrey Chew*, World Scientific, Singapore (1985), 17-22.

21. S. Mallat, Multiresolution approximations and wavelet orthonormal bases of $L^2(R)$, *Trans. Amer. Math. Soc.*, 315 (1989), 69-88.

22. ______, A theory for multiresolution signal decomposition: the wavelet representation, *IEEE Trans. PAMI*, 11 (1989), 674-693.

23. Y. Meyer, *Ondelettes ét Opérateurs, I: Ondelettes*, Hermann, Paris, (1990).

24. ______, Principe d'incertitude bases hilbertiennes et algébres d'opérateurs, Séminaire Bourbaki, 662 (1985-1986).

25. Z. Nashed and G. Walter, General sampling theorems for functions in reproducing kernel Hilbert spaces, *Math. Control, Signals, Systems*, 4 (1991), 367-390.

26. G. Nelson, L. Pfeifer, and R. Wood, High speed octave band digital filtering, *IEEE Trans. Audio Electroacoustics*, AU-20, 1 (1972), 58-65.

27. M. Rieffel, Von Neumann algebras associated with pairs of lattices in lie groups, *Math. Ann.*, 257 (1981), 403-418.

28. W. Schempp, Rader ambiguity functions, the Heisenberg group, and holomorphic theta series, *Proc. Amer. Math. Soc.*, 92 (1984), 103-110.

29. Ph. Tchamitchian, Biorthogonalité et théorie des opérateurs, *Rev. Math. Iberoamer.*, 3 (1987), 163-189.

30. G. Walter, A sampling theorem for wavelet subspaces, *IEEE Inform. Theory*, 38 (1992), 881-884.

31. Young, *An Introduction to Nonharmonic Fourier Series*, Academic Press, New York (1980).

32. J. Zak, Finite translations in solid state physics, *Phys. Rev. Lett.*, 19 (1967), 1385-1397.

AUTHOR INDEX

SUBJECT INDEX

S